Plant Genetic Resources In Vitro Culture: Present Situation and Prospects for Propagation, Conservation and Sustainable Use

Plant Genetic Resources In Vitro Culture: Present Situation and Prospects for Propagation, Conservation and Sustainable Use

Guest Editors

Anna De Carlo
Waed Tarraf

MDPI

Basel • Beijing • Wuhan • Barcelona • Belgrade • Novi Sad • Cluj • Manchester

Guest Editors
Anna De Carlo
Institute of BioEconomy-IBE
National Research Council-CNR
Florence
Italy

Waed Tarraf
Institute of BioEconomy-IBE
National Research Council-CNR
Florence
Italy

Editorial Office
MDPI AG
Grosspeteranlage 5
4052 Basel, Switzerland

This is a reprint of the Special Issue, published open access by the journal *Plants* (ISSN 2223-7747), freely accessible at: www.mdpi.com/journal/plants/special_issues/Plant_Genetic_Resources_In_Vitro_Culture.

For citation purposes, cite each article independently as indicated on the article page online and using the guide below:

Lastname, A.A.; Lastname, B.B. Article Title. *Journal Name* **Year**, *Volume Number*, Page Range.

ISBN 978-3-7258-2668-1 (Hbk)
ISBN 978-3-7258-2667-4 (PDF)
https://doi.org/10.3390/books978-3-7258-2667-4

Cover image courtesy of Anna De Carlo

Contents

Preface . . . . **vii**

Waed Tarraf and Anna De Carlo
In Vitro Biotechnology for Conservation and Sustainable Use of Plant Genetic Resources
Reprinted from: *Plants* **2024**, *13*, 1897, https://doi.org/10.3390/plants13141897 . . . . **1**

Carla Benelli, Waed Tarraf, Tolga İzgü, Monica Anichini, Cecilia Faraloni and Maria Cristina Salvatici et al.
Long-Term Conservation for the Safeguard of *Abies nebrodensis*: An Endemic and Endangered Species of Sicily
Reprinted from: *Plants* **2024**, *13*, 1682, https://doi.org/10.3390/plants13121682 . . . . **6**

Bibiana Tirado, Víctor Manuel Gómez-Rodríguez, Carlos Iván Cruz-Cárdenas, Lily Xochilt Zelaya-Molina, Humberto Ramírez-Vega and Gabriela Sandoval-Cancino
In Vitro Conservation of Mexican Garlic Varieties by Minimal Growth
Reprinted from: *Plants* **2023**, *12*, 3929, https://doi.org/10.3390/plants12233929 . . . . **31**

Meimei Tan, Ruoxin Chen, Xingran Chen, Muhammad Qasim Shahid, Xiangdong Liu and Jinwen Wu
In Vitro Induction of Interspecific Hybrid and Polyploidy Derived from *Oryza officinalis* Wall
Reprinted from: *Plants* **2023**, *12*, 3001, https://doi.org/10.3390/plants12163001 . . . . **45**

Soumaya El Merzougui, Carla Benelli, Rachida El Boullani and Mohammed Amine Serghini
The Cryopreservation of Medicinal and Ornamental Geophytes: Application and Challenges
Reprinted from: *Plants* **2023**, *12*, 2143, https://doi.org/10.3390/plants12112143 . . . . **62**

Jayanthi Nadarajan, Azadeh Esfandiari, Liya Mathew, Jasmine Divinagracia, Claudia Wiedow and Ed Morgan
Development, Management and Utilization of a Kiwifruit (*Actinidia* spp.) In Vitro Collection: A New Zealand Perspective
Reprinted from: *Plants* **2023**, *12*, 2009, https://doi.org/10.3390/plants12102009 . . . . **93**

Nóra Mendler-Drienyovszki and Katalin Magyar-Tábori
Response of Rowan Berry (*Sorbus redliana*) Shoot Culture to Slow Growth Storage Conditions
Reprinted from: *Plants* **2023**, *12*, 1287, https://doi.org/10.3390/plants12061287 . . . . **108**

Virginia Sarropoulou, Eleni Maloupa and Katerina Grigoriadou
Cretan Dittany (*Origanum dictamnus* L.), a Valuable Local Endemic Plant: In Vitro Regeneration Potential of Different Type of Explants for Conservation and Sustainable Exploitation
Reprinted from: *Plants* **2023**, *12*, 182, https://doi.org/10.3390/plants12010182 . . . . **124**

Carla Benelli, Waed Tarraf, Tolga Izgu and Anna De Carlo
In Vitro Conservation through Slow Growth Storage Technique of Fruit Species: An Overview of the Last 10 Years
Reprinted from: *Plants* **2022**, *11*, 3188, https://doi.org/10.3390/plants11233188 . . . . **140**

Naseem Ahmad, Nigar Fatima, Mohammad Faisal, Abdulrahman A. Alatar and Ranjith Pathirana
Photosynthetic Parameters and Oxidative Stress during Acclimation of Crepe-Myrtle (*Lagerstroemia speciosa* (L.) Pers.) in a *meta*-Topolin-Based Micropropagation System and Genetic Fidelity of Regenerated Plants
Reprinted from: *Plants* **2022**, *11*, 1163, https://doi.org/10.3390/plants11091163 . . . . **158**

Nikola Škoro, Suzana Živković, Slaana Jevremović and Nevena Puač
Treatment of Chrysanthemum Synthetic Seeds by Air SDBD Plasma
Reprinted from: *Plants* **2022**, *11*, 907, https://doi.org/10.3390/plants11070907 . . . . **175**

Preface

Plant genetic resources comprise a wide range of plant species, such as herbs, shrubs, and trees, that are of great importance for ensuring food, nutrition, and health security. Biotic or abiotic stresses, human activities, and climate change have long threatened the biodiversity in their natural habitat, causing an exponential loss in plant germplasm throughout the world. Thus, there is an urgent need to slow down genetic erosion and support their conservation and sustainable utilization. Advances in biotechnology can generate new opportunities for the rational use and regeneration of valuable plant material by adopting techniques such as in vitro culture and in vitro conservation. This reprint discussed novel approaches to in vitro culture that can positively influence the use of plant genetic resources for biomass or secondary metabolite production, as well as efficient in vitro storage methods and cryopreservation technologies applied to ensure a sustainable supply of high-quality plants as raw materials for different uses.

Anna De Carlo and Waed Tarraf
Guest Editors

plants

Editorial

In Vitro Biotechnology for Conservation and Sustainable Use of Plant Genetic Resources

Waed Tarraf * and Anna De Carlo

Institute of BioEconomy, National Research Council (CNR/IBE), Sesto Fiorentino, 50019 Florence, Italy; anna.decarlo@ibe.cnr.it
* Correspondence: waed.tarraf@ibe.cnr.it

1. Introduction

This Special Issue contains novel contributions related to the role and use of in vitro culture in the conservation and sustainable use of plant genetic resources. This area is of increasing interest in the global community to guarantee adequate supplies of food for future generations. All cultivated, wild relatives of cultivated species, traditional cultivars, landraces, and advanced breeding lines of plants are included in plant genetic resources [1]. Indeed, the increase in world population growth has led to significant pressure being exerted on many valuable plant species, which, in turn, has led to the genetic erosion of important germplasms from their natural habitats [2]. However, the loss of biodiversity due to the overexploitation of natural populations, random harvest at different levels from the wild, natural hazards, population growth, and climate change are considered great threats to plant genetic resources [3]. This risk can be overcome by adopting biotechnological techniques, including rapid and large-scale propagation [4] and genetic improvement in plants [5], to achieve sustainable agricultural production and enhance food security. There are two strategies to conserve biodiversity: (i) in situ conservation, where the plant species are conserved where they are found and are maintained in their original location [6] with minimum interventions from humans, and (ii) ex situ conservation, where plant materials are preserved outside their natural habitats [7]. However, for a more effective ex situ conservation program, plant tissue culture has emerged as an interesting approach to maintaining the germplasm in healthy and controlled conditions [8]. Among these ex situ conservation approaches, micropropagation offers a rapid and efficient method for the large-scale production of economically important plants in a short time and limited space [9]. Additionally, in vitro conservation effectively maintains plant germplasm, especially rare and endangered species, and recalcitrant seed and vegetatively propagated species [10].

This Special Issue comprises ten contributions focused on the role of in vitro culture for the propagation, conservation, and sustainable use of plant genetic resources.

Citation: Tarraf, W.; De Carlo, A. In Vitro Biotechnology for Conservation and Sustainable Use of Plant Genetic Resources. *Plants* **2024**, *13*, 1897. https://doi.org/10.3390/plants13141897

Received: 2 July 2024
Accepted: 9 July 2024
Published: 10 July 2024

2. Novel Approaches for the Conservation of Plant Genetic Resources

The International Union for Conservation of Nature (IUCN) estimates the global conservation status of species, and according to the latest updated data, 26,276 plant species out of 66,536 species, where they evaluated 425,035 plants, are classified as threatened and have been put on the red list of threatened species [11]. Therefore, cultivation of these species is highly recommended to limit the overexploitation practices that seriously threaten the sustainable use of these resources. Jain et al. [12] demonstrated the effectiveness of biotechnological methods for germplasm conservation, including in vitro propagation, genetic transformation, DNA banks, and cryopreservation. These techniques allow for preserving pathogen-free material, elite plants, and genetic diversity in the short-, medium-, and long-term.

Compared to conventional propagation methods, synthetic seeds can offer an efficient alternative tool for the propagation and conservation of valuable plant species that are difficult to propagate [13]. Combining synseed technology with micropropagation represents perfect biotechnology that could be convenient in the agriculture sector to improve the production of year-round plants. Many studies have proven the role of plasma treatments in improving seed germination and accelerating the conversion of plantlets into whole plants [14,15]. For example, seed treatment with high-frequency atmospheric-pressure plasma, often called "cold plasma", can positively affect seed germination and subsequently ensure plant development of conventional seeds without adding chemicals that are harmful to the environment and human health [16]. In this SI, Škoro et al. [17] applied a novel approach to obtain artificially encapsulated *Chrysanthemum* shoot tips before sowing in the soil. They showed the high efficiency of the surface dielectric barrier discharge air cold plasma treatment in directly developing plasma-treated synthetic seeds into entire plantlets. This treatment significantly improved the agronomic seed quality by minimizing contamination and promoting a considerable regrowth and conversion of *Chrysanthemum* synseeds into whole plantlets, either in vitro (under aseptic) or ex vitro (non-aseptic) conditions.

Many innovations in in vitro culture protocols for the propagation and breeding program of many species were reported in the present SI. For example, the use of meta-Topolin to improve the micropropagation of the *Lagerstroemia speciosa* [18]. This substance could successfully replace several traditional cytokinins [19,20] and possess an advantage for the in vitro propagation of true-to-type plant material appropriate for market needs, conservation strategies, and pharmaceutical purposes. Particularly, obtaining formulations and industrial products derived from micropropagated *L. speciosa* plants characterized by a higher level of corosolic acid (the future anti-diabetic drug).

This Special Issue included the indirect organogenesis protocol of *Origanum dictamnus* by Sarropoulou et al. [21], which was obtained for the first time. Based on the IUCN classification, it has been evaluated as near-threatened; therefore, the in vitro regeneration of this valuable endemic plant is highly recommended to decrease its overharvesting in nature. The authors developed an efficient protocol for the in vitro indirect regeneration testing of the potential of different plant tissues and organs to regenerate shoots or roots.

Tan et al. [22] created an interspecific hybrid of *Oryza officinalis* and cultivated rice, verifying the fertility of its pollen and embryo sac using cytological analysis. They obtained an optimal protocol to induce polyploidy by producing high-quality callus followed by colchicine treatment at 400 mg$\cdot$L^{-1} for 2 days. This work can provide a solution for cross-compatibility in *O. officinalis* and may pave the way for further improvement programs in rice.

Benelli et al. [23] underlined the importance of the slow growth storage (SGS) technique as an efficient in vitro approach for the preservation of many fruit species by controlling their growth under in vitro conditions. It is known as "minimal growth storage" due to changes applied to some physical, chemical, or nutritional factors to decrease the growth of plantlets. Also, it is called "cold storage" when low temperatures replace the standard growth conditions. In this review, the effect of many factors on the SGS of shoot cultures from temperate and tropical species has been deeply discussed and supported by published works during the last ten years.

In the context of SGS, Mendler-Drienyovszki and Magyar-Tábori [24] investigated the cold storage of the genetic resources of endangered species from the genus *Sorbus*. They reported for the first time the cold storage ability of in vitro *S. redliana* shoots at 4 °C under dark conditions. After 52 weeks, the stored shoot cultures maintained 100% survival with an efficient multiplication rate when regrown under normal culture conditions. This result allowed for plant materials without at least three subcultures to be maintained compared to the periodical subculture of four weeks, leading to a reduction in costs and saving working time.

Also, Tirado et al. [25] evaluated the in vitro conservation of Mexican garlic varieties by minimal growth. The results underlined a maximum storage of one year at 5 °C on a Murashige and Skoog medium supplied with a combination of sucrose and sorbitol.

Therefore, optimizing a protocol can be applied to other garlic varieties for medium-term storage in germplasm banks.

Cryopreservation as another biotechnological tool is applied to rescue plant genetic resources that are often at risk of loss due to overexploitation, sometimes combined with other biotic and abiotic stresses, leading to a decline in their natural habitats. The review of El Merzougui et al. [26] presented an update on the application of cryopreservation to medicinal and ornamental geophytes over the last 20 years. Indeed, several factors that limit the success of bulbous germplasm conservation were reported. Such a review will highly help the biologists and cryobiologists in their further research to optimize the cryopreservation protocols of geophytes.

The use of cryopreservation for the preservation of kiwifruit is reported in this SI by Nadarajan et al. [27]. The authors outlined the current status of the kiwifruit collection in New Zealand and detailed the ongoing development of in vitro collection for germplasm conservation. As a result of the spread of *Pseudomonas syringae* pv. *Actinadiae*-biovar 3 in 2010, which had a destructive effect on the health of field collections, almost all the *Actinidia* accessions maintained in field collections are introduced under in vitro conditions. These collections hold about 450 genotypes from many species, so managing such a huge collection requires appropriate protocols. The authors discussed the methods applied for the medium-term storage and long-term conservation of accessions of *Actinidia*.

Moreover, this Special Issue presented unique research on the long-term conservation of conifers that are under immediate risk of extinction. Benelli et al. [28] developed and validated cryopreservation protocols for pollen, zygotic embryos, and embryogenic callus to establish a cryobank for *Abies nebrodensis* Thereby, the authors suggested a strategy that allows for the safe preservation of the remaining population of *A. nebrodensis* and further opens the door for similar initiatives for other critically endangered conifers.

3. Conclusions and Future Perspectives

This Special Issue highlighted ex situ conservation as a highly efficient strategy for germplasm preservation. Many novel approaches are discussed, such as synseed technology and cold plasma treatment, the use of meta-Topolin in multiplication medium to replace many traditional cytokinins, and indirect organogenesis of near-threatened species. Also, the Special Issue included different research on in vitro culture technology, in vitro preservation, and cryopreservation. Further efforts are required to promote the in vitro conservation of plant germplasm collection and extend the applications of obtained protocols to other plant species, in particular rare and under-threat species, to ensure the sustainable use of the current plant genetic resources for the next generations.

Author Contributions: Conceptualization, W.T. and A.D.C.; writing—original draft preparation, W.T. and A.D.C.; writing—review and editing, W.T. and A.D.C. All authors have read and agreed to the published version of the manuscript.

Funding: This research received no external funding.

Data Availability Statement: Not applicable.

Acknowledgments: The Guest Editors thank all the contributors who shared their research via this Special Issue and reviewers for their valuable comments and suggestions for improving the manuscripts. Also, we appreciate the help of the Editorial staff for their kind support and assistance in managing this Special Issue.

Conflicts of Interest: The authors declare no conflicts of interest.

References

1. Ulukan, H. Plant genetic resources and breeding: Current scenario and future prospects. *Int. J. Agric. Biol.* **2011**, *13*, 447–454.
2. Salgotra, R.K.; Chauhan, B.S. Genetic diversity, conservation, and utilization of plant genetic resources. *Genes* **2023**, *14*, 174. [CrossRef]

3. Kardos, M. Conservation genetics. *Curr. Biol.* **2021**, *31*, R1185–R1190. [CrossRef]
4. Jan, N.; Qazi, H.A.; Ramzan, S.; John, R. Developing stress-tolerant plants through in vitro tissue culture: Family Brassicaceae. In *Biotechnologies of Crop Improvement*; Gosal, S., Wani, S., Eds.; Springer: Cham, Switzerland, 2018; Volume 1, pp. 327–372.
5. Purwantoro, A.; Purwestri, Y.A.; Lawrie, M.D.; Semiarti, E. Genetic transformation via plant tissue culture techniques: Current and future approaches. In *Advances in Plant Tissue Culture: Current Developments and Future Trends*; Rai, A.C., Kumar, A., Modi, A., Singh, M., Eds.; Elsevier: London, UK, 2022; pp. 131–156.
6. Pandotra, P.; Gupta, S. Biotechnological approaches for conservation of plant genetic resources and traditional knowledge. In *Plant Genetic Resources and Traditional Knowledge for Food Security*; Salgotra, R., Gupta, B., Eds.; Science and Business Media: Singapore, 2016; pp. 121–135.
7. Rajpurohit, D.; Jhang, T. In situ and ex situ conservation of plant genetic resources and traditional knowledge. In *Plant Genetic Resources and Traditional Knowledge for Food Security*; Salgotra, R., Gupta, B., Eds.; Springer Science and Business Media: Singapore, 2016; pp. 137–162.
8. Abdalla, N.; El-Ramady, H.; Seliem, M.K.; El-Mahrouk, M.E.; Taha, N.; Bayoumi, Y.; Shalaby, T.A.; Dobránszki, J. An academic and technical overview on plant micropropagation challenges. *Horticulturae* **2022**, *8*, 677. [CrossRef]
9. Teraiya, S.; Nirmal, D.; Joshi, P. Potential scope and prospects of plant growth-promoting microbes (PGPMs) in micropropagation technology. In *Plant-Microbe Interaction—Recent Advances in Molecular and Biochemical Approaches: Volume 2: Agricultural Aspects of Microbiome Leading to Plant Defence*; Meena, M., Swapnil, P., Eds.; Elsevier: Amsterdam, The Netherlands, 2023; Volume 2, pp. 249–277.
10. Engelmann, F. Use of biotechnologies for the conservation of plant biodiversity. *Vitr. Cell. Dev. Biol.-Plant* **2011**, *47*, 5–16. [CrossRef]
11. IUCN Plant Species in the World and Its Annual Increase. 2023. Available online: https://nc.iucnredlist.org/redlist/content/attachment_files/2023-1_RL_Table_1a.pdf (accessed on 9 July 2024).
12. Jain, M.; Johnson, T.S.; Krishnan, P. Biotechnological approaches to conserve the wealth of nature: Endangered and rare medicinal plant species. *J. Nat. Remedies* **2012**, *12*, 97–107.
13. Saxena, A.; Shukla, M.; Saxena, P. Synthetic seeds: Relevance to endangered germplasm conservation in vitro. In *Synthetic Seeds: Germplasm Regeneration, Preservation and Prospects*; Faisal, M., Alatar, A., Eds.; Springer: Cham, Switzerland, 2019; pp. 21–60.
14. Puač, N.; Gherardi, M.; Shiratani, M. Plasma agriculture: A rapidly emerging field. *Plasma Process. Polym.* **2018**, *15*, 1700174. [CrossRef]
15. Attri, P.; Ishikawa, K.; Okumura, T.; Koga, K.; Shiratani, M. Plasma agriculture from laboratory to farm: A review. *Processes* **2020**, *8*, 1002. [CrossRef]
16. Adhikari, B.; Adhikari, M.; Park, G. The effects of plasma on plant growth, development, and sustainability. *Appl. Sci.* **2020**, *10*, 6045. [CrossRef]
17. Škoro, N.; Živković, S.; Jevremović, S.; Puač, N. Treatment of *Chrysanthemum* synthetic seeds by Air SDBD Plasma. *Plants* **2022**, *11*, 907. [CrossRef]
18. Ahmad, N.; Fatima, N.; Faisal, M.; Alatar, A.A.; Pathirana, R. Photosynthetic parameters and oxidative stress during Acclimation of crepe-myrtle parameters and oxidative stress during acclimation of crepe-myrtle (*Lagerstroemia speciosa* (L.) Pers.) in a meta-Topolin-based micropropagation system and genetic fidelity of regenerated plants. *Plants* **2022**, *11*, 1163.
19. Erişen, S.; Kurt-Gür, G.; Servi, H. In vitro propagation of *Salvia sclarea* L. by meta-Topolin, and assessment of genetic stability and secondary metabolite profiling of micropropagated plants. *Ind. Crops Prod.* **2020**, *157*, 112892. [CrossRef]
20. Saeiahagh, H.; Mousavi, M.; Wiedow, C.; Bassett, H.B.; Pathirana, R. Effect of cytokinins and sucrose concentration on the efficiency of micropropagation of 'Zes006' *Actinidia* chinensis var. chinensis, a red-fleshed kiwifruit cultivar. *Plant Cell. Tissue Organ Cult.* **2019**, *138*, 1–10. [CrossRef]
21. Sarropoulou, V.; Maloupa, E.; Grigoriadou, K. Cretan dittany (*Origanum dictamnus* L.), a valuable local endemic plant: In vitro regeneration potential of different type of explants for conservation and sustainable exploitation. *Plants* **2023**, *12*, 182. [CrossRef]
22. Tan, M.; Chen, R.; Chen, X.; Shahid, M.Q.; Liu, X.; Wu, J. In vitro induction of interspecific hybrid and polyploidy derived from Oryza officinalis Wall. *Plants* **2023**, *12*, 3001. [CrossRef]
23. Benelli, C.; Tarraf, W.; Izgu, T.; De Carlo, A. In vitro conservation through slow growth storage technique of fruit species: An overview of the last 10 years. *Plants* **2022**, *11*, 3188. [CrossRef] [PubMed]
24. Mendler-Drienyovszki, N.; Magyar-Tábori, K. Response of rowan berry (*Sorbus redliana*) Shoot culture to slow growth storage conditions. *Plants* **2023**, *12*, 1287. [CrossRef] [PubMed]
25. Tirado, B.; Gómez-Rodríguez, V.M.; Cruz-Cárdenas, C.I.; Zelaya-Molina, L.X.; Ramírez-Vega, H.; Sandoval-Cancino, G. In vitro conservation of Mexican garlic varieties by minimal growth. *Plants* **2023**, *12*, 3929. [CrossRef]
26. El Merzougui, S.; Benelli, C.; El Boullani, R.; Serghini, M.A. The cryopreservation of medicinal and ornamental geophytes: Application and challenges. *Plants* **2023**, *12*, 2143. [CrossRef]

27. Nadarajan, J.; Esfandiari, A.; Mathew, L.; Divinagracia, J.; Wiedow, C.; Morgan, E. Development, management and utilization of a kiwifruit (*Actinidia* spp.) In vitro collection: A New Zealand perspective. *Plants* **2023**, *12*, 2009. [CrossRef]
28. Benelli, C.; Tarraf, W.; Izgü, T.; Anichini, M.; Faraloni, C.; Salvatici, M.C.; Antonietta, M.; Danti, R.; Lambardi, M. Long-term conservation for the safeguard of *Abies nebrodensis*: An endemic and endangered species of Sicily. *Plants* **2024**, *13*, 1682. [CrossRef] [PubMed]

plants

Article

Long-Term Conservation for the Safeguard of *Abies nebrodensis*: An Endemic and Endangered Species of Sicily

Carla Benelli [1], Waed Tarraf [1,*], Tolga İzgü [1], Monica Anichini [1], Cecilia Faraloni [1], Maria Cristina Salvatici [2], Nourhene Jouini [3], Maria Antonietta Germanà [3], Roberto Danti [4] and Maurizio Lambardi [1]

1 Institute for BioEconomy (IBE), National Research Council (CNR), Via Madonna del Piano 10, 50019 Sesto Fiorentino, Florence, Italy; carla.benelli@ibe.cnr.it (C.B.); tolga.izgu@ibe.cnr.it (T.İ.); monica.anichini@ibe.cnr.it (M.A.); cecilia.faraloni@ibe.cnr.it (C.F.); maurizio.lambardi@ibe.cnr.it (M.L.)
2 Institute of Chemistry of Organometallic Compounds (ICCOM)-Electron Microscopy Centre (Ce.M.E.), National Research Council (CNR), Via Madonna del Piano 10, 50019 Sesto Fiorentino, Florence, Italy; salvatici@ceme.fi.cnr.it
3 Department of Agricultural, Food and Forestry Sciences, University of Palermo, 90128 Palermo, Italy; nourhene.jouini@unipa.it (N.J.); mariaantonietta.germana@unipa.it (M.A.G.)
4 Institute for Sustainable Plant Protection (IPSP), National Research Council (CNR), Via Madonna del Piano 10, 50019 Sesto Fiorentino, Florence, Italy; roberto.danti@ipsp.cnr.it
* Correspondence: waed.tarraf@ibe.cnr.it

Abstract: The combined approaches between ex situ and in situ conservation are of great importance for threatened species in urgent need of protection. This study aims to develop concrete actions to preserve the relic of 30 adult trees of the Sicilian fir (*Abies nebrodensis*) from extinction using long-term germplasm conservation in liquid nitrogen (LN, −196 °C). Pollen grains were collected, and their moisture content (MC) was measured. Then, viability (2,3,5-tryphenyl tetrazolium chloride, TTC), in vitro germinability, and enzymatic antioxidant activity (ascorbate peroxidase, APX; catalase, CAT) were evaluated before and after cryopreservation. Seeds collected from mature cones underwent X-ray analysis, and only full seeds were used to excise the zygotic embryos (ZEs) for cryopreservation. The MC percentage of ZEs was determined, and then they were plunged in LN with (+PVS2) or without (−PVS2) Plant Vitrification Solution 2; untreated ZEs were used as a control. Viability (TTC test) and in vitro germination were assessed for all ZEs (+PVS2, −PVS2, and control). Embryogenic callus (EC) lines obtained from mature ZEs were cryopreserved applying the 'encapsulation-dehydration' technique. This study has allowed, after optimizing cryopreservation protocols for pollen, ZEs, and EC of *A. nebrodensis*, to establish the first cryobank of this endangered species in Polizzi Generosa (Palermo, Italy), inside the 'Madonie Regional Park'. The strategy developed for Sicilian fir conservation will pave the way for similar initiatives for other critically endangered conifer species.

Keywords: Sicilian fir; cryopreservation; cryobank; pollen; zygotic embryos; embryogenic callus

check for updates

Citation: Benelli, C.; Tarraf, W.; İzgü, T.; Anichini, M.; Faraloni, C.; Salvatici, M.C.; Jouini, N.; Germanà, M.A.; Danti, R.; Lambardi, M. Long-Term Conservation for the Safeguard of *Abies nebrodensis*: An Endemic and Endangered Species of Sicily. *Plants* **2024**, *13*, 1682. https://doi.org/10.3390/plants13121682

Academic Editor: Carolina Sánchez-Romero

Received: 12 March 2024
Revised: 12 June 2024
Accepted: 13 June 2024
Published: 18 June 2024

1. Introduction

Abies nebrodensis, commonly known as the Sicilian fir, is a critically endangered conifer species endemic to the Madonie Regional Park in the north of Sicily. This species is subject to significant threats caused by genetic erosion, fragmentation, poor natural regeneration, and a high rate of empty seeds leading to low seed germination. The remaining natural population is alarmingly small, consisting of only 30 adult trees, underscoring the urgency for effective conservation strategies [1–3]. According to the International Union for Conservation of Nature (IUCN), *A. nebrodensis* is classified as CR-D in the Red List of Endangered Species (https://top50.iucn-mpsg.org/species/1; accessed on 12 March 2024).

Historical exploitation and habitat degradation have contributed to the species decline, with its wood once highly valued for construction due to its elasticity and resistance. The past few decades have seen various projects aimed at both in situ and ex situ conservation, including experimental plantations and protective measures like fencing to support natural regeneration. Studies have highlighted *A. nebrodensis* as a species highly variable in seed production and the vital role of embryo presence in seeds for germination [4]. In addition, this species is characterized by a very limited natural regeneration due to the high rate of empty seeds, slow growth of seedlings, rocky soils, and grazing by wild herbivores.

The LIFE4FIR project (http://www.life4fir.com/it/), started in 2019, aims to develop an effective strategy to improve the conservation status of *A. nebrodensis* (Figure 1a) through a comprehensive approach that includes different actions: protecting the relic population, increasing genetic diversity by controlled cross-pollination, breeding selected outbred seedlings, and reforesting with selected seedlings to create new cores of the species within the Madonie Regional Park [1]. The project strategy is also based on ex situ conservation strategies, like establishing a seed bank and a cryobank for long-term germplasm conservation. Among the conservation techniques, cryopreservation stands out as a particularly promising method for the safe, long-term maintenance of *A. nebrodensis*.

Figure 1. *Abies nebrodensis* tree (**a**); male cones (**b**); pollen (**c**).

The current conservation efforts greatly focus on biotechnological approaches, and cryopreservation has emerged as a crucial tool for the long-term maintenance of germplasm [5,6], allowing for the storage of plant material at ultra-low temperatures, effectively halting any enzymatic or chemical activity that could damage the biological material [7]. Under these conditions, any enzymatic or chemical activity is stopped, preventing structural and physiological damage to the stored biological material. This method allows for the indefinite storage of valuable plant material, making it a widely used tool for ex situ conservation of a variety of crops, fruit, and forest trees [8–11]. At present, plant cryopreservation techniques are available for storage of explants derived from in vivo and in vitro conditions [6,7,11], such as meristem and shoot tips, root tips, callus, pollen, seeds, cyanobacteria, algae, bryophytes, ferns, suspension cultures, embryogenic cultures of both zygotic and somatic embryos, and dormant buds [11–15].

This biotechnological approach for long-term conservation represents a safe and cost-effective tool where samples are stored in small volumes, protected from contamination, and without the need for laborious management. It should be noted that using desiccation, vitrification, and encapsulation-based methods for the cryopreservation of in vitro-derived shoot tips, several cryobanks of different economically important crops were established [16–20]. Similar efforts have been directed at the cryopreservation of some endangered species [21,22]. For example, several strategies were developed for threatened plants such as lilies, orchids, and redwood [23] as well as for the conservation of seeds and shoot tips of Canadian cherry birch [24].

In addition to shoot tips, several other explant types have been used in cryopreservation, such as pollen, zygotic embryos (ZEs), and embryogenic callus (EC). Particularly, pollen conservation is an important tool for the maintenance of plant genetic resources and can promote improved efficiency in breeding programs as well as germplasm conservation and exchange [25]. Cryopreservation of embryogenic tissue enables large-scale propagation and preservation of forestry resources [26]. Zygotic embryos, embryonic axes, and somatic embryos of different temperate and tropical species, including crops, fruit, and forest trees, have been effectively cryopreserved [27].

Within the genus *Abies*, the species subjected to cryopreservation include *A. cephalonica* L. [28,29], *A. nordmanniana* [30,31], *A. alba* [32], and some fir hybrids [33]. Currently, methodologies of conifer breeding integrate cryopreservation with somatic embryogenesis, extensively evaluated for *A. alba* [32].

For the above reasons, this study was focused on finding the optimal protocols for cryopreservation of pollen, ZEs, and EC to establish a cryobank for the critically endangered *A. nebrodensis*.

2. Materials and Methods

2.1. Plant Material

Mature cones of *Abies nebrodensis* trees were collected during October 2020 in their natural habitat, located in the north-west of Sicily, in the Madonie Regional Park, Palermo (Italy), considering all the relevant guidelines and regulations to avoid any damage to the relict trees. Seeds extracted from a large part of the 30 residual trees were packed in sealable paper bags, signed, and transferred to the laboratory of the CNR-IBE, where they were stored at 4 °C until further use. In May 2022, the male cones were gently removed from adult trees. It was not possible to sample all 30 trees due to either a lack of seed production from some trees (only 24 are mature) or difficulties in collecting cones from trees located in impervious sites.

2.2. X-ray Analysis

To select full seeds (i.e., seeds containing well-formed embryos), all seeds extracted from mature cones from trees numbers 6, 8, 10, 12, 13, 19, 21, 22, and 27 were X-ray radiographed through the X-ray apparatus "Gilardoni radio light" (Lecco, Italy) [2] set up as follows: 25 kV, 3 mA (soft X-rays), a focus-film distance of 45 cm, and a 2-min time of exposure. Seeds from each tree were placed in plastic square well plates (20 × 20 cm), 100 at a time, each seed in a separate well. The seeds were in direct contact with a Carestream X-ray film during the exposure. After X-ray application, the films were washed in development solution (4 min) and then in fixation solution (3 min) on a shaker (20 rpm). The films were washed under tap water (3–5 s) before further examination by the film viewer screen.

2.3. Procedures for the Conservation in Liquid Nitrogen (LN, 196 °C)

2.3.1. Pollen

Male cones were collected in mid-May 2022 from trees N° 1, 2, 6, 7, 8, 9, 10, 11, 12, 13, 14, 15, 16, 17, 18, 19, 21, 22, 23, 24, 27, and 29. After extraction from cones, the pollen was cleaned and sieved to remove all the impurities (Figure 1b,c) and maintained in paper bags at room temperature for 2 days. The moisture content (MC) of fresh pollen from different trees was measured using the moisture analyzer (Mettler-Toledo AG, Laboratory & Weighing Technologies, Greifensee, Switzerland). Measurements of MC were performed with 0.1 g of pollen per replicate and repeated three times for each tree.

Morphological pollen observations were carried out at the CEME-Centro di Microscopie Elettroniche "Laura Bonzi"-CNR Research Area (Florence, Italy), using a Gaia 3 (Tescan s.r.o., Brno, Czech Republic) FIB-SEM (Focused Ion Beam-Scanning Electron Microscope) electron beam for scanning electron microscope (SEM) imaging with a voltage of 2 kV, operating in high vacuum mode, and a secondary electron (in-beam SE) detector. Samples were deposited on a stub and then coated with an ultrathin coating of gold. Pollen grains were photographed, and SEM measurements of the polar axis (P) were performed. At least 70 fresh pollen grains were examined to ensure a complete morphological analysis. Furthermore, the pollen observations were complemented by a stereomicroscope (Zeiss Stemi 2000 C, Jena, Germany) and an optical microscope (Leica DM-500, Heerbrugg, Switzerland) for counting viable and germinated pollen grains.

Viability Test

To evaluate the viability of pollen before and after LN immersion, the 2,3,5-tryphenyl tetrazolium chloride (TTC) test was used. A TTC 1% solution (purity 99%, Sigma-Aldich, Buchs, Switzerland) was prepared by adding 200 mg of TTC and 12 g of sucrose in 20 mL of distilled water. Two drops of this mixture were dropped on a microscope slide; the pollen was dusted over it, covered with a coverslip, and kept at room temperature for 24–48 h in the dark [34]. Following incubation, 400 pollen grains were randomly counted under the optical microscope (Leica DM-500). Pollen grains stained red were categorized as "viable", whereas colorless-stained pollen was "non-viable". The estimation of pollen vitality was expressed as the percentage of stained grains in the total number of grains counted. The TTC test was performed in two replicates for each tree. Three microscopic field views were observed per replicate, containing a minimum of 300 pollen grains/replication.

In Vitro Germinability Test

The assessment of in vitro pollen germinability, both prior to and after cryopreservation, was carried out by placing pollen grains on a semisolid medium and evaluating the elongation of the pollen tube. After 48 h of incubation, the pollen tube that achieved a length at least three times the diameter of the pollen grain was considered germinated [35]. The composition of the germination medium consisted of boric acid (50 mg L^{-1}), sucrose (15 g L^{-1}), and plant agar (6 g L^{-1}; Duchefa Biochemie, Haarlem, The Netherlands). The pollen was maintained in the dark at 25 °C as an optimal temperature for in vitro germination assays of most species [36]. For each plant, two replicates were conducted, and within each replicate, three microscopic fields were randomly selected for observation (at least 300 grains/replicate). The percentage of germination represented the number of germinated pollen grains in relation to the total number of grains counted. The number of germinated pollen grains was counted using a Leica DM-500 optical microscope.

Pollen Catalase and Ascorbate Peroxidase

The determination of catalase (CAT) and ascorbate peroxidase (APX) activity was carried out for pollen samples of trees 6, 8, 9, 11, 13, 14, 22, 24, and 27, according to Ren et al. [37]. For both CAT and APX activity measurement, three replicates of 0.03 g pollen for each tree, before and after immersion in LN, were used for homogenization and extraction. For CAT activity, the absorbance, after the addition of 30% H_2O_2, was measured at a wavelength of 240 nm. For APX activity, the absorbance, after the addition of 30% H_2O_2, was measured at a wavelength of 290 nm. Each analysis was repeated three times. The results are expressed as U/g protein.

Immersion, Storage in Liquid Nitrogen, and Recovery of Pollen

Samples of pollen were transferred in 2-mL cryovials. For each tree, two cryovials (0.30 g/cryovial) were immersed directly into LN. After storage in LN, for at least 1 h, the cryovials containing the pollen were thawed under a laminar flow cabinet for 2 h at room temperature. Subsequently, following the procedure described above, the viability (TTC) and in vitro germinability of cryopreserved pollen (Figure 2) were tested.

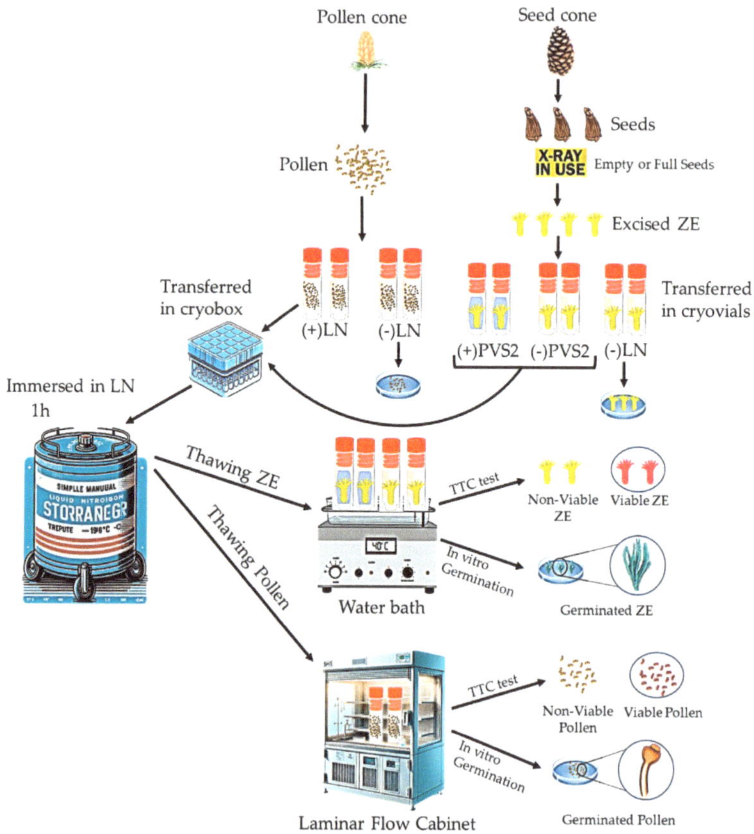

Figure 2. Steps of cryopreservation process for pollen and zygotic embryos (ZE) from *A. nebrodensis*. TTC: tryphenyl tetrazolium chloride; PVS2: Plant Vitrification Solution 2; LN: liquid nitrogen.

2.3.2. Excised Zygotic Embryos

Under laminar flow, X-rayed seeds containing mature embryos were treated with 70% EtOH (5 min), rinsed with sterile distilled water 3 times, treated with sodium hypochlorite (20% v/v) with a few drops of Tween 20 solution (20 min), and rinsed again 3 times with sterile distilled water. Finally, seeds were imbibed in water for 48 h under sterile conditions and then opened to excise the ZEs.

Viability (TTC Test) and In Vitro Germination Test of ZEs

The viability of ZEs was determined by the TTC. Briefly, ZEs were soaked in TTC solution 0.1% (w/v) in 50 mM Tris-HCl buffer (pH 7.6) for 24 h, in total darkness at 30 °C. After staining, the ZEs were placed on moist filter paper to observe their viability. The red color of embryonic tissues was the main indicator of ZEs viability.

For the germination test, before and after the LN, excised ZEs were cultured in vitro on hormone-free Murashige and Skoog (MS; Sigma-Aldrich, St. Louis, MO, USA) [38] medium (MS-HF), containing sucrose (20 g L^{-1}) and agar (7 g L^{-1}) at pH 5.8. All ZEs were maintained at 24 °C under a 16-h photoperiod (60 µmol $m^{-2}s^{1}$ of photosynthetically active radiation). After 3 weeks of culture, the germination rate was evaluated.

Immersion in LN and Recovery of ZEs

Prior to the immersion in LN, the MC of ZEs (approx. 15 ZEs = 0.1 g/tree) was determined under the sterile laminar airflow following the method of Ayala et al. [39].

From each tree, a total number of 108 ZEs was divided into three groups; each group included three replicates: (1) submerged in Plant Vitrification Solution 2 (+PVS2; 30% glycerol, 15% ethylene glycol, 15% dimethyl sulfoxide in MS, and 0.4 M sucrose [40]), (2) without Plant Vitrification Solution 2 treatment (−PVS2), and (3) untreated control (neither PVS2 nor LN). The cryovial (2 mL) containing 4 ZEs from the +PVS2 and −PVS2 groups were immersed in LN for at least 1 h. Thereafter, the cryovials were taken out of the LN and thawed in a water bath (40 °C) for 1 min. Under laminar flow, cryopreserved ZEs were washed with liquid MS medium containing 1.2 M sucrose (washing solution) for 20 min. All ZEs cryopreserved (+PVS2 and −PVS2) or not cryopreserved (control) were subjected to viability and in vitro germination tests (Figure 2), as described above. Both tests were carried out with three replicates per tree (6 zygotic embryos/replicate).

2.3.3. Embryogenic Callus

The EC was obtained following a protocol developed by Jouini et al. [2]. Briefly, ECs obtained from ZEs were developed and subcultured every 4 weeks on Schenk and Hildebrandt (SH; Sigma-Aldrich, St. Louis, MO, USA) [41] media supplemented with 1 mg L^{-1} 6-benzyladenine (BA; Sigma-Aldrich, St. Louis, MO, USA), 1 g L^{-1} casein (Sigma-Aldrich, St. Louis, MO, USA), 500 mg L^{-1} glutamine (Duchefa Biochemie, Haarlem, The Netherlands), 20 g L^{-1} sucrose, and 7 g L^{-1} agar at 24 °C in dark conditions. After eight subcultures, the EC cryopreservation was carried out using the encapsulation-dehydration technique. Portions of EC obtained from the ZE of tree N°10 were encapsulated in Ca-Alginate beads as described by Standardi and Micheli [42]. Then, the beads were transferred into sterilized filter paper inside the Petri dishes (60 Ø mm) and placed inside a glass jar (500 cc) containing 150 g of silica gel previously sterilized in an oven at 105 °C for 16 h. The jars were sealed with tape, and the beads were exposed to silica gel at room temperature for 1, 2, 3, 4, and 5 h. A control treatment of encapsulated ECs without desiccation was also included. The desiccated beads were placed in 2 mL cryovials (5 per each), replicated three times, and plunged directly into LN for at least 1 h. Thereafter, cryopreserved encapsulated ECs in cryovials were thawed in a water bath at 40 °C for 2 min and treated with a washing solution for 20 min. Beads were placed on fresh SH regrowth medium, supplemented with 1 mg L^{-1} BA, 1 g L^{-1} casein, 500 mg L^{-1} glutamine, 20 g L^{-1} sucrose, and 7 g L^{-1} agar. The regrowth ability of encapsulated ECs was determined when the callus broke through the gel of the bead. The same protocol was applied for the other ECs obtained from trees N° 7, 8, 21, and 22.

2.4. Statistical Analysis

The primary cause of the variation in the amount of zygotic embryos and pollen collected from *A. nebrodensis* trees depends on the production of trees, which was extremely variable from one year to the next. All cryopreservation experiments followed the randomized block trial design. The moisture content of pollen, the viability and germination of pollen, and ZEs, both before and after cryopreservation, were expressed as percentages. All percentages of data were subjected to an arcsine transformation. Means were differentiated through analysis of variance (ANOVA), followed by the least significant difference (LSD) post-hoc test to evaluate the differences in germination and viability of pollen and ZEs, and CAT and APX data analysis activities within *A. nebrodensis* trees. For the moisture content of the pollen, Duncan's multiple range test was used. Data analysis was conducted using JMP® software (SAS Institute, Cary, NC, USA) version 5.00, with the significance level at $p \leq 0.001$. The Pearson correlation analysis between the germination of pollen and its viability, as well as the germination of ZEs and its viability, was computed using the corrplot package in the R programming language (version 4.3.1).

3. Results and Discussion

The *A. nebrodensis* species is affected by a very low seed germination rate, mainly depending on a large number of empty seeds [4]. Moreover, Sicilian fir is characterized by abundant seed production every 3–4 years, similar to other conifer species [43]. For this reason, the quantity of explants and the number of trees analyzed varied in the current study.

3.1. Evaluation of X-ray Analysis

Based on the X-ray image, two groups of seeds were distinguished: (1) viable and able to germinate (Figure 3a), and (2) unviable, either empty or with an undeveloped embryo and endosperm (Figure 3b). It was observed that a significant proportion of the seeds, varying by *A. nebrodensis* tree, were devoid of ZEs [2]. The validation of X-ray analysis was confirmed through the dissection and direct observation of the X-rayed seeds, along with an assessment of their germination potential (Figure 3c,d).

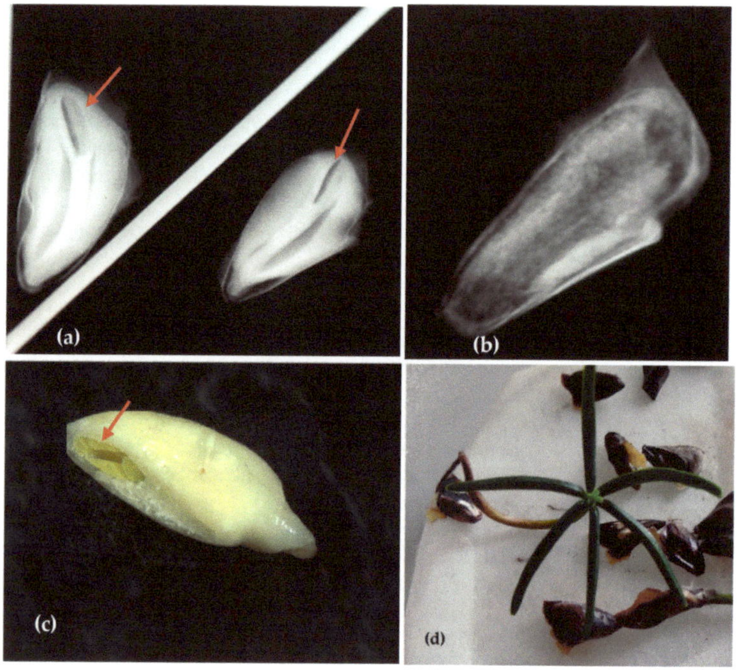

Figure 3. Seeds after X-ray exposure: full seed with embryo ((**a**); red arrow), empty seed (**b**) and validation of X-ray analysis: presence of embryo ((**c**); red arrow), and seed germination (**d**).

The X-ray technique was essential for verifying the presence of the ZEs within the seeds [44]. This approach allowed us to save time in the excision of the ZEs used for cryopreservation purposes and EC induction, as empty seeds were discarded. Earlier, Fedorkov [45] confirmed the advantage of this method to optimize seed storage and improve the production of plant material. Although radiographic inspection is potentially harmful, seeds are exposed to a non-lethal dose during the test; hence, no damage was expected to occur or adversely affect germination [46].

Furthermore, X-ray analysis is a non-destructive method that safely allows the selection of high-quality seeds from deteriorating and dead seeds [47]. Therefore, X-ray tests have been widely applied to evaluate the quality of conifer seeds, such as in *Cupressus sempervirens*, *C. arizonica* [48], *Pinus sylvestris* [49–51], *P. sibirica*, and *P. koraiensis* [52] to separate the empty seeds from seed samples.

3.2. Conservation in Liquid Nitrogen (LN, −196 °C)

3.2.1. Pollen

Morphological Characteristics of Pollen

The pollen of *A. nebrodensis* was observed by a scanning electron microscope (Figure 4), a Leica stereomicroscope, and a Leica Optical Microscope (Figure 5), and these examinations revealed that the pollen had the typical characteristics of the genus *Abies* [53–55]. Pollen grains resulted isodiametric with an elliptical central body with two lateral air sacs (bisaccate) and one aperture leptoma. The air sacs are clearly protruding from the body.

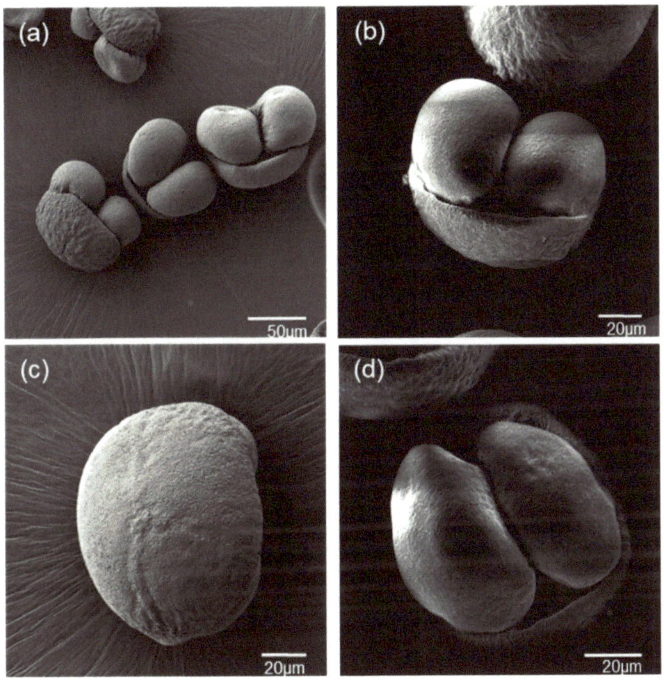

Figure 4. Pollen grains of *Abies nebrodensis* under SEM. (**a**) Pollen grains in the polar and equatorial views; (**b**) two sacci in equatorial views; (**c**) two sacci in polar distal view; (**d**) polar proximal view.

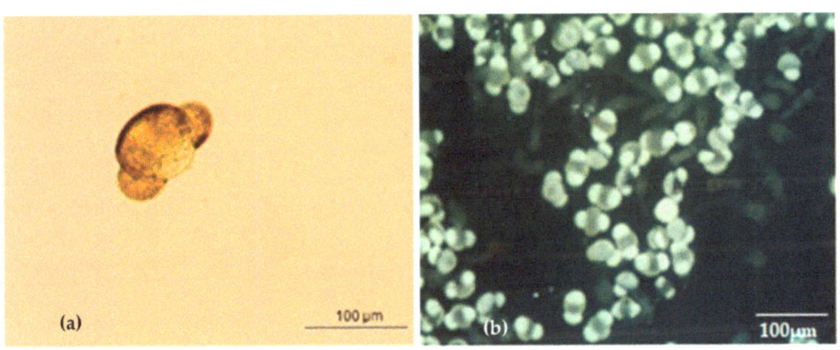

Figure 5. Pollen grain morphology of *A. nebrodensis* observed under optical microscope ((**a**); from Frascella et al. [1]), stereomicroscope (**b**).

Measurements of the polar axis (P) of *A. nebrodensis* pollen grains exhibited a range from 68.11 to 92.34 μm, with an average of 83.47 μm in accordance with the study of

Wrońska-Pilarek et al. [55] on *A. alba*. Most of the observed pollen grains showed a large size, over 80 µm (73%).

Evaluation of Pollen MC

The MC in explants used in cryopreservation is a crucial factor for the success of this conservation method. For this reason, before the cryopreservation process, the MC content of pollen samples from different *A. nebrodensis* trees was recorded. Figure 6 shows that MC significantly varied among trees and was restricted in a percentage ranging between 6.8 and 11.3%. Trees 19 and 21 had the lowest percentage, while tree 6 exhibited the highest MC.

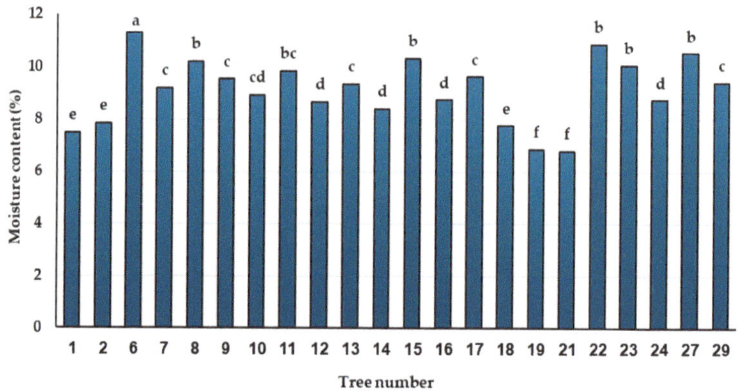

Figure 6. Moisture content of pollen in tested trees of *A. nebrodensis* before cryopreservation. Values with the same letters are not significantly different at the 0.05 level (Duncan's multiple range test).

Many factors influence the viability and germinability of cryopreserved pollen, mainly the MC, which is critical to avoid the formation of deleterious ice crystals breaking the cell membranes [12] and also the developmental stage of pollen grains (bi- or tricellular) [25,56,57].

Different methods are applied for pollen dehydration to adjust the moisture content [58–60]. The pollen MC from 8 to 10% avoids tissue damage during the freezing process, regardless of the final cold storage method [61]. Connor and Towill [62] reported the success of long-term conservation of pollen with moisture contents between 7 and 20% by applying −80 to −196 °C. Moreover, in some conifer species, MC percentages of pollen of 9.8–10.1 were recorded before immersion in LN [63]. In the case of *A. nebrodensis*, the MC of pollen maintained for two days at room temperature was suitable (about 10%) for direct immersion in LN without compromising its viability and germinability. This aspect is notable because it makes the cryopreservation procedure easier and faster.

Weatherhead et al. [64] obtained successful pollen cryopreservation of potato (*Solanum* spp.) without desiccation for 9 months, with no significant reduction in pollen viability. Moreover, Anushma et al. [65] immersed the pollen of nine wild *Solanum* spp. directly in LN without dehydration; pollen viability and germinability were retained in all species without any significant reduction during the cryostorage for 36 weeks. Similarly, the cryopreservation of non-dehydrated pollen from *Diospyros* spp. and *Olea europaea* maintained both viability and germination capability for up to 360 days of storage in LN [66,67].

Viability and In Vitro Pollen Germinability

Pollen samples from different trees of *A. nebrodensis* were subjected to a TTC test to assess their viability in fresh and cryopreserved samples (Figure 7a–c). Results of the TTC test showed (Table 1) that pollen was viable both in the control group (−LN) and in the group that was stored in LN. Significant differences were observed in terms of tree and treatment, and the interaction between tree and treatment was also significant. In the control, the highest viability rates were recorded for trees N° 8 and 23 (98.88%), while after

LN treatment, the highest viability was observed in trees 6 (95.08%) and 18 (95.52%). Trees N° 2, 19, 21, and 29 exhibited low viability rates in both the control (from 3.88 to 9.26%) and cryopreserved pollen (from 6.45 to 33.29%). However, the pollen from all trees responded favorably to the application of LN in terms of viability. The effect of the *A. nebrodensis* tree on pollen viability, both before and after cryopreservation, was significant. The highest mean viability, regardless of control and LN treatments, was recorded in trees N° 8, 17, 18, and 24, ranging from 93% to 95% (Table 1).

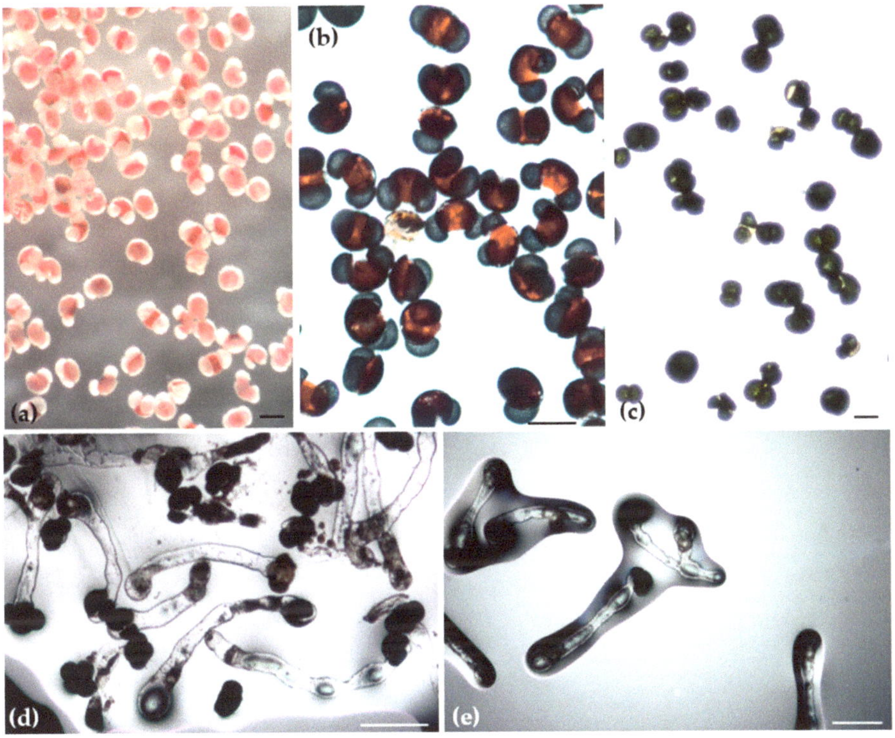

Figure 7. Cryopreserved pollen grains of *Abies nebrodensis*: viable pollen grains after TTC under stereomicroscope (**a**) and under microscope (**b**); non-viable pollen grains under optical microscope (**c**); in vitro germination of pollen grains under optical microscope (**d**,**e**) (Bars, 100 µm).

For the in vitro germinability test, the control and cryopreserved pollen samples (Figure 7d,e) were placed on semisolid medium and incubated at a constant temperature of 25 °C for 48 h. Observations conducted post-incubation identified the presence of pollen tube formation in both groups. The in vitro germinability outcomes indicated that the tree, treatment, and the interaction between the tree and treatment were significant (Table 2). In the control group (−LN), the germinability rate for the tree-by-treatment interaction varied from a low of 9.16% to a high of 99.72%. The highest germinability rate in the control was observed in trees N°15 (99.72%) and 14 (99.44%). Following immersion in LN, the germination capacity of the pollen from the twenty-two trees ranged from 12.66 to 99.33%. Trees N°23 and 24 showed the highest germinability, with 99.33% and 99%, respectively, whereas no germinability was observed in the cryopreserved pollen of tree 19. Generally, the treatment with LN was found to enhance pollen germination (mean 81.95%) compared to the control (mean 79.16%).

Table 1. Viability percentage of pollen before and after cryopreservation.

N° Tree (T)	Treatment (Tr)	Tr xT	T Mean	Tr Mean
1	Control +LN	77.77 h–l 74.51 i–m	76.14 GHI	
2	Control +LN	9.16 n 6.45 n	7.81 LM	
6	Control +LN	90.00 b–f 95.08 abc	92.54 ABC	
7	Control +LN	88.89 c–g 70.14 klm	79.51 FGH	
8	Control +LN	98.88 a 91.69 b–f	95.29 A	
9	Control +LN	79.16 g–k 92.71 bcde	85.94 DEF	
10	Control +LN	88.16 b–f 93.72 bcd	90.94 ABCD	
11	Control +LN	90.55 b–f 93.66 bcd	92.10 ABCD	Control 68.93 B
12	Control +LN	65.88 lm 82.27 f–j	74.08 GHI	
13	Control +LN	90.44 bcde 89.33 b–g	89.89 BCD	+LN 73.22 A
14	Control +LN	74.99 i–m 73.14 i–m	74.07 HI	
15	Control +LN	65.83 lm 70.70 klm	68.27 IJ	
16	Control +LN	90.27 b–f 84.02 e–i	87.14 DEF	
17	Control +LN	93.61 bcd 93.99 bcd	93.80 AB	
18	Control +LN	96.39 ab 95.52 abc	95.95 AB	
19	Control +LN	3.88 n 7.69 n	5.79 M	
21	Control +LN	4.99 n 33.29 m	34.68 K	
22	Control +LN	33.05 m 86.65 d–h	59.85 J	
23	Control +LN	98.88 a 66.38 lm	82.63 CDE	
24	Control +LN	94.44 bcd 93.28 bcd	93.86 AB	
27	Control +LN	71.94 jklm 92.24 b–f	82.09 EFG	
29	Control +LN	9.26 n 24.43 m	16.85 L	
p value		0.0002 ***	0.0001 ***	0.0001 ***

LSD_{tree}: 5.81 ***. $LSD_{treatment}$: 1.75 ***. $LSD_{tree \times treatment}$: 8.22 ***. *** $p < 0.001$; +LN: Immersed in liquid nitrogen; −LN: Control not immersed in liquid nitrogen. The values of Tr xT are the mean of two replicates/tree (at least 300 pollen grains/replicate). Different letters within a column indicate significantly different means (LSD test). Lowercase letters indicate significant differences between *A. nebrodensis* trees and treatments. Uppercase letters indicate significant differences among total trees and the mean of treatments.

Table 2. Germinability percentage of pollen before and after cryopreservation.

N° Tree (T)	Treatment (Tr)	TrxT	T Mean	Tr Mean
1	Control +LN	84.72 l–p 82.66 op	83.69 G	
2	Control +LN	22.62 qr 41.66 q	20.83 I	
6	Control +LN	88.89 i–p 86.66 k–p	87.77 EFG	
7	Control +LN	90.00 h–p 80.00 p	85.00 FG	
8	Control +LN	91.11 h–p 92.00 g–o	91.55 DEF	
9	Control +LN	87.77 j–p 89.00 i–p	88.38 EFG	
10	Control +LN	98.61 abcd 93.00 f–o	95.80 ABC	
11	Control +LN	93.89 d–l 91.00 h–p	92.44 CDE	
12	Control +LN	92.22 g–o 90.66 e–m	91.44 CDE	Control 79.16 B
13	Control +LN	89.44 i–p 96.33 a–h	92.88 CDE	+LN
14	Control +LN	99.44 ab 91.33 e–n	95.38 ABC	81.95 A
15	Control +LN	99.72 a 93.66 b–j	96.69 AB	
16	Control +LN	90.83 h–p 97.66 a–g	94.25 BCD	
17	Control +LN	85.83 l–p 98.33 a–e	92.08 CDE	
18	Control +LN	83.61 n–p 84.66 m–p	84.14 FG	
19	Control +LN	23.89 qr 0.00 t	11.94 I	
21	Control +LN	30.28 q 94.00 c–k	62.14 H	
22	Control +LN	91.66 g–o 93.33 e–m	92.50 CDE	
23	Control +LN	98.05 a–f 99.33 a	98.69 A	
24	Control +LN	98.89 abcd 99.00 abc	98.94 A	
27	Control +LN	90.83 h–p 96.00 a–i	93.41 CDE	
29	Control +LN	9.16 s 12.66 rs	10.91 I	
p value		0.0001 ***	0.0009 ***	0.0001 ***

LSD_{tree}: 6.69 ***. $LSD_{treatment}$: 2.01 ***. $LSD_{tree\times treatment}$: 9.47 ***. *** $p < 0.001$; +LN: Immersed in liquid nitrogen; −LN: Control not immersed in liquid nitrogen. The values of Tr xT are the mean of two replicates/tree (at least 300 pollen grains/replicate). Different letters within a column indicate significantly different means (LSD test). Lowercase letters indicate significant differences between *A. nebrodensis* trees and treatments. Uppercase letters indicate significant differences among total trees and the mean of treatments.

The results of the correlation analysis revealed a significant correlation between the viability and germinability percentages of pollen before and after liquid nitrogen treatment (Figure 8). This correlation was positive either in pollen cryopreserved or not cryopreserved reporting values of $R^2 = 0.78$ and $R^2 = 0.83$, respectively.

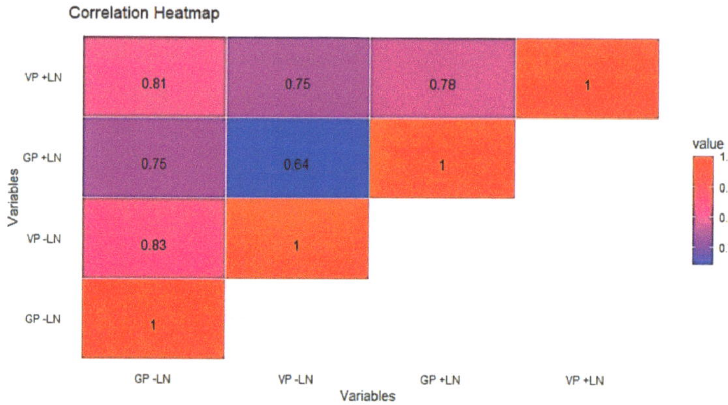

Figure 8. Heatmap for Pearson's linear correlation coefficients between viability (VP) and germinability (GP) of pollen. −LN: control; +LN: immersion in liquid nitrogen.

Pollen Catalase and Ascorbate Peroxidase Activities

The effects of storage temperature of LN on enzymatic activities, CAT and APX, were determined for pollen samples of nine trees (Figure 9). Changes in CAT and APX activity after freezing were different among trees. Indeed, freezing induced a reduction in CAT activity in all the trees compared with the activity measured in crude extracts from fresh pollen, with the exception of trees N°8 and 9, where this reduction was not significant. In trees 14, 24, and 27, a drastic decline was detected, with activity almost absent (about 98% of the decrease), while for the other trees, a reduction of 20–55% was observed.

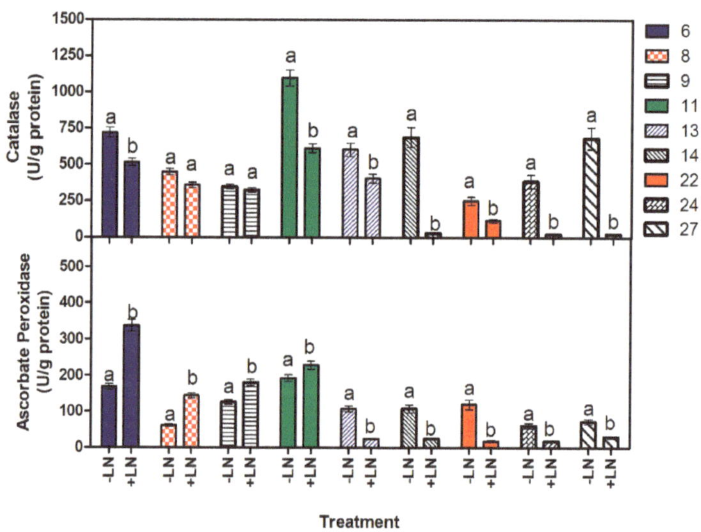

Figure 9. Changes of CAT and APX activity before (−LN) and after liquid nitrogen (+LN) in pollen from the tested *A. nebrodensis* trees. Different letters indicate significant differences between the values for each tree. The data are the means of three replicates.

Concerning the APX activity, in cryopreserved pollen of five trees out of nine examined (13, 14, 22, 24, and 27), the values decreased, and the differences compared to fresh pollen resulted in a range of 24% to 79%. The major reduction in activity was detected in the range of 70–79% for the trees N°13, 14, 22, and 24, while it was 58% for the trees N°27. In the other trees, the APX activity, after LN, significantly increased to 18% and 43% in trees 11 and 9, respectively, and almost doubled in trees 6 and 8. Changes in antioxidant enzyme activities before and after LN were significantly different for each tree, except for CAT in trees 9 and 8. Interestingly, the trees showing an increase in APX activity after LN (6, 8, 9, and 11) had the lowest decrease in CAT activity, indicating that in these trees the stress due to freezing was less damaging.

Similar to other environmental stresses, freezing and cold stress can affect plant metabolism, causing the generation of several kinds of active oxygen molecules. In response to oxidative stress, plants develop antioxidant defense systems via enzymatic and non-enzymatic antioxidant reactions to maintain normal metabolisms and functions in the cell, thus protecting themselves against oxidative stress [37,68–71].

CAT and APX are two significant antioxidant enzymes. Among the tested trees in this study, only three (14, 24, and 27) showed a strong decrease in CAT activity after immersion in LN. In general, an increase in antioxidant activities is normally promoted under stressful conditions to react to the stress, generating free radicals. However, according to the conditions, the response to stress can be different. The decrease in CAT activity after freezing has already been observed in frozen apple tissue [72]. This decline was attributed to damage occurring at the tetrameric enzyme structure and to the loss of the total CAT protein amount. Other findings reported that the decrease in CAT activity may be linked to the conversion of the superoxide radical to hydrogen peroxide, resulting in excessive oxidative stress [73]. Indeed, Jia et al. [74] and Jiang et al. [75] reported that the application of CAT to the LN-stored pollen could significantly improve the viability of pollen after cryopreservation.

However, it is important to note that a considerable APX activity was present in the above-mentioned trees; it may indicate a positive response to freezing stress.

The CAT activity declined in the most cryopreserved trees, while N°6, 8, 9, and 11 showed higher levels of both antioxidant enzyme activity (CAT and APX), and they could be considered more tolerant to the freezing condition. Despite the freezing stress, the cryopreserved pollen in the current study responded well in terms of viability and germinability, proving to be an adequate propagule for the cryopreservation of *A. nebrodensis*.

The importance of pollen cryopreservation for the long-term conservation of genetic resources has been highlighted in many plant species, including ornamental, fruit, and forest [76–79]. Pollen grains are suitable genetic materials for preservation for their small size and desiccation tolerance, and their conservation is beneficial in breeding programs overcoming seasonal and geographical restrictions [25]. Pollen from *Prunus persica* [80] and *Carya illinoensis* [81] was preserved in LN for 10 years and for 5 years in *Vitis vinifera* [82]. Zhang et al. [83] successfully cryopreserved pollen grains from 51 cultivars of *Prunus mume* for the establishment of the cryobank.

In *A. nebrodensis*, pollen preservation is fundamental to carrying out controlled cross-breeding among trees of this species to counteract the high level of inbreeding due to fragmentation and cross-contamination with other fir species present near its natural area, such as *A. alba* and *A. cephalonica*.

Pollen from conifer trees for germplasm conservation has been successfully preserved in LN. Dehydrated pollen of *Picea pungens* and *Pinus ponderosa* showed germination rates of 84% after 6 months of cryopreservation [59,62]; while for *Pseudotsuga menziesii*, the pollen cryopreserved for 1 year showed a germination rate of 81%, exhibiting a fertility nearly as high as that of fresh pollen when applied for controlled pollinations [84]. These data were in accordance with our findings on *A. nebrodensis*, where the maximum pollen germinability was recorded at 99% after cryopreservation. Lanteri et al. [63] reported good results in terms of pollen germinability after preservation at −196 °C for *Picea abies* (78.2%) and *Pinus nigra* (65.8%), without significant differences between cryopreserved and fresh pollen. In the same study, all the assessed conifer species responded well to the cryopreservation process, and it was notable that in *Picea abies*, *P. sylvestris*, and *P. uncinata*, the pollen germinability rate increased in the first month of LN storage.

In tropical forest trees, pollen cryopreservation was also reported for teak and sandalwood [85,86]. In the former species (*Tectona grandis*), after 2 days of cryopreservation, the pollen showed a very slight decrease in germination (52.18%) compared to the fresh pollen (58.25%), while in *Santalum album* L, where dried anthers with pollen were preserved in LN, no reduction was observed in the germination of cryopreserved pollen (84.20%) compared to fresh dehydrated pollen (84.50%).

The pollen viability and germinability evaluation is an essential step in verifying the success of its storage conditions. In several cryopreserved *A. nebrodensis* trees, the viability and germinability increased with respect to the control. This event was reported for other species such as *Prunus nume* [83], pineapple [78], *Solanum* spp. [65], pecan [87], mango, and litchi pollen [88]. This increase can be attributed to the very low temperature or/and the dehydration that, in combination or individually, contributed to the breakdown of the grain dormancy with an improvement in the percentage of vitality and germination [78,83]. Moreover, the viability and germinability of pollen before and after LN can be dependent on other factors such as initial grain water content and reserve compounds, external temperature, relative humidity, and gamete maturity. Indeed, Anushma et al. [65] reported that the variable response to cryopreservation observed in wild species of *Solanum* pollen could be due to inherent genetic variations and differences in the level of maturity. In Serbian spruce (*Picea omorika*), Batos and Miljkovic [89] indicated that climate conditions preceding pollen maturation influenced the pollen quality and vitality after conservation at −20 °C for 14 years. Pollen production and its subsequent performance depend on various factors, with wide variability between and within species. In particular, environmental conditions can influence pollen physiological processes such as carbohydrate storage reserves [90]. Indeed, carbohydrates, water content, and pollen viability and longevity are closely related [91,92]. In *A. nebrodensis*, the different conditions in the location of the plants (soil, wind exposure, altitudes) could influence the germination of pollen storage, even if our findings demonstrated a clear effect of the genotype.

3.2.2. Zygotic Embryos

Viability and In Vitro Germination of ZEs

The applicability of a simplified cryopreservation protocol, with or without PVS2, for the long-term storage of ZEs was investigated for *A. nebrodensis*, and viability and germination tests were applied. Prior to cryopreservation, the MC of the non-dehydrated ZEs (Figure 10a) was 58% and decreased under the sterile laminar airflow to achieve 25–27% after 2 h [39].

The results of the viability test (Figure 10b,c) revealed that the treatment and the interaction between treatment and tree were not statistically significant (Table 3). However, the differences in tree averages were found to be statistically significant. Notably, after cryopreservation, trees N°8, 10, 13, and 27 exhibited 100% viability in the group of + PVS2 treatment. The presence of PVS2 has a protective role on plant tissues [13,27,40]. Moreover, it was interesting that only ZEs from tree N°10 maintained 100% viability after LN application, both with PVS2 and without PVS2, compared with the control. In the group −PVS2+LN treatment, viability rates of 100% were observed only in the ZEs of trees N° 6 and 10. Considering either the groups (+PVS2+LN or −PVS2+LN) the viability fluctuated between 66% and 100%.

Table 3 demonstrated that in the control group, the percentage of viability (75.69%) was lower than those observed in both the −PVS2+LN (81.94%) and +PVS2+LN (92.01%) treatments.

The investigation further demonstrated that the viability rates were influenced by the tree effect, as statistical analysis showed a significant variation among *A. nebrodensis* trees. These findings underscore the critical role of genetic factors in determining the susceptibility of ZEs to the cryopreservation process.

Figure 10. Untreated zygotic embryo (**a**), cryopreserved zygotic embryo after TTC test: non-viable (**b**) and viable (**c**); cryopreserved germinated zygotic embryo after 7 (**d**) and 20 days (**e**).

Table 3. Viability percentage of zygotic embyos before and after cryopreservation.

N° Tree (T)	Treatment (Tr)		TrxT	T Mean	Tr Mean
6	Control	UN	33.33 a	74.07 BC	
	+LN	−PVS2	100 a		
		+PVS2	88.88 a		
8	Control	UN	83.33 a	90.74 ABC	
	+LN	−PVS2	88.88 a		
		+PVS2	100 a		
10	Control	UN	100 a	100 A	Control 75.69 A
	+LN	−PVS2	100 a		
		+PVS2	100 a		
12	Control	UN	66.66 a	66.66 C	
	+LN	−PVS2	66.66 a		
		+PVS2	66.66 a		−PVS2 +LN
13	Control	UN	88.89 a	85.18 ABC	81.94 A
	+LN	−PVS2	66.66 a		
		+PVS2	100 a		
21	Control	UN	100 a	88.88 ABC	+PVS2 +LN 92.01 A
	+LN	−PVS2	77.77 a		
		+PVS2	88.88 a		
22	Control	UN	33.33 a	65.74 C	
	+LN	−PVS2	72.22 a		
		+PVS2	91.66 a		
27	Control	UN	100 a	94.44 AB	
	+LN	−PVS2	83.33 a		
		+PVS2	100 a		
	p value		0.2503	0.0498 *	0.1126

LSD_{tree}: 22.91 *. $LSD_{treatment}$: Not Significant. $LSD_{tree \times treatment}$: Not Significant. * $p < 0.05$. UN: untreated Control, neither PVS2 nor LN. The values of Tr xT are the mean of three replicates per tree (6 zygotic embryos/replicate). Different letters within a column indicate significantly different means (LSD test). Lowercase letters indicate significant differences between *A. nebrodensis* trees and treatments. Uppercase letters indicate significant differences among total trees and the mean of treatments.

The in vitro germination test, conducted on ZEs (Figure 10d,e), revealed that the tree, treatment, and tree-treatment interaction were all statistically significant (Table 4). Germination rates exhibited variation among trees across the control group, as well as after LN applications, both with and without the submerge in PVS2. In particular, tree N°10 achieved a 100% germination rate across all three treatment conditions (control, −PVS2+LN, and +PVS2+LN). Generally, a significant reduction in germination rates for ZEs of all other trees was observed post-cryopreservation. In tree N°6, the control recorded a higher germination rate (88.89%), while, following LN treatment, the germination decreased to 16.66% without PVS2 or to 0% with PVS2. Also, the germination of trees N° 12 and 27 was null following the LN application with PVS2. These results highlight that the ZEs of the control group exhibited superior in vitro germination rates compared to the ZEs cryopreserved.

Table 4. Germination percentage of zygotic embryos before and after cryopreservation.

N° Tree (T)	Treatment (Tr)		TrxT	T Mean	T Tr Mean
	Control	UN	88.89 ab		
6	+LN	−PVS2	16.66 fghi	35.18 B	
		+PVS2	0.00 i		
	Control	UN	83.33 abc		
8	+LN	−PVS2	58.33 cde	50.00 B	
		+PVS2	8.33 ghi		
	Control	UN	100 a		
10	+LN	−PVS2	100 a	100 A	Control 65.97 A
		+PVS2	100 a		
	Control	UN	33.33 efgh		
12	+LN	−PVS2	33.33 efgh	14.81 C	
		+PVS2	0.00 i		−PVS2 +LN 39.79 B
	Control	UN	77.78 abc		
13	+LN	−PVS2	58.33 cde	50.37 B	
		+PVS2	15 fghi		
	Control	UN	75 bcd		
21	+LN	−PVS2	35 defg	46.66 B	+PVS2 +LN 25.11 B
		+PVS2	30 d-f		
	Control	UN	36.11 defg		
22	+LN	−PVS2	5.55 hi	29.76 BC	
		+PVS2	47.61 cdef		
	Control	UN	33.33 efgh		
27	+LN	−PVS2	11.11 ghi	14.81 C	
		+PVS2	0.00 i		
	p value		0.0003 ***	0.0001 ***	0.0001 ***

LSD_{tree}: 15.79 ***. $LSD_{treatment}$: 11.63 ***. $LSD_{tree \times treatment}$: 27.36 ***. *** $p < 0.001$; UN: untreated Control, neither PVS2 nor LN. The values of Tr xT are the mean of three replicates per tree (6 zygotic embryos /replicate). Different letters within a column indicate significantly different means (LSD test). Lowercase letters indicate significant differences between *A. nebrodensis* trees and treatments. Uppercase letters indicate significant differences among total trees and the mean of treatments.

Additionally, the tree was shown to play a significant role in in vitro germination (Table 4). Tree N°10 showed the highest germination response, while in the other trees, it ranged from 14.81 to 50.37%. This underscores the significance of trees effects on the success of in vitro germination in pre- and post-cryopreservation.

A correlation analysis was conducted to explore the relationship between viability and germination in ZEs (Figure 11). The analysis revealed a non-significant correlation between viability and germination across all treatments. This finding suggests an absence of a relation between the germination and viability of ZEs, both before and after cryopreservation treatments, indicating that the processes influencing viability may not directly correspond to those affecting the germination of ZEs in the context of cryopreservation.

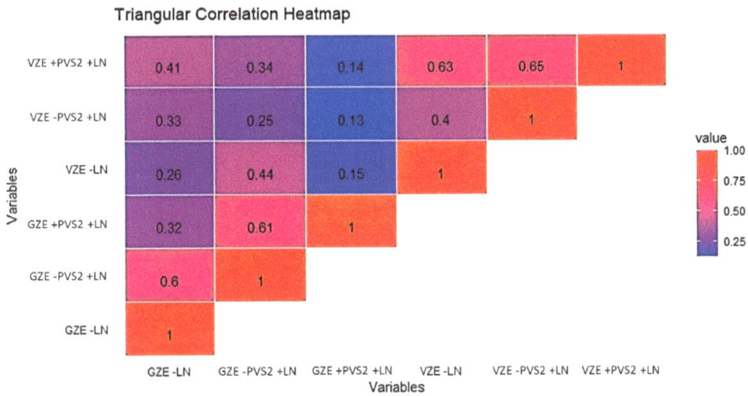

Figure 11. Heatmap for Pearson's linear correlation coefficients between viability (VZE) and germination (GZE) of zygotic embryos. −LN: control; +PVS2+LN: with PVS2 and immersion in liquid nitrogen; −PVS2+LN: without PVS2 and immersion in liquid nitrogen.

In conifers, Ayala et al. [39] investigated the cryopreservation of mature ZEs from the interspecific hybrid *Pinus elliottii* var. elliottii x *Pinus caribaea* var. hondurensis, achieving a 100% survival rate after 120 min of desiccation, leading to a moisture content of 23.7%. This precise desiccation treatment facilitated a high regeneration rate, with 86.7% of the surviving explants producing an average of 16.1 adventitious buds. This success, achieved with a simple desiccation step followed by direct immersion in LN, provides valuable insights into effective cryopreservation strategies. This finding is in accordance with our results, where ZEs were only desiccated and not immersed in PVS2 (−PVS2) before cryopreservation. In *Castanea sativa*, Corredoira et al. [93] reported a high survival rate of 93–100% in the embryonic axes after desiccation in a laminar flow cabinet to 20–24% MC. Following the LN storage, about 63% of these axes developed into whole plants. For the cryopreservation of embryonic axes of several *Quercus* spp. by the drying method, a recovery range of 10–80% was reported, while the treatment with 15% dimethyl sulfoxide (DMSO) solution or the application of the encapsulation-dehydration method resulted in a recovery of 40% and 12%, respectively [94]. In another study on *Citrus x aurantiifolia*, excised embryonic axes showed survival rates of 60% to 67% when undergoing dehydration and LN, significantly higher than the 10% to 13% survival following cryopreservation-vitrification with PVS2 [95]. Our investigation revealed a wide range of germination rates (0 to 100%) in post-cryopreservation, with a significant treatment-x-tree interaction, regardless of the presence of PVS2. The significant variability among trees observed in *A. nebrodensis* for some trees, demonstrating exceptionally high viability and germination rates, reflected the importance of genetic factors (the genotype effect) in determining cryopreservation success. Thus, the current cryopreservation protocol may offer a better protective effect for ZEs in some trees. Likewise, in ZEs of oil palm, a high success rate of cryopreservation was obtained, achieving up to 96.67% viability and 90.88% germination through an easy air-drying procedure without chemical pretreatment [96]. Pinto et al. [97] demonstrated that dehydration to 23% of MC achieved after 60 min, followed by osmotic solution rehydration, was optimal for cryopreservation of *Coffea arabica* ZEs, with a significant increase in germination rates. Moreover, the dehydration time significantly influenced both the viability and germination percentage of coffee ZEs, highlighting the delicate balance between removing sufficient moisture to prevent ice crystal formation and retaining enough to ensure cellular integrity. In this study, the optimal ZEs desiccation (25–27%) was obtained after 120 min under laminar airflow.

Therefore, dehydration of ZEs prior to their introduction into LN is a method chosen for the cryopreservation of species producing recalcitrant seeds [98]. Recently, an innovative approach was applied to cryopreserve the embryonic axes of *Syzygium maire*, a species challenged by recalcitrant seeds, using a metal-mesh vacuum infiltration method with PVS2, resulting in a post-cryopreservation survival of 19% following a 20-min PVS2 incubation [99]. The difference in recovery rates among different studies highlights the species-specific and technique-specific challenges and successes inherent to cryopreservation efforts.

3.2.3. Embryogenic Callus

The capability to induce EC from mature embryos of *A. nebrodensis* was demonstrated by Jouini et al. [2], and it could be a crucial step towards ex situ conservation and large-scale propagation of this endangered fir, although it was not possible to induce EC from all the fertile trees. Indeed, the ECs were obtained from a few trees, namely N°7, 8, 10, 21, and 22, due to the variation in the response of mature ZEs to the induction medium, and even when ECs were produced, their growth was very slow.

Immature or mature embryos were widely employed to establish embryogenic cultures from coniferous species [100]. To reduce costs and time spent on periodical subculture and to minimize the occurrence of somaclonal variation, somatic embryogenesis combined with cryopreservation is needed to maintain the juvenility and regenerability of lines [101]. In this study, the possibility of cryopreserving EC obtained from ZEs was investigated by applying a procedure of encapsulation-dehydration of callus samples (Figure 12).

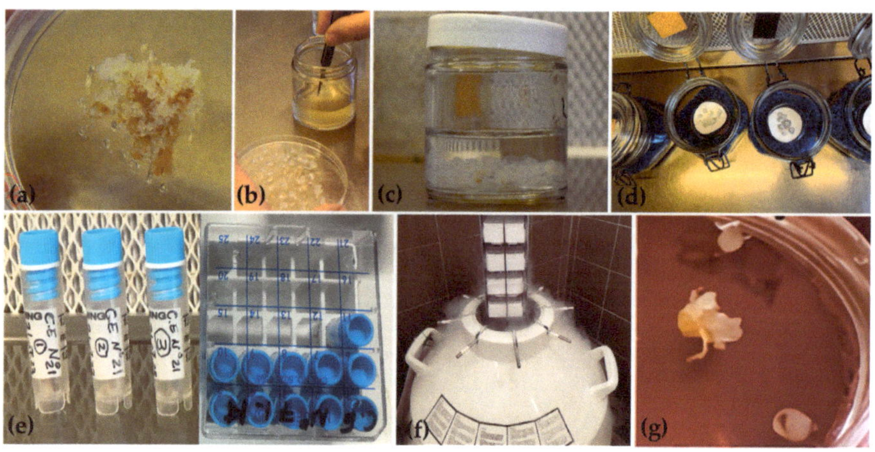

Figure 12. Flowchart depicts the major steps of cryopreservation for embryogenic callus of *A. nebrodensis*. Embryogenic callus-EC (**a**); preparation of EC beads (**b**,**c**); EC beads dehydration on silica gel (**d**); dehydrated EC beads in cryovials and cryoboxes (**e**); immersion in liquid nitrogen (**f**); regrowth of cryopreserved EC (**g**).

Before freezing in LN, encapsulated EC beads were gradually dehydrated over silica gel. The initial MC of non-dehydrated beads was 91%, and after 5 h of desiccation, it decreased to 23% (Figure 13). This percentage was suitable for the cryopreservation of EC beads from tree N°10. Generally, the most preferred MC for encapsulated explants, either dehydrated under a laminar airflow cabinet or over silica gel, is around 20% on a fresh weight basis [27]. The optimal MC of alginate beads, prior to direct immersion in LN, is largely dependent on the plant species, e.g., 39.50% for *Phoenix dactylifera* [102], 35% for *Quercus suber* [103] and 22% for *Thymus moroderi* [104]. For *A. nebrodensis*, acceptable survival was observed for EC alginate beads with a MC of 23%. The first evidence of regrowth was visible after 4 weeks on SH medium, with 35% of EC beads developing after 8 weeks. Then, following the same procedure of encapsulation-dehydration, ECs obtained from the other lines were cryopreserved, reaching a regrowth rate of 35% in tree N°7, 30% in N°8, 32% in N°21, and 40% in N°22 by the end of 8 weeks. The recovery period, including the latency phase, is affected by the *A. nebrodensis* tree effect [105]. In our protocol, with the application of the encapsulation-dehydration technique, no cryoprotectant was used [100] to minimize the water content of calli beads, thus avoiding the toxic effect of cryoprotectants such as PVS2 [106] and lessening the risk of tissue damage.

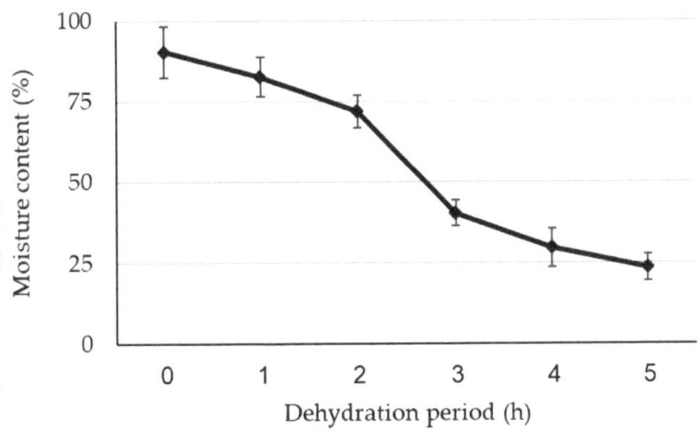

Figure 13. Moisture content of *A. nebrodensis* embryogenic callus beads after silica gel dehydration at different times.

The regeneration percentage in different conifer species determined the success of cryopreservation protocols. In the case of *A. nebrodensis*, with the encapsulation-vitrification method, 80% of recovery was achieved from non-cryopreserved EC beads, while no proliferation was obtained from those immersed in LN [2]. Our results exhibited for the first time the possibility to reach 30–40% of regrowth following the encapsulation-dehydration method after the freezing in LN, although the results depend strictly on the *A. nebrodensis* tree. Application of the slow-freezing cryopreservation protocol in hybrid firs *Abies alba* × *A. cephalonica* and *Abies alba* × *A. numidica* led to 37.5%–100% regrowth of cryopreserved embryogenic tissues [33]. Similarly, applying the same method to embryogenic lines from *A. alba* enabled regeneration ranging from 91.66 to 100% [107]. Slow-freezing is an effective method for the cryopreservation of EC, but it is not simple and requires specific programmable apparatus for slow-cooling before the sample immersion into LN. A proper protocol with an 80% survival by vitrification method was optimized to cryopreserve embryogenic cell lines of *Larix kaempefri* × *Larix gmelinii*, involving pre-culturing of embryogenic tissue on a medium containing an osmotic agent, followed by cryoprotectant treatment [108].

Survival rates of 84.4% and 86.7% were reached by the stepwise dehydration method (sucrose steps from 0.25 to 1.0 M for 7 days, followed by desiccation over silica gel) of embryogenic tissues from *A. alba* × *A. numidica* and *P. nigra*, respectively, 28 and 35 days after recovery from LN [109]. In another study with the vitrification method, Latutrie and Aronen [110] observed 80–93% recovery for *P. sylvestris* embryogenic lines.

Regrowth with our encapsulation-dehydration protocol was reliable but lower than for other conifer species, regardless of cryopreservation technique. The preliminary results obtained from the current study demonstrate that the encapsulation-dehydration method can be successfully used for the cryopreservation of *A. nebrodensis* ECs. Further investigation is needed to improve the regrowth percentage after immersion in LN of EC of

Sicilian fir, such as pre-culture of encapsulated ECs in medium supplemented with osmotic agents for a specific time or modification of the regrowth medium to optimize the recovery percentage. To the best of our knowledge, this is the first report on the application of this technique for *A. nebrodensis*, offering an alternative method for the efficient and long-term storage of its germplasm in LN.

4. Conclusions

This study, developed in the framework of the LIFE4FIR Project, financed by the European Union, represents a qualifying example of actions aimed at safeguarding the Sicilian fir, which is listed by the International Union for Conservation of Nature (IUCN) as an endangered species due to serious genetic erosion. For this reason, the plan of protection of the 30 relict trees involved a series of actions combining two strategies: in situ, by monitoring their health status and protecting their survival in the natural habitat, and *ex situ*, through the conservation of the seeds and the cryopreservation of organs and tissues taken from the relict trees. *A. nebrodensis* is a conifer characterized by the presence of a low percentage of full seeds with an unknown life span in terms of years, even in the sporadic years of abundant seed production. Therefore, in addition to the establishment of a 'Seed Bank', the project, through a cryopreservation approach, has obtained effective conservation of pollen, zygotic embryos, and lines of embryogenic callus in liquid nitrogen. In this study, the cryopreservation protocols for several explants of *A. nebrodensis* were optimized and validated, reaching a satisfactory recovery percentage of pollen, ZEs, and EC after LN. Therefore, the obtained results enabled the secure preservation of the germplasm of the remaining population and the establishment of a cryobank inside the 'MAN/Museum of the *A. nebrodensis*' located near the 'Madonie Regional Park' in the Municipality of Polizzi Generosa.

To our knowledge, the Sicilian fir 'Cryobank' is currently a unique example for the application of cryopreservation technique to a conifer at immediate risk of extinction, can pave the way for similar strategies with other conifers subjected to advanced genetic erosion, and also serves as a critical model to preserve biodiversity in facing the challenges of environmental changes.

Author Contributions: Conceptualization, W.T., M.L. and C.B.; methodology, W.T., T.İ., C.F., M.C.S., N.J., M.A. and C.B.; data curation, M.A., T.İ. and W.T.; writing—original draft preparation, W.T., T.İ., C.F. and C.B.; writing—review and editing, W.T., T.İ., M.A.G. and C.B.; supervision, C.B., R.D. and M.L.; funding acquisition, R.D. and M.L. All authors have read and agreed to the published version of the manuscript.

Funding: This research was funded by the European Union by the LIFE4FIR project, grant number LIFE18 NAT/IT/000164.

Data Availability Statement: Data are contained within the article.

Acknowledgments: The authors want to thank Gaetano La Placa and Giuseppe Di Noto for the collection of plant material used in this study.

Conflicts of Interest: The authors declare no conflict of interest. The funders had no role in the design of the study; in the collection, analyses, or interpretation of data; in the writing of the manuscript; or in the decision to publish the results.

References

1. Frascella, A.; Della Rocca, G.; Barberini, S.; Emiliani, G.; Secci, S.; Lambardi, M.; Benelli, C.; Tarraf, W.; Izgu, T.; Schicchi, R.; et al. Innovative in situ and ex situ conservation strategies of the madonie fir Abies nebrodensis. *Sustainability* **2022**, *14*, 12643. [CrossRef]
2. Jouini, N.; Yahyaoui, E.; Tarraf, W.; İzgü, T.; Benelli, C.; Lambardi, M.; Germanà, M.A. Somatic embryogenesis in Abies nebrodensis, an endangered Sicilian fir. *Plant Cell Tissue Organ Cult.* **2023**, *152*, 393–404. [CrossRef]
3. Tarraf, W.; Izgu, T.; Jouini, N. Strategies for the conservation by biotechnological approaches of Abies nebrodensis, a relict conifer of Sicily. *Acta Hortic.* **2023**, *1359*, 215–221. [CrossRef]
4. Scialabba, A. Seed germination in Abies nebrodensis (Pinaceae). In *Mediterranean Plant Germination Reports*; Magrini, S., Salmeri, C., Eds.; Fl. Medit: Palermo, Italy, 2019; Volume 29, pp. 272–276.

5. Benelli, C. Plant Cryopreservation: Look at the present and the future. *Plants* **2021**, *10*, 2744. [CrossRef]
6. Nagel, M.; Pence, V.; Ballesteros, D.; Lambardi, M.; Popova, E.; Panis, B. Plant cryopreservation: Principles, applications, and challenges of banking plant diversity at ultralow temperatures. *Annu. Rev. Plant Biol.* **2024**, *75*, 7.1–7.28. [CrossRef]
7. Engelmann, F. Plant cryopreservation: Progress and prospects. *Vitr. Cell. Dev. Biol.—Plant* **2004**, *40*, 427–433. [CrossRef]
8. Engelmann, F. Importance of cryopreservation for the conservation of plant genetic resources. In *Cryopreservation of Tropical Plant Germplasm. Current Research Progress and Application*; Engelmann, F., Takagi, H., Eds.; Japan International Research Center for Agricultura: Tsukuba, Japan; International Plant Genetic Resources Institute: Rome, Italy, 2000; pp. 8–20.
9. Gonzalez Benito, M.E.; Clavero-Ramirez, I.; López-Aranda, J.M. The use of cryopreservation for germplasm conservation of vegetatively propagated crops. *Span. J. Agric. Res.* **2004**, *2*, 341–351. [CrossRef]
10. Benelli, C.; De Carlo, A.; Engelmann, F. Recent advances in the cryopreservation of shoot-derived germplasm of economically important fruit trees of Actinidia, Diospyros, Malus, Olea, Prunus, Pyrus and Vitis. *Biotechnol. Adv.* **2013**, *31*, 175–185. [CrossRef] [PubMed]
11. Wang, M.R.; Lambardi, M.; Engelmann, F.; Pathirana, R.; Panis, B.; Volk, G.M.; Wang, Q.C. Advances in cryopreservation of in vitro-derived propagules: Technologies and explant sources. *Plant Cell Tissue Organ Cult.* **2021**, *144*, 7–20. [CrossRef]
12. Benson, E.E. Cryopreservation theory. In *Plant Cryopreservation: A Practical Guide*; Reed, B.M., Ed.; Springer: New York, NY, USA, 2008; pp. 15–32.
13. Reed, B.M. *Plant Cryopreservation: A Practical Guide*; Springer: New York, NY, USA, 2008.
14. Kulus, D.; Zalewska, M. Cryopreservation as a tool used in long-term storage of ornamental species—A review. *Sci. Hortic.* **2014**, *168*, 88–107. [CrossRef]
15. Li, H.H.; Lu, J.L.; Lo, H.E.; Tsai, S.; Lin, C. Effect of cryopreservation on proteins from the ubiquitous marine dinoflagellate *Breviolum* sp. (family symbiodiniaceae). *Plants* **2021**, *10*, 1731. [CrossRef] [PubMed]
16. Ruta, C.; Lambardi, M.; Ozudogru, E.A. Biobanking of vegetable genetic resources by in vitro conservation and cryopreservation. *Biodivers. Conserv.* **2020**, *29*, 3495–3532. [CrossRef]
17. Höfer, M.; Hanke, M.V. Cryopreservation of fruit germplasm. *Vitr. Cell. Dev. Biol.—Plant* **2017**, *53*, 372–381. [CrossRef]
18. Keller, E.R.J.; Senula, A.; Grübe, M.; Diekmann, K.; Dehmer, K.J. Fifteen years of cryopreservation in the IPK genebank—Experience, conclusions and outlook. *Acta Hortic.* **2014**, *1039*, 249–264. [CrossRef]
19. Jenderek, M.M.; Reed, B.M. Cryopreserved storage of clonal germplasm in the USDA National Plant Germplasm System. *Vitr. Cell. Dev. Biol.—Plant* **2017**, *53*, 299–308. [CrossRef]
20. Panis, B.; Nagel, M.; Van den Houwe, I. Challenges and Prospects for the Conservation of Crop Genetic Resources in Field Genebanks, in In Vitro Collections and/or in Liquid Nitrogen. *Plants* **2020**, *9*, 1634. [CrossRef] [PubMed]
21. Pence, V.C. In vitro methods and cryopreservation: Tools for endangered exceptional species preservation and restoration. *Acta Hortic.* **2014**, *1039*, 73–80. [CrossRef]
22. Pence, V.C. Tissue cryopreservation for plant conservation: Potential and challenges. *Int. J. Plant Sci.* **2014**, *175*, 40–45. [CrossRef]
23. Kaviani, B.; Kulus, D. Cryopreservation of endangered ornamental plants and fruit crops from tropical and subtropical regions. *Biology* **2022**, *11*, 847. [CrossRef]
24. Rathwell, R.; Popova, E.; Shukla, M.R.; Saxena, P.K. Development of cryopreservation methods for cherry birch (*Betula lenta* L.), an endangered tree species in Canada. *Can. J. For. Res.* **2016**, *46*, 1284–1292. [CrossRef]
25. Volk, G.M. Collecting pollen for genetic resources conservation. In *Collecting Plant Genetic Diversity: Technical Guidelines—2011 Update*; Guarino, L., Rao, V.R., Reid, R., Eds.; CAB International: Wallingford, UK, 2011; pp. 1–10.
26. Lelu-Walter, M.A.; Thompson, D.; Harvengt, L.; Sanchez, L.; Toribio, M.; Pâques, L.E. Somatic embryogenesis in forestry with a focus on Europe: State-of-the-art, benefits, challenges and future direction. *Tree Genet. Genomes* **2013**, *9*, 883–899. [CrossRef]
27. Engelmann, F. Cryopreservation of Embryos: An Overview. In *Plant Embryo Culture. Methods in Molecular Biology*; Springer: New York, NY, USA, 2011; Volume 710, pp. 155–184.
28. Aronen, T.S.; Krajnakova, J.; Häggman, H.M.; Ryynänen, L.A. Genetic fidelity of cryopreserved embryogenic cultures of open-pollinated *Abies cephalonica*. *Plant Sci.* **1999**, *142*, 163–172. [CrossRef]
29. Krajňáková, J.; Sutela, S.; Aronen, T.; Gömöry, D.; Vianello, A.; Häggman, H. Long-term cryopreservation of Greek fir embryogenic cell lines: Recovery, maturation and genetic fidelity. *Cryobiology* **2011**, *63*, 17–25. [CrossRef] [PubMed]
30. Nörgaard, J.V.; Baldursson, S.; Krogstrup, P. Genotypic differences in the ability of embryogenic Abies nordmanniana cultures to survive cryopreservation. *Silvae Genet.* **1993**, *42*, 93–97.
31. Misson, J.P.; Druart, P.; Panis, B.; Watillon, B. Contribution to the study of the maintenance of somatic embryos of *Abies nordmanniana* LK: Culture media and cryopreservation method. *Propag. Ornam. Plants* **2006**, *6*, 17–23.
32. Krajňáková, J.; Bertolini, A.; Gömöry, D.; Vianello, A.; Häggman, H. Initiation, long-term cryopreservation, and recovery of *Abies alba* Mill. embryogenic cell lines. *Vitr. Cell. Dev. Biol.—Plant* **2013**, *49*, 560–571. [CrossRef]
33. Salaj, T.; Matusikova, I.; Panis, B.; Swennen, R.; Salaj, J. Recovery and characterisation of hybrid firs (*Abies alba* x *A. cephalonica*, *Abies alba* x *A. numidica*) embryogenic tissues after cryopreservation. *CryoLetters* **2010**, *31*, 206–217.
34. Buchner, L.; Eisen, A.K.; Šikoparija, B.; Jochner-Oette, S. Pollen viability of Fraxinus excelsior in storage experiments and investigations on the potential effect of long-range transport. *Forests* **2022**, *13*, 600. [CrossRef]
35. Pinney, K.; Polito, V.S. Olive pollen storage and in vitro germination. *Acta Hortic.* **1990**, *286*, 207–210. [CrossRef]

36. Ferri, A.; Giordani, E.; Padula, G.; Bellini, E. Viability and in vitro germinability of pollen grains of olive cultivars and advanced selections obtained in Italy. *Adv. Hort. Sci.* **2008**, *22*, 116–122.
37. Ren, R.; Li, Z.; Zhang, L.; Zhou, H.; Jiang, X.; Liu, Y. Enzymatic and nonenzymatic antioxidant systems impact the viability of cryopreserved Paeonia suffruticosa pollen. *Plant Cell Tissue Organ Cult.* **2021**, *144*, 233–246. [CrossRef]
38. Murashige, T.; Skoog, F. A revised medium for rapid growth and bioassays with tobacco tissue cultures. *Physiol. Plant* **1962**, *15*, 473–497. [CrossRef]
39. Ayala, L.P.E.; Luna, C.V.; Brugnoli, E.A.; Espasandin, F.D.; Duarte, M.J.; González, A.M.; Gauchat, M.E.; Moncaleán Guillén, P.; Sansberro, P.A. Cryopreservation of mature zygotic embryos, shoot bud regeneration, and field establishment of *Pinus elliottii* var. elliottii x *Pinus caribaea* var. hondurensis in vitro-derived plants. *Trees—Struct. Funct.* **2023**, *37*, 417–433. [CrossRef]
40. Sakai, A.; Kobayashi, S.; Oiyama, I. Cryopreservation of nucellar cells of navel orange (*Citrus sinensis* Osb. var. *brasiliensis Tanaka) by vitrification. Plant Cell Rep.* **1990**, *9*, 30–33. [PubMed]
41. Schenk, R.O.Y.U.; Hildebrandt, A.C. Medium and techniques for induction and growth of monocotyledonous and dicotyledonous plant cell cultures. *Can. J. Bot.* **1972**, *50*, 199–204. [CrossRef]
42. Standardi, A.; Micheli, M. Protocols for Micropropagation of Selected Economically-Important Horticultural Plants. In *Protocols for Micropropagation of Selected Economically-Important Horticultural Plants. Methods in Molecular Biology*; Springer: New York, NY, USA, 2012; Volume 994, pp. 397–418.
43. Owens, J.N.; Blake, M.D. Forest tree Seed Production. In *A Review of the Literature and Recommendations for Future Research*; Petawawa National Forestry Institute: Chalk River, ON, Canada, 1985; Volume PI-X-53.
44. Skrzyszewska, K.; Chłanda, J. A study on the variation of morphological characteristics of silver fir (*Abies alba* Mill.) seeds and their internal structure determined by X-ray radiography in the beskid sâdecki and beskid niski mountain ranges of the Carpathians (southern Poland). *J. For. Sci.* **2009**, *55*, 403–414. [CrossRef]
45. Fedorkov, A. Climatic adaptation of seed maturity in scots pine and Norway spruce populations. *Silva Fenn.* **2001**, *35*, 119–123. [CrossRef]
46. Gagliardi, B.; Marcos-Filho, J. Germination and seed structure assessed by the X-ray test Relationship between germination and bell pepper seed structure assessed by the X-ray test. *Sci. Agric.* **2011**, *68*, 411–416. [CrossRef]
47. Dell'Aquila, A. Development of novel techniques in conditioning, testing and sorting seed physiological quality. *Seed Sci. Technol.* **2009**, *37*, 608–624. [CrossRef]
48. Chavagnat, A.; Bastien, J.C. Determination of the quality of *Cupressus sempervirens* and *Cupressus arizonica* seeds by X-ray radiography. *Seed Sci. Technol.* **1991**, *19*, 139–146.
49. Bye, H.R.; Edvardsen, Ø.M. The use of X-ray for prediction of seed viability in cones" The Norwegian Forest Seed Station. In Proceedings of the ISTA Congress, Budapest, Hungary, 17–19 May 2004.
50. Nygren, M.; Himanen, K.; Ruhanen, H. Viability and germination of Scots pine seeds after freezing of harvested cones in vitro. *Can. J. For. Res.* **2016**, *46*, 1035–1041. [CrossRef]
51. Tuomainen, T.V.; Himanen, K.; Helenius, P.; Kettunen, M.I.; Nissi, M.J. Quantitative magnetic resonance imaging of scots pine seeds and the assessment of germination potential. *Can. J. For. Res.* **2022**, *52*, 685–695. [CrossRef]
52. Arkhipov, M.V.; Priyatkin, N.S.; Gusakova, L.P.; Karamysheva, A.V.; Trofimuk, L.P.; Potrakhov, N.N.; Bessonov, V.B.; Shchukina, P.A. Microfocus X-ray method for detecting hidden defects in seeds of woody forest species and other types of vascular plants. *Tech. Phys.* **2020**, *65*, 324–332. [CrossRef]
53. Bagnell, J.R.; Robert, C. Species distinction among pollen grains of Abies, Picea, and Pinus in the Rocky Mountain area (a scanning electron microscope study). *Rev. Palaeobot. Palynol.* **1975**, *19*, 203–220. [CrossRef]
54. Saito, T.; Tsuchida, K. Pollen morphology of the genus. *Jpn. J. Palynol.* **1992**, *38*, 158–171.
55. Wrońska-Pilarek, D.; Dering, M.; Bocianowski, J.; Lechowicz, K.; Kowalkowski, W.; Barzdajn, W.; Hauke-Kowalska, M. Pollen morphology and variability of *Abies alba* Mill. genotypes from South-Western Poland. *Forests* **2020**, *11*, 1125. [CrossRef]
56. Sidhu, R.K. Pollen storage in vegetable crops: A review. *J. Pharmacogn. Phytochem.* **2019**, *8*, 599–603.
57. Rajasekhraran, P.E. *ICAR Short Course on In Vitro Conservation and Cryopreservation-New Options to Conserve Horticultural Genetic Resources*; Indian Institute of Horticultural Research: Hessaraghatta Lake PO Bangalore, India, 2006; pp. 1–199.
58. Wang, L.; Wu, J.; Chen, J.; Fu, D.; Zhang, C.; Cai, C.; Ou, L. A simple pollen collection, dehydration, and long-term storage method for litchi (*Litchi chinensis* Sonn.). *Sci. Hortic.* **2015**, *188*, 78–83. [CrossRef]
59. Dinato, N.B.; Imaculada Santos, I.R.; Zanotto Vigna, B.B.; de Paula, A.F.; Fávero, A.P. Pollen cryopreservation for plant breeding and genetic resources conservation. *CryoLetters* **2020**, *41*, 115–127.
60. Araújode Oliveira, A.C.; da Silva Lédo, A.; Polek, M.; Krueger, R.; Shepherd, A.; Volk, G.M. Optimization of in vitro germination and cryopreservation conditions for preserving date palm pollen in the USDA National Plant Germplasm System. *Plant Cell Tissue Organ Cult.* **2021**, *144*, 223–232. [CrossRef]
61. Sprague, J.R.; Johnson, V.W. Extraction and storage of loblolly pine (*Pinus taeda*) pollen [Germination]. In Proceedings of the Proceedings-Southern Conference on Forest Tree Improvement (USA), Gainesville, FL, USA, 14–16 June 1977.
62. Connor, K.F.; Towill, L.E. Pollen-handling protocol and hydration/dehydration characteristics of pollen for application to long-term storage. *Euphytica* **1993**, *68*, 77–84. [CrossRef]
63. Lanteri, S.; Belletti, P.; Lotito, S. Storage of pollen of Norway spruce and different pine species. *Silvae Genet.* **1992**, *42*, 104–109.

64. Weatherhead, M.A.; Grout, B.W.W.; Henshaw, G.G. Advantages of storage of potato pollen in liquid nitrogen. *Potato Res.* **1978**, *21*, 331–334. [CrossRef]
65. Anushma, P.L.; Rajasekharan, P.E.; Singh, T.H. Pollen cryobanking in wild solanums for conservation and utilization of nuclear genetic diversity. *Genet. Resour. Crop Evol.* **2023**, 1–11. [CrossRef]
66. Ferri, A.; Giordani, E.; Benelli, C. Viability and in vitro germination rate of pollen grains of *D. kaki* Thunb., *D. lotus* L. and *D. virginiana* L. in relation to storage time and temperatures. *Acta Hortic.* **2013**, *996*, 97–102. [CrossRef]
67. Benelli, C.; Giordani, E.; Bonetti, D.; Nin, S. Effect of storage temperature on pollen viability and in vitro germinability of fruit. *Acta Italus Hortus* **2024**, *40*, 114–118.
68. Gill, S.S.; Tuteja, N. Reactive oxygen species and antioxidant machinery in abiotic stress tolerance in crop plants. *Plant Physiol. Biochem.* **2010**, *48*, 909–930. [CrossRef] [PubMed]
69. Valizadeh-Kamran, R.; Toorchi, M.; Mogadam, M.; Mohammadi, H.; Pessarakli, M. Effects of freeze and cold stress on certain physiological and biochemical traits in sensitive and tolerant barley (*Hordeum vulgare*) genotypes. *J. Plant Nutr.* **2018**, *41*, 102–111. [CrossRef]
70. Balestrasse, K.B.; Tomaro, M.L.; Batlle, A.; Noriega, G.O. The role of 5-aminolevulinic acid in the response to cold stress in soybean plants. *Phytochemistry* **2010**, *71*, 2038–2045. [CrossRef]
71. Taiz, L.; Zeiger, E.; Moller, I.M.; Murphy, A. *Plant Physiology and Development*, 6th ed.; Sinauer Associates, Inc.: Sunderland, MA, USA, 2015.
72. Gong, Y.; Toivonen, P.M.A.; Wiersma, P.A.; Lu, C.; Lau, O.L. Effect of freezing on the activity of catalase in apple flesh tissue. *J. Agric. Food Chem.* **2000**, *48*, 5537–5542. [CrossRef]
73. Poobathy, R.; Sinniah, U.R.; Xavier, R.; Subramaniam, S. Catalase and superoxide dismutase activities and the total protein content of protocorm-like bodies of Dendrobium Sonia-28 subjected to Vitrification. *Appl. Biochem. Biotechnol.* **2013**, *170*, 1066–1079. [CrossRef]
74. Jia, M.X.; Jiang, X.R.; Xu, J.; Di, W.; Shi, Y.; Liu, Y. CAT and MDH improve the germination and alleviate the oxidative stress of cryopreserved Paeonia and Magnolia pollen. *Acta Physiol. Plant.* **2018**, *40*, 37. [CrossRef]
75. Jiang, M.X.; Li, Z.D.; Ren, R.F.; Xu, J.; Li, B.L.; Liu, Y.; Jiang, X.R.; Di, W. MDH and CAT increase the germination of cryopreserved Paeonia pollen by regulating the ROS and apoptosis-like events. *Acta Hortic.* **2019**, *1234*, 105–112. [CrossRef]
76. Xu, J.; Li, B.; Liu, Q.; Shi, Y.; Peng, J.; Jia, M.; Liu, Y. Wide-scale pollen banking of ornamental plants through cryopreservation. *CryoLetters* **2014**, *35*, 312–319.
77. Borghezan, M.; Clauman, A.D.; Steinmacher, D.A.; Guerra, M.P.; Orth, A.I. In vitro viability and preservation of pollen grain of kiwi (*Actinidia chinensis* var. deliciosa (A. Chev.) A. Chev). *Crop Breed. Appl. Biotechnol.* **2011**, *11*, 338–344. [CrossRef]
78. da Silva, R.L.; de Souza, E.H.; de Jesus Vieira, L.; Pelacani, C.R.; Souza, F.V.D. Cryopreservation of pollen of wild pineapple accessions. *Sci. Hortic.* **2017**, *219*, 326–334. [CrossRef]
79. Rajasekharan, P.E.; Rohini, M.R. Pollen cryopreservation: Advances and prospects. In *Pollen Cryopreservation Protocols*; Springer Protocols Handbooks; Rajasekharan, P.E., Rohini, M.R., Eds.; Humana Press: New York, NY, USA, 2023; pp. 1–18.
80. Omura, M.; Akiham, T.A. Pollen preservation of fruit trees for gene banks in Japan. *Plant Genet. Res. Newsl.* **1980**, *43*, 28–31.
81. Shang, X. Studies on Cryopreservation of Peony Pollen. Master's Thesis, Beijing Forestry University Library, Beijing, China, 2005.
82. Ganeshan, S.; Alexander, M. Fertilizing ability of cryopreserved grape, *Vitis vinifera* L. pollen. *Vitis* **1990**, *29*, 145–150.
83. Zhang, Y.-L.; Chen, R.-D.; Huang, C.-J.; Liu, Y. Cryo-banking of *Prunus mume* pollen and its application in cross-breeding. *Cryo Letters* **2009**, *30*, 165–170.
84. Copes, D.L. Fertility of Douglas-fir Pollen After One Year of Storage in Liquid Nitrogen. *For. Sci.* **1985**, *31*, 569–574.
85. Anilkumar, G.S.; Rajasekharan, P.E. Cryopreservation of Sandalwood (*Santalum album* L.) Pollen. In *Pollen Cryopreservation Protocols*; Springer Protocols Handbooks; Rajasekharan, P., Rohini, M., Eds.; Humana Press: New York, NY, USA, 2023; pp. 601–607.
86. Anilkumar, G.S.; Rajasekharan, P.E. Cryopreservation of Pollen for Long-Term Storage in Teak (*Tectona grandis*). In *Pollen Cryopreservation Protocols*; Springer Protocols Handbooks; Rajasekharan, P., Rohini, M., Eds.; Humana Press: New York, NY, USA, 2023; pp. 609–616.
87. Sparks, D.; Yates, I.E. Pecan pollen stored over a decade retains viability. *HortScience* **2002**, *37*, 176–177. [CrossRef]
88. Chaudhury, R.; Malik, S.K.; Rajan, S. An improved pollen collection and cryopreservation method for highly recalcitrant tropical fruit species of mango (*Mangifera indica* L.) and litchi (*Litchi chinensis* Sonn.). *Cryo Letters* **2010**, *31*, 268–278.
89. Batos, B.; Miljković, D. The vitality of the Serbian spruce (*Picea omorika*) pollen during the long-term cryopreservation. *Grana* **2019**, *58*, 433–446. [CrossRef]
90. Franchi, G.G.; Bellani, L.; Nepi, M.; Pacini, E. Types of carbohydrate reserves in pollen: Localization, systematic distribution and ecophysiological significance. *Flora* **1996**, *191*, 143–159. [CrossRef]
91. Nepi, M.; Franchi, G.G.; Pacini, E. Pollen hydration status at dispersal: Cytophysiological features and strategies. *Protoplasma* **2001**, *216*, 171–180. [CrossRef]
92. Pacini, E.; Guarnieri, M.; Nepi, M. Pollen carbohydrates and water content during development, presentation, and dispersal: A short review. *Protoplasma* **2006**, *228*, 73–77. [CrossRef]
93. Corredoira, E.; San-José, M.C.; Ballester, A.; Vieitez, A.M. Cryopreservation of zygotic embryonic axes and somatic embryos of european chestnut. *CryoLetters* **2004**, *25*, 33–42. [PubMed]

94. Gonzalez-Benito, M.E.; Martín, C., III. 9 Cryopreservation of Quercus (Oak) Species. In *Cryopreservation of Plant Germplasm II*; Towill, L.E., Bajaj, Y.P.S., Eds.; Springer: Berlin/Heidelberg, Germany, 2013; pp. 312–322.
95. Camellia, N.A.; Muhaimin, A.K.; Shukri, M.A. Desiccation and cryopreservation of key lime (*Citrus* × *aurantiifolia*) seed and zygotic embryonic axes. *Res. Crop.* **2020**, *21*, 268–275.
96. Wei, Q.; Shi, P.; Khan, F.S.; Htwe, Y.M.; Zhang, D.; Li, Z.; Wei, X.; Yu, Q.; Zhou, K.; Wang, Y. Cryopreservation and cryotolerance mechanism in zygotic embryo and embryogenic callus of oil palm. *Forests* **2023**, *14*, 966. [CrossRef]
97. de Pinto, M.S.; Paiva, R.; da Silva, D.P.C.; Santos, P.A.A.; de Freitas, R.T.; Silva, L.C. Cryopreservation of coffee zygotic embryos: Dehydration and osmotic rehydration. *Cienc. Agrotecnologia* **2016**, *40*, 380–389. [CrossRef]
98. Berjak, P.; Walker, M.; Mycock, D.J.; Wesley-Smith, J.; Watt, P.; Pammenter, N.W. Cryopreservation of recalcitrant zygotic embryos. In *Cryopreservation of Tropical Plant Germplasm*; Engelmann, F., Takagi, H., Eds.; International Plant Genetic Resources Institute: Rome, Italy, 2000; pp. 140–155.
99. van der Walt, K.; Nadarajan, J.; Mathew, L.; Bettoni, J.C.; Souza, J.A. Advances in cryopreservation of Syzygium maire (swamp maire, maire tawake) zygotic embryos, a critically endangered tree species endemic to New Zealand. *Front. Conserv. Sci.* **2023**, *4*, 1269881. [CrossRef]
100. Häggman, H.M.; Aronen, T.S.; Ryynänen, L.A. Cryopreservation of Embryogenic Cultures of Conifers. In *Somatic Embryogenesis in Woody Plants*; Springer: Dordrecht, The Netherlands, 2000.
101. Ballesteros, D.; Martínez, M.T.; Sánchez-Romero, C.; Montalbán, I.A.; Sales, E.; Moncaleán, P.; Arrillaga, I.; Corredoira, E. Current status of the cryopreservation of embryogenic material of woody species. *Front. Plant Sci.* **2024**, *14*, 1337152. [CrossRef] [PubMed]
102. Metwali, E.M.R.; Kadasa, N.M.S.; Soliman, H.I.A.; Almaghrabi, O.A.; Fuller, M.P. Cryopreservation of embryogenic callus of date palm (*Phoenix dactylifera*) cv. Magdoul through encapsulation-dehydration technology. *Int. J. Agric. Biol.* **2020**, *24*, 1762.
103. Fernandes, P.; Rodriguez, E.; Pinto, G.; Roldán-Ruiz, I.; De Loose, M.; Santos, C. Cryopreservation of *Quercus suber* somatic embryos by encapsulation-dehydration and evaluation of genetic stability. *Tree Physiol.* **2008**, *28*, 1841–1850. [CrossRef]
104. Marco-Medina, A.; Casas, J.L.; Elena González-Benito, M. Comparison of vitrification and encapsulation-dehydration for cryopreservation of Thymus moroderi shoot tips. *CryoLetters* **2010**, *31*, 301–309. [PubMed]
105. Lineros, Y.; Balocchi, C.; Muñoz, X.; Sánchez, M.; Ríos, D. Cryopreservation of *Pinus radiata* embryogenic tissue: Effects of cryoprotective pretreatments on maturation ability. *Plant Cell Tissue Organ Cult.* **2018**, *135*, 357–366. [CrossRef]
106. Volk, G.M.; Harris, J.L.; Rotindo, K.E. Survival of mint shoot tips after exposure to cryoprotectant solution components. *Cryobiology* **2006**, *52*, 305–308. [CrossRef]
107. Salaj, T.; Panis, B.; Klubicova, K.; Salaj, J. Cryopreservation of *Abies alba* embryogenic tissues by slow-freezing method. *Not. Bot. Horti Agrobot. Cluj-Napoca* **2022**, *50*, 12770. [CrossRef]
108. Ma, M.; Wang, X.; Zhang, C.; Pak, S.; Wu, H.; Yang, J.; Li, C. Enhancing the cryopreservation system of larch embryogenic culture by optimizing pre-culture, osmoprotectants, and rapid thawing. *Forests* **2023**, *14*, 1621. [CrossRef]
109. Hazubska-Przybył, T.; Wawrzyniak, M.K.; Obarska, A.; Salaj, T. Cryopreservation of *Abies alba* × *A. numidica* and *Pinus nigra* embryogenic tissues by stepwise dehydration method. *Plant Methods* **2024**, *20*, 10. [CrossRef]
110. Latutrie, M.; Aronen, T. Long-term cryopreservation of embryogenic *Pinus sylvestris* cultures. *Scand. J. For. Res.* **2013**, *28*, 103–109. [CrossRef]

plants

MDPI

Article

In Vitro Conservation of Mexican Garlic Varieties by Minimal Growth

Bibiana Tirado [1], Víctor Manuel Gómez-Rodríguez [1], Carlos Iván Cruz-Cárdenas [2], Lily Xochilt Zelaya-Molina [2], Humberto Ramírez-Vega [1] and Gabriela Sandoval-Cancino [2,*]

[1] Centro Universitario de los Altos, Universidad de Guadalajara, Av. Rafael Casillas Aceves 1200, Tepatitlán de Morelos 47620, Jal, Mexico; bibiana.tirado9313@alumnos.udg.mx (B.T.); victor.gomez@cualtos.udg.mx (V.M.G.-R.); humberto.rvega@academicos.udg.mx (H.R.-V.)

[2] Centro Nacional de Recursos Genéticos-INIFAP, Av. Boulevard de la Biodiversidad 400, Tepatitlán de Morelos 47600, Jal, Mexico; cruz.ivan@inifap.gob.mx (C.I.C.-C.); zelaya.lily@inifap.gob.mx (L.X.Z.-M.)

* Correspondence: sandoval.gabriela@inifap.gob.mx

Abstract: Garlic (*Allium sativum* L.) is one of the 30 crops that are essential for world food; therefore, its conservation should be considered a priority. There are two main plant conservation strategies: in situ and ex situ conservation. Both strategies are important; nevertheless, ex situ field conservation is affected by biotic and abiotic factors. A complementary strategy to preserve garlic germplasm in the medium term is through in vitro culture by minimal growth. The aim of this study was to evaluate the in vitro conservation of three Mexican garlic varieties by minimal growth. Garlic plants obtained from in vitro garlic bulbs were preserved in six culture media at 25, 18, and 5 °C. A randomized design was used and an analysis of the variance of the survival, contamination, and shoot height of the explants was performed at 30, 60, 90, 180, 270, and 365 days of culture. The results showed that the in vitro conservation of Pebeco, Tacátzcuaro Especial, and Huerteño garlic varieties was optimally obtained for one year at 5 °C in a basal Murashige and Skoog (MS) culture medium with 68.46 g L^{-1} sucrose and 36.43 g L^{-1} sorbitol. Thus, the achieved protocol can be adapted to other varieties of garlic for medium-term storage in germplasm banks.

Keywords: garlic; genetic resources; in vitro conservation; osmotic agents

Citation: Tirado, B.; Gómez-Rodríguez, V.M.; Cruz-Cárdenas, C.I.; Zelaya-Molina, L.X.; Ramírez-Vega, H.; Sandoval-Cancino, G. In Vitro Conservation of Mexican Garlic Varieties by Minimal Growth. *Plants* **2023**, *12*, 3929. https://doi.org/10.3390/plants12233929

Academic Editors: Anna De Carlo and Waed Tarraf

Received: 19 September 2023
Revised: 9 November 2023
Accepted: 13 November 2023
Published: 22 November 2023

1. Introduction

Garlic (*Allium sativum* L.) is a vegetable native to Central Asia and is one of the most important species of the *Allium* genus due to its particular taste and smell, as well as its beneficial properties for human health [1]. Garlic bulbs contain organosulforates, flavanols, saponins, and sapogenins—substances with great potential in medicine and agriculture [2,3]. These compounds have antimicrobial, antioxidant, and flavor-enhancer effects, among others [4].

There are 910 varieties of garlic grown in 35 countries [5], but some of these are only conserved under field conditions and, therefore, are at risk of loss due to biotic and abiotic factors, even during post-harvest [6]. For that reason, it is important to carry out garlic preservation and its genetic resources [7].

Germplasm banks are essential to preserve the genetic diversity of species of agri-food importance, such as garlic. There are 1750 germplasm banks around the world that preserve different species in medium-term and long-term conservation, including elite and local varieties and wild relatives [8]. The preservation of genetic resources is stipulated in international treaties such as the Convention on Biological Diversity and the International Treaty on Plant Genetic Resources for Food and Agriculture [9].

Conservation of sub-orthodox and recalcitrant seeds and vegetatively propagated species, such as garlic, is carried out by in vitro culture techniques [10]. In vitro plant culture enables cells, tissues, and organs to grow in aseptic culture media [11].

Minimal growth is an in vitro culture technique for medium-term plant conservation, from a few months to more than a year, according to the species [12]. Plant growth is slowed by chemical and physical factors. Some of these factors are the decrease of culture medium nutrients, the addition of osmotic agents, growth retardants, mineral oils, low conservation temperature, low light intensity, and shorter photoperiod [13].

This medium-term plant conservation technique has been applied to different species using specific protocols for each one and even for each genotype due to genetic diversity [14]. Regarding garlic, Benke et al. [15] reported 70% survival of the Bhima variety after five months in Murashige and Skoog (MS) culture medium supplemented with sucrose (68.46 g L^{-1}) and sorbitol (36.43 g L^{-1}). Likewise, Pardo et al. [16] obtained 73% survival of the Boconó clone after seven months of storage in MS medium (25%) containing 45 g L^{-1} of sucrose. Both investigations were performed under standard light and temperature conditions (16 h light photoperiod, 25 μm m^{-2} s^{-1}, and 25 ± 1 °C). On the other hand, Hassan et al. [17] preserved Balady and Seds 40 varieties for 15 months at 4 °C in MS medium with sorbitol (72.87 g L^{-1}) in dark conditions and obtained survival rates of 35.7% and 90% for each variety, respectively.

Different garlic varieties and their wild relatives are in medium-term and long-term preservation worldwide [18]. However, in most cases, there is a cultivar- or genotype-specific response. In Mexico, there are garlic varieties of agri-food and breeding importance for which there are no in vitro conservation protocols, and they are not stored in a germplasm bank. In this sense, it is vital to generate in vitro conservation protocols because the current conservation of garlic germplasm depends on field regeneration and noncontrolled storage conditions where it is exposed to biotic and abiotic threats, such as unfavorable environmental conditions, pillage, pests, and diseases. Minimal growth provides a medium-term in vitro conservation alternative that allows to safely preserve important germplasm under controlled conditions, efficiently using space and resources. Due to this, the objective of this work was the in vitro preservation of three Mexican garlic varieties by minimal growth.

2. Results

2.1. Shoot Height

In the case of Pebeco's explant shoot height, significant differences related to temperature were observed at all periods except at 90 days ($p < 0.05$). In culture media, significant differences were present in this variable at 30, 60, and 365 days ($p < 0.05$). At 365 days, the explant length was shorter in culture medium M1 formulated with Dunstan and Short (BDS) salts and 100 g L^{-1} sucrose (67.54 mm) (Table 1).

On the other hand, significant differences in shoot height of Tacátzcuaro Especial related to conservation temperature were observed at all cultivation periods ($p < 0.05$). Specifically, at 365 days, a shorter length of 89.17 mm was recorded at 25 °C. Also, the effect of culture media was present with reports of significant differences at all cultivation periods ($p < 0.05$). Less growth (78.61 mm) was obtained in culture medium M1 at 365 days (Table 2).

For Huerteño's, significant differences in shoot height between the tested temperatures were observed at 30, 60, and 90 days with reports of less growth at 5 °C ($p < 0.05$). Also, significant differences in culture media were noted, particularly a shorter length of 49.41 mm in culture medium M1 was recorded at 365 days (Table 3).

Table 1. Effect of storage temperature and culture media on shoot height (mm) of the Pebeco garlic variety.

	Days					
Temperature	**30**	**60**	**90**	**180**	**270**	**365**
5 °C	21.08 ± 1.04 b	55.04 ± 4.59 b	75.04 ± 6.33 a	106.37 ± 5.95 a	108.28 ± 6.02 a	106.47 ± 6.23 a
18 °C	59.10 ± 4.42 a	81.39 ± 5.83 a	82.69 ± 5.94 a	82.79 ± 6.12 b	81.94 ± 6.20 b	81.97 ± 6.20 b
25 °C	58.50 ± 6.66 a	71.29 ± 7.79 ab	65.70 ± 7.00 a	69.11 ± 7.46 b	68.99 ± 7.41 b	69.49 ± 7.60 b
Culture Medium						
M1: BDS + 100.0 g L^{-1} suc	40.37 ± 4.33 abc	57.74 ± 5.41 b	61.60 ± 5.57 a	71.52 ± 8.04 a	69.53 ± 7.86 a	67.54 ± 7.24 b
M2: MS (25%) + 45.0 g L^{-1} suc	48.66 ± 7.16 abc	64.56 ± 6.44 ab	72.45 ± 7.18 a	81.45 ± 8.79 a	76.54 ± 8.10 a	73.96 ± 7.97 ab
M3: MS + 15.0 g L^{-1} mannitol + 15.0 g L^{-1} suc	37.45 ± 5.35 bc	62.91 ± 8.06 ab	70.69 ± 8.83 a	85.44 ± 10.79 a	86.08 ± 10.99 a	86.77 ± 11.14 ab
M4: MS + 72.87 g L^{-1} sorbitol	31.60 ± 5.34 c	60.18 ± 10.76 ab	69.53 ± 11.20 a	88.41 ± 10.74 a	95.60 ± 11.26 a	96.85 ± 11.55 ab
M5: MS + 68.46 g L^{-1} suc + 36.43 g L^{-1} sorbitol	61.80 ± 10.05 a	94.47 ± 9.81 a	91.20 ± 8.70 a	106.41 ± 7.73 a	106.67 ± 7.88 a	105.95 ± 8.08 a
M6: MS + 30.0 g L^{-1} suc	57.47 ± 10.49 ab	75.58 ± 10.72 ab	81.38 ± 11.66 a	83.32 ± 11.84 a	84.00 ± 11.79 a	84.80 ± 11.90 ab

Values of mean ± SE with the same letters per column do not present a significant difference (Tukey $p < 0.05$). MS: Murashige and Skoog [19], BDS: Dunstan and Short [20], suc: sucrose. For each temperature, the value is the media obtained from all six culture media, while for each culture medium, the value is the media obtained from the three tested temperatures.

Table 2. Effect of storage temperature and culture media on shoot height (mm) of Tacátzcuaro Especial garlic variety.

	Days					
Temperature	**30**	**60**	**90**	**180**	**270**	**365**
5 °C	22.37 ± 1.03 b	50.30 ± 3.49 c	78.73 ± 5.16 b	109.36 ± 5.05 a	115.20 ± 4.98 a	115.85 ± 5.06 a
18 °C	88.22 ± 4.98 a	116.09 ± 3.65 a	112.76 ± 4.43 a	115.54 ± 3.49 a	116.86 ± 3.57 a	116.57 ± 3.54 a
25 °C	79.95 ± 6.52 a	92.05 ± 6.48 b	89.50 ± 6.35 b	90.36 ± 6.82 b	87.13 ± 6.66b	89.17 ± 6.70 b
Culture Medium						
M1: BDS + 100.0 g L^{-1} suc	57.46 ± 8.26 b	69.63 ± 7.85 b	75.32 ± 7.83 c	79.58 ± 7.61 b	77.27 ± 7.65 b	78.61 ± 7.35 b
M2: MS (25%) + 45.0 g L^{-1} suc	74.82 ± 9.29 ab	89.12 ± 7.75 ab	99.16 ± 6.15 abc	112.70 ± 6.68 a	113.25 ± 5.88 a	112.92 ± 5.90 a
M3: MS + 15.0 g L^{-1} mannitol + 15.0 g L^{-1} suc	54.23 ± 7.73 bc	80.19 ± 9.22 ab	87.21 ± 8.52 abc	98.98 ± 8.99 ab	100.18 ± 9.14 ab	101.29 ± 9.27 ab
M4: MS + 72.87 g L^{-1} sorbitol	34.77 ± 4.99 c	73.63 ± 10.13 b	80.66 ± 9.40 bc	98.58 ± 7.85 ab	106.71 ± 8.31 a	107.42 ± 8.42 a
M5: MS + 68.46 g L^{-1} suc + 36.43 g L^{-1} sorbitol	80.96 ± 9.50 a	102.02 ± 6.54 a	106.66 ± 6.07 ab	123.00 ± 3.21 a	124.06 ± 3.14 a	124.66 ± 3.12 a
M6: MS + 30.0 g L^{-1} suc	78.84 ± 10.19 a	102.27 ± 8.44 a	112.97 ± 7.62 a	117.67 ± 7.71 a	116.91 ± 7.89 a	118.29 ± 7.91 a

Values of mean ± SE with the same letters per column do not present a significant difference (Tukey $p < 0.05$). MS: Murashige and Skoog [19], BDS: Dunstan and Short [20], suc: sucrose. For each temperature, the value is the media obtained from all six-culture media, while for each culture medium, the value is the media obtained from the three tested temperatures.

Table 3. Effect of storage temperature and culture media on shoot height (mm) of Huerteño garlic variety.

	Days					
Temperature	**30**	**60**	**90**	**180**	**270**	**365**
5 °C	23.26 ± 1.61b	40.13 ± 3.47 b	67.44 ± 6.18 b	89.23 ± 7.62 a	88.26 ± 7.47 a	87.96 ± 7.56 a
18 °C	69.81 ± 7.23 a	101.32 ± 27.29 a	75.64 ± 7.53 ab	73.42 ± 7.65 a	74.66 ± 7.48 a	74.18 ± 7.45 a
25 °C	87.87 ± 6.82 a	92.62 ± 6.68 ab	90.21 ± 6.55 a	89.99 ± 6.73 a	87.19 ± 6.59 a	88.25 ± 6.79 a
Culture Medium						
M1: BDS + 100.0 g L^{-1} suc	39.62 ± 7.77 c	44.45 ± 7.75 b	47.85 ± 7.65 b	51.80 ± 8.93 b	49.24 ± 8.31 b	49.41 ± 8.43 b
M2: MS (25%) + 45.0 g L^{-1} suc	61.03 ± 9.01 abc	69.07 ± 8.93 ab	77.04 ± 8.90 ab	86.85 ± 9.89 ab	84.71 ± 9.58 ab	83.20 ± 9.75 ab
M3: MS + 15.0 g L^{-1} mannitol + 15.0 g L^{-1} suc	50.95 ± 9.64 bc	59.20 ± 9.65 ab	64.87 ± 9.72 ab	76.88 ± 10.95 ab	74.85 ± 10.67 ab	75.42 ± 10.91 ab
M4: MS + 72.87 g L^{-1} sorbitol	52.57 ± 9.39 bc	69.38 ± 10.36 ab	80.28 ± 9.61 ab	90.46 ± 10.45 a	89.77 ± 10.21 a	91.25 ± 10.37 a
M5: MS + 68.46 g L^{-1} suc + 36.43 g L^{-1} sorbitol	72.65 ± 11.06 ab	84.42 ± 10.37 ab	97.56 ± 10.30 a	94.04 ± 10.97 a	94.69 ± 10.20 a	95.16 ± 10.40 a
M6: MS + 30.0 g L^{-1} suc	85.06 ± 10.23 a	141.62 ± 52.62 a	98.99 ± 8.53 a	105.26 ± 8.76 a	106.95 ± 8.69 a	106.33 ± 8.63 a

Values of mean ± SE with the same letters per column do not present a significant difference (Tukey $p < 0.05$). MS: Murashige and Skoog [19], BDS: Dunstan and Short [20], suc: sucrose. For each temperature, the value is the media obtained from all six culture media, while for each culture medium, the value is the media obtained from the three tested temperatures.

In addition, Figure 1 shows the shoot height development of the three garlic varieties studied during 365 days under minimal growth conditions, where the shoots of Tacátzcuaro Especial showed greater height over time compared with Pebeco and Huerteño.

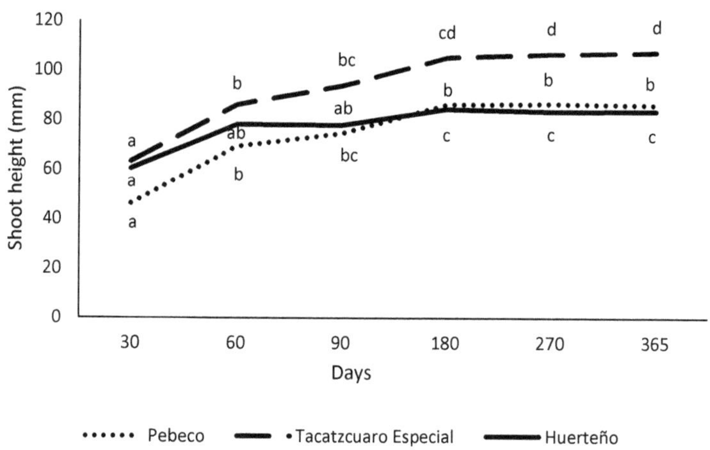

Figure 1. Shoot height development of three garlic varieties during 365 days under minimal growth conditions. The mean values in each evaluation period for each variety that shares the same letter did not differ significantly ($p < 0.05$) according to the Tukey HSD test.

2.2. Survival and Contamination

In the case of Pebeco´s explant survival and contamination, significant differences ($p < 0.05$) linked to temperature and culture medium were observed. At 365 days, higher survival was observed at 5 °C (72.2%) and in culture medium M5 (66.7%). At the same evaluation period, there was less contamination in explants incubated at 5 °C (47.2%) and planted in M4 culture medium (38.9%) (Table 4).

Table 4. Survival and contamination (%) of three garlic varieties at 365 days under minimal growth conditions.

	Pebeco		Tacátzcuaro Especial		Huerteño	
Temperature	**Survival**	**Contam.**	**Survival**	**Contam.**	**Survival**	**Contam.**
5 °C	72.2 ± 7.6 a	47.2 ± 8.4 c	95.8 ± 2.9 a	14.6 ± 5.1 a	58.3 ± 7.2 a	43.8 ± 7.2 b
18 °C	13.9 ± 5.8 b	100.0 ± 0.0 a	81.3 ± 5.7 ab	27.1 ± 6.5 a	33.3 ± 6.9 b	81.3 ± 5.7 a
25 °C	22.2 ± 7.0 b	75.0 ± 7.3 b	68.1 ± 6.9 b	29.8 ± 6.7 a	25.0 ± 6.3 b	43.8 ± 7.2 b
Culture Medium						
M1: BDS + 100.0 g L^{-1} suc	22.2 ± 10.1 b	83.3 ± 9.0 a	75.0 ± 9.0 a	33.3 ± 9.9 a	16.7 ± 7.8 a	54.2 ± 10.4 a
M2: MS (25%) + 45.0 g L^{-1} suc	16.7 ± 9.0 b	88.9 ± 7.62 a	88.0 ± 6.9 a	29.1 ± 9.5 a	33.3 ± 9.8 a	66.7 ± 9.8 a
M3: MS + 15.0 g L^{-1} mannitol + 15.0 g L^{-1} suc	22.2 ± 10.1 b	83.3 ± 9.0 a	79.2 ± 8.5 a	16.7 ± 7.8 a	33.3 ± 9.8 a	62.5 ± 10.1 a
M4: MS + 72.87 g L^{-1} sorbitol	44.4 ± 12.1 ab	38.9 ± 11.8 b	75.0 ± 9.0 a	20.8 ± 8.5 a	54.2 ± 10.4 a	50.0 ± 10.4 a
M5: MS + 68.46 g L^{-1} suc + 36.43 g L^{-1} sorbitol	66.7 ± 11.4 a	72.2 ± 10.9 ab	91.3 ± 6.0 a	26.1 ± 9.4 a	54.2 ± 10.4 a	54.2 ± 10.4 a
M6: MS + 30.0 g L^{-1} suc	44.4 ± 12.1 ab	77.8 ± 10.1 a	83.3 ± 7.8 a	16.7 ± 7.8 a	41.7 ± 10.3 a	50.0 ± 10.4 a

Values of mean ± SE with the same letters per column do not present a significant difference (Tukey $p < 0.05$). MS: Murashige and Skoog [19], BDS: Dunstan and Short [20], suc: sucrose, contam.: contamination. For each temperature, the value is the media obtained from all six culture media, while for each culture medium, the value is the media obtained from the three tested temperatures.

The survival of Tacátzcuaro Especial´s explants showed only significant differences between temperature treatments after 365 days of conservation with reports of a greater rate (95.8%) at 5 °C. In culture media, significant differences were also observed in this variable at all culture periods, except after 365 days ($p < 0.05$). At 365 days, the survival range was 75 to 91% in different culture media (Table 4). Regarding Tacátzcuaro Especial contamination, significant differences linked to storage temperature were observed only at 180 and 270 days ($p < 0.05$). Contamination at 365 days showed no significant differences at 5, 18, and 25 °C (14.6, 27.1, and 29.8%, respectively), although less contamination was present at 5 °C. In culture media, no significant differences were observed for this variable at all periods. At 365 days, the contamination ranged between 16.7 and 33.3% in different culture media, and explants in culture media M3 and M6 showed less contamination (~17%) (Table 4).

Significant differences in the survival of Huerteño's explants were observed at different temperatures after 365 days. Higher survival of 58.3% was reached at 5 °C compared with 18 and 25 °C (33.3 and 25.0%, respectively). In culture media, significant differences were noted only at 180 days, which coincides with a significant survival reduction of Huerteño's explants (Figure 2). At 365 days, the plant survival range in different culture media was from 16.7 to 54.2% (Table 4).

In the case of shoot contamination, significant differences related to conservation temperature were observed for all cultivation periods of Huerteño plants. At 365 days, less contamination was reported at 5 and 25 °C (43.8%) while higher contamination of 81.3% was reported at 18 °C. In culture media, contamination observed in different culture media was between 50.0 and 66.7% after 365 days of storage under minimal growth conditions (Table 4).

The change in the survival of the three Mexican garlic varieties over time was studied (Figure 2). TacátzcuaroEspecial's survival was higher at all tested periods compared with the other varieties. Although the obtained results were different between varieties, the survival was constant until culture day 180, when the survival of Pebeco and Huerteño varieties significantly decreased regardless of the temperature and culture medium used.

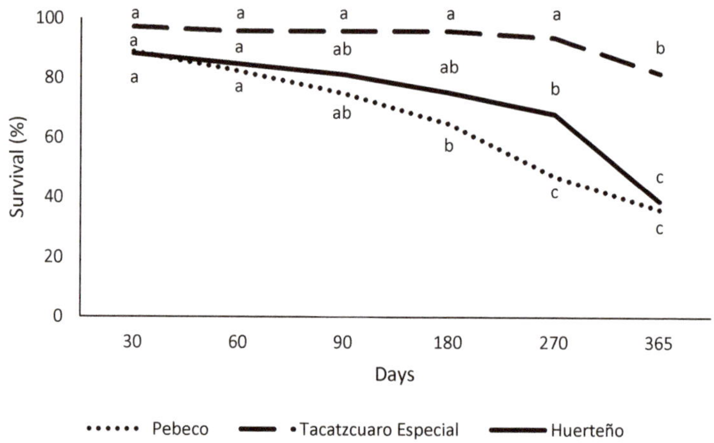

Figure 2. Survival development of three garlic varieties during 365 days under minimal growth conditions. The mean values in each evaluation period for each variety that shares the same letter did not differ significantly ($p < 0.05$) according to the Tukey HSD test.

On the other hand, Figure 3 shows general aspects of the minimal growth morphological effect on three garlic varieties in six culture media at 25, 18, and 5 °C after 30 days of culture. It was notable that plants at 5 °C remained smaller compared with those conserved at the other temperatures.

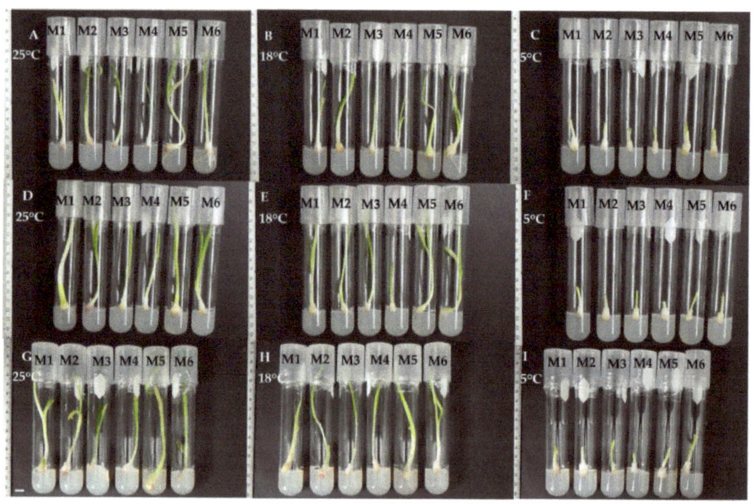

Figure 3. General aspects of minimal growth of garlic in six culture media at 30 days of culture at 25, 18, and 5 °C. (**A–C**) Pebeco; (**D–F**) Tacátzcuaro Especial; (**G–I**) Huerteño. M1: BDS + 100.0 g L^{-1} suc; M2: MS (25%) + 45.0 g L^{-1} suc; M3: MS + 15.0 g L^{-1} mannitol + 15.0 g L^{-1} suc; M4: MS + 72.87 g L^{-1} sorbitol; M5: MS + 68.46 g L^{-1} suc + 36.43 g L^{-1} sorbitol; M6: MS + 30.0 g L^{-1} suc. MS: Murashige and Skoog [19], BDS: Dunstan and Short [20], suc: sucrose. The bar indicates 10 mm.

The in vitro garlic plants of three varieties were kept under minimal growth conditions (six culture media stored at 25, 18, and 5 °C) for 365 days. After this period, better morphological characteristics were observed in plants incubated at 5 °C, such as a bright green color, adequate tissue turgency, and in some cases, bulbification (Figure 4).

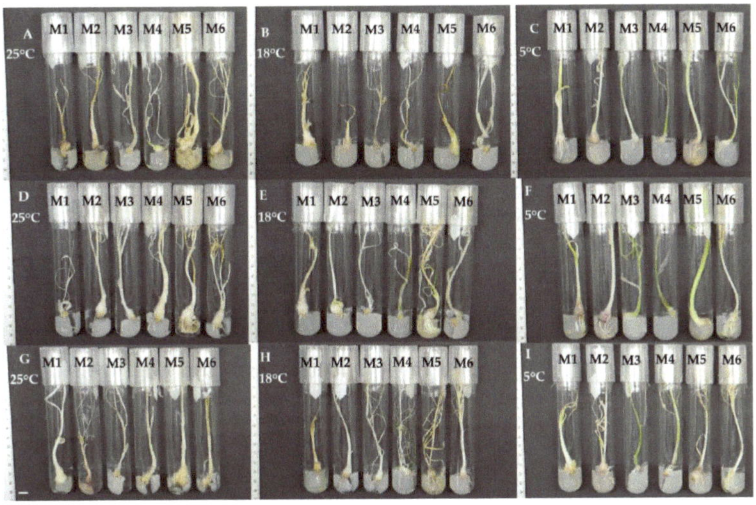

Figure 4. General aspects of minimal growth of garlic in six culture media at 365 days of culture at 25, 18, and 5 °C. (**A–C**) Pebeco; (**D–F**) Tacátzcuaro Especial; (**G–I**) Huerteño. M1: BDS + 100.0 g L^{-1} suc; M2: MS (25%) + 45.0 g L^{-1} suc; M3: MS + 15.0 g L^{-1} mannitol + 15.0 g L^{-1} suc; M4: MS + 72.87 g L^{-1} sorbitol; M5: MS + 68.46 g L^{-1} suc + 36.43 g L^{-1} sorbitol; M6: MS + 30.0 g L^{-1} suc. MS: Murashige and Skoog [19], BDS: Dunstan and Short [20], suc: sucrose. The bar indicates 10 mm.

2.3. Regeneration and Recovery

After 365 days of conservation at 25, 18, and 5 °C in six different culture media, plants stored at 5 °C showed a higher regeneration percentage and recovered their normal growth (Table 5, Figure 5). Specifically, the garlic variety Tacátzcuaro Especial had the highest values of regeneration at 5 °C (Table 5). In the case of all three garlic varieties, regenerated aseptic plants were sufficient as starting material for a future micropropagation step.

Table 5. Plant regeneration of different garlic varieties explants after 365 days under minimal growth conditions.

	Pebeco		Tacátzcuaro Especial		Huerteño	
Temperature	**Plant Regen. (%)**	**Number of RegeneratedPlants**	**Plant Regen. (%)**	**Number of RegeneratedPlants**	**Plant Regen. (%)**	**Number of RegeneratedPlants**
5 °C	44.4 ± 8.4 a	16	60.4 ± 7.1 a	29	33.3 ± 6.9 a	16
18 °C	0 ± 0 b	0	14.6 ± 5.1 b	7	2.1 ± 2.1 b	1
25 °C	5.5 ± 3.9 b	2	2.1 ± 2.1 b	1	0 ± 0 b	0
Culture Medium						
M1: BDS + 100.0 g L^{-1} suc	16.7 ± 9.0 a	3	20.8 ± 8.4 ab	5	8.3 ± 5.8 a	2
M2: MS (25%) + 45.0 g L^{-1} suc	11.1 ± 7.6 a	2	41.7 ± 10.3 a	10	16.7 ± 7.8 a	4
M3: MS + 15.0 g L^{-1} mannitol + 15.0 g L^{-1} suc	16.7 ± 9.03 a	3	29.1 ± 9.5 ab	7	12.6 ± 6.9 a	3
M4: MS + 72.87 g L^{-1} sorbitol	22.2 ± 10.0 a	4	20.8 ± 8.5 ab	5	12.6 ± 6.9 a	3
M5: MS + 68.46 g L^{-1} suc + 36.43 g L^{-1} sorbitol	16.7 ± 9.0 a	3	39.1 ± 10.4 ab	9	12.6 ± 6.9 a	3
M6: MS + 30.0 g L^{-1} suc	16.7 ± 9.0 a	3	4.1 ± 4.1 b	1	8.3 ± 0.8 a	2

Values of mean ± SE with the same letters per column do not present a significant difference (Tukey $p < 0.05$). MS: Murashige and Skoog [19], BDS: Dunstan and Short [20], suc: sucrose, regen.: regeneration. For each temperature, the value is the media obtained from all six culture media, while for each culture medium, the value is the media obtained from the three tested temperatures.

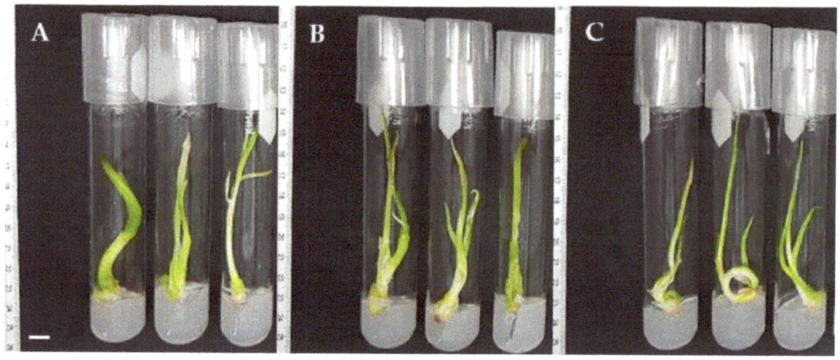

Figure 5. Garlic plants after minimal growth recovery phase (Murashige and Skoog medium with 30 g L^{-1} of sucrose and incubated at 25 °C). (**A**) Pebeco, (**B**) Tacátzcuaro Especial and (**C**) Huerteño. The bar indicates 10 mm.

3. Discussion

Preservation of plant genetic resources is essential to guarantee the well-being of the world's population with increasing agri-food needs [21]. Also, plant genetic resource conservation contributes to ensuring agricultural and food security for many generations, either, through the use of these resources, to restore specific plant populations or to perform genetic improvement and generate varieties better adapted to the challenges ahead [22].

In this sense, genetic diversity preserved in gene banks must be immediately available [23]. Therefore, minimal growth is an in vitro conservation technique that facilitates the availability, accessibility, and multiplication of plant germplasm [18].

In this study, 5 °C allowed the plant's conservation time to reach up to 365 days in the Pebeco, Tacátzcuaro Especial, and Huerteño varieties. These results are the first report of a medium-term conservation protocol for these varieties. Cold is one of the main factors that slows plant growth and development [24]. Low temperatures reduce the photosynthetic rate; however, in some cold-sensitive species, they cause depolymerization of chloroplast microtubules [25].

The conservation period of these three varieties in the present study was longer than reported by Benke et al. [15] who preserved the garlic variety Bhima for five months and obtained similar survival with the same culture medium (M5) under standard incubation conditions. Although the M5 culture medium formulated with basal MS medium, sucrose, and sorbitol did not limit the growth of the explants, it favored their survival.

On the other hand, sucrose, sorbitol, and sugars, in general, at high concentrations can cause osmotic stress in plant cells [26]. Osmotic stress affects cell division, morphogenesis, and survival [27]. Gelmesa et al. [28] reported that 0.2 M sorbitol reduced in vitro growth to different degrees in three genotypes of potato (*Solanum tuberosum* L.). In this study, most garlic plants formed bulbs at the three conservation temperatures and mainly at M1 culture medium (100 g L^{-1} sucrose), probably due to a plant survival mechanism strategy induced by osmotic stress. For instance, Alexopoulos and colleagues [29] found that a higher concentration of 8% sucrose in the MS culture medium contributed to enhanced in vitro bulb production in sea lilies (*Pancratium maritimum*). Similarly, in a separate study, the inclusion of 7% sucrose in the MS culture medium was shown to promote bulb formation in elephant garlic (*Allium ampeloprasum* L.) [30]. Moreover, in another investigation conducted by Youssef et al. [31], the use of 9% and 12% sucrose in the MS culture medium resulted in a notable increase in bulb formation for lilies (*Lilium orientalis* cv. "Starfighter").

Other factors that influence growth and survival during in vitro culture are the explant type, its phenological state, the size and volume of the culture container, and the volume of culture media used [32]. In this sense, during medium-term in vitro conservation, one of the most important variables is shoot height since it is one of the variables that allows choosing the best conditions to extend subculture time [33]. In this study, garlic plants showed their maximum growth in a relatively short period at 25 °C compared with other cultivated species like sweet potato and native species like *Alcantarea nahoumii* (Bromeliaceae) (Leme) J. R. Grant, which were conserved for 4 and even 24 months without significant growth, respectively [34,35]. Nevertheless, shoot growth did stabilize in time, which indicates that all variables and their effects on variety conservation should be considered and evaluated during an appropriate period. Although low temperatures did not limit initial rapid plant growth, they did extend subculture time with higher survival rates in all cases with better morphological characteristics.

Another aspect to take into account when establishing a conservation protocol such as minimal growth is the effect of genotype under specific in vitro conditions. In the present study, different responses can be observed between the three garlic varieties evaluated, whose survival ranges oscillate between 95.8 and 54.2%. This coincides with what was found by Benke et al. [15] who obtained survival percentages between 20 and 70 depending on the garlic genotype evaluated. Therefore, it is necessary to perform the necessary research in order to generate species-specific and sometimes, as in the case of garlic, variety-specific conservation protocols.

One of the main obstacles in in vitro culture is endogenous contamination of the explant [36]. Bacterial growth in the initial stages of in vitro culture is frequent; however, bacterial contamination has also been observed during the multiplication and acclimatization stages [37].

In this study, more bacterial contamination was observed at 18 °C in culture medium M2 (MS medium at 25% supplemented with 45 g L^{-1} sucrose). This could be attributed to the fact that the optimum temperature range for the development of the garlic plant is 12 to 24 °C, and this could also benefit the growth of the natural microbiota of garlic [38].

Regarding culture media, sucrose is the main carbon source in plants in vitro [39]. In case of minimal growth, over time the carbon source is gradually reduced in the culture medium, and the remaining components in it concentrate due to medium dehydration caused by water evaporation. These phenomena cause greater plant stress due to the eventual change in culture medium components availability and osmotic potential. Osmotic stress could be the reason for a greater presence of endogenous garlic microbiota over time. Also in bacteria, sucrose operons are expressed in the presence of sucrose and when other preferred carbon sources are depleted [40].

On the other hand, endophytic microorganisms colonize the vascular tissues of plants and, in most cases, survive surface chemical disinfection [41] and, therefore, can appear months after in vitro establishment, as is the case in the present study. These microorganisms can have a negative or positive effect on plants. In this sense, a bacteriostatic effect has been observed in endophytes isolated from garlic bulbs [42], and endophytic bacteria from garlic roots promote plant growth in vitro [43]. Furthermore, garlic bulbs and garlic oil have different components with antifungal effects [3]. This could be one of the reasons why only major bacterial contamination was observed in two of the three varieties evaluated. Likewise, this could favor the survival of contaminated explants. However, contaminated living plants were discarded for evaluation purposes, and after 365 days of evaluation, they were preserved by minimal growth.

In the present study, plant regeneration after 365 days in conservation by minimal growth was lower than plant survival for all three garlic varieties. Nevertheless, Tacátzcuaro Especial has the highest number of regenerated plants. These results are in accordance with other minimal growth reports in which lower explant regeneration than explant survival at the end of the conservation period was obtained [44,45]. This suggests that plant survival is not sufficient to choose an adequate number of initial plants for medium-term conservation.

The capacity of plant cells to regrow after a medium or long conservation period depends on many factors, like the establishment of adequate regrowth conditions, species, the status of the donor plant or initial material, explant type, and the in vitro conservation technique [22].

4. Materials and Methods

4.1. Plant Material

Healthy garlic bulbs (*Allium sativum* L.) of Pebeco, Tacátzcuaro Especial, and Huerteño varieties were collected from the field collection in Bajío Experimental Field-INIFAP (National Institute of Agricultural and Livestock Forestry Research), Celaya, Guanajuato, Mexico. These garlic varieties were generated by INIFAP following a traditional breeding program that used the "Taiwan "commercial garlic variety, introduced in 1978 [46].

4.2. Initial Explants and Culture Conditions

Garlic bulbs (Figure 6A,D,G) were divided into individual cloves, which were peeled and washed with commercial detergent and water for 10 min. Garlic cloves were placed in fungicide (1 mL L^{-1}, Bravo® 720, Syngenta, Cartagena, Colombia) and fungicide–bactericide (6.25 g L^{-1}, Agri-Mycin®500, Zoetis, Mexico City, Mexico) solutions for 20 min each with constant stirring. Subsequently, they were disinfected with ethanol at 70% (*v*/*v*) for one minute and with a solution with 1.5% NaOCI for 20 min under a laminar flow cabinet (Labconco, Missouri, USA). After each solution, the cloves were rinsed three times with sterile distilled water and cut into 15 mm long and 10 mm wide pieces. Explants were placed in hormone-free semi-solid Murashige and Skoog (MS) medium (PhytoTech Labs, Kansas, USA) [46] supplemented with 30 g L^{-1} sucrose (PhytoTech Labs, Kansas, USA) and solidified with 9 g L^{-1} agar (PhytoTech Labs, Kansas, USA). The pH medium was adjusted to 5.8 $\pm$ 0.1 with sodium hydroxide (NaOH; 1N) before sterilization in an autoclave at 121 °C and 1.3 atm pressure for 20 min. The plant cultures were kept for 30 days in a growth room at a temperature of 25 $\pm$ 1 °C, photoperiod of 16 h light/8 h dark, and a photosynthetic photon flux density of 25 µmol m^{-2} s^{-1} provided by white fluorescent light (Figure 6B,E,H).

Figure 6. Initial explants and minimal growth experiments. (**A**) Pebeco, (**D**) Tacátzcuaro Especial and (**G**) Huerteño garlic bulbs. (**B**) Pebeco, (**E**) Tacátzcuaro Especial and (**H**) Huerteño plants in MS culture medium supplemented with 30 g L^{-1} sucrose 30 days after in vitro establishment. (**C**) Pebeco, (**F**) Tacátzcuaro Especial and (**I**) Huerteño initial explants in different culture media on day 0 in minimal growth. The bar indicates 10 mm.

4.3. Minimal Growth Experiment and Data Collection

Aseptic in vitro garlic plants were cut to measure 10 mm in length and 5 mm in diameter and placed on six different culture media with a pH of 5.8 and 9 gL^{-1} of agar (Table 6, Figure 6C,F,I). The initial explants were cultivated for 365 days in growth rooms at a temperature of 5, 18, and 25 ± 1 °C, a photoperiod of 16 h light/8 h dark, and a photosynthetic photon flux density of 25 µmol m^{-2} s^{-1} provided by white fluorescent light.

Table 6. Culture media composition used for garlic minimal growth.

Culture Media	Mannitol (g L^{-1})	Sucrose (g L^{-1})	Sorbitol (g L^{-1})
M1: BDS	0.0	100.0	0.0
M2: 25% MS	0.0	45.0	0.0
M3: 100% MS	15.0	15.0	0.0
M4: 100% MS	0.0	0.0	72.87
M5: 100% MS	0.0	68.46	36.43
M6: 100% MS (control)	0.0	30.0	0.0

MS: Murashige and Skoog [19]; BDS: Dunstan and Short [20].

The experiment was performed using a randomized design, and the experimental unit was an explant in a test tube (25 × 150 mm) containing 10 mL of culture medium. Shoot height, survival, and contamination (visual assessment of bacteria presence) were evaluated in explants at 30, 60, 90, 180, 270, and 365 days. In the case of each temperature treatment, the value is the media obtained from six culture media while for each culture medium, the value is the media obtained from the three tested temperatures. For each treatment, six repetitions were used for Pebeco and eight for Tacátzcuaro Especial and Huerteño varieties, respectively. After 365 days of culture, aseptic surviving plants were removed from the minimum growth conditions and cultured for 30 days for recovery in test tubes containing 10 mL of MS medium supplemented with 0.5 mg L^{-1} of benzyladenine (BA), 30 g L^{-1} of sucrose, and 9 g L^{-1} of agar. The pH medium was adjusted to 5.8 ± 0.1 with NaOH (1N) before sterilization in an autoclave at 121 °C and 1.3 atm pressure for 20 min. Plant cultures were kept for 30 days in a growth room at a temperature of 25 ± 1 °C, photoperiod of 16 h light/8 h dark, and a photosynthetic photon flux density of 25 µmol m^{-2} s^{-1} provided by white fluorescent light.

4.4. Statistical Analysis

Statistical analysis of the data was carried out using one-way analysis of variance (ANOVA) and followed by Tukey´s test to reveal significant differences between the means at a level of $p < 0.05$ using the statistical software Statgraphics Centurion, Version 17, Virginia, USA.

5. Conclusions

In the present study, optimal conditions for minimal growth conservation of Pebeco, Tacátzcuaro Especial and Huerteño garlic varieties for a one-year period were established. Basal MS culture medium supplemented with sucrose and sorbitol at 5 °C led to the best response for the conservation of three Mexican garlic varieties. With these results, a new alternative was generated for the in vitro preservation of important garlic varieties that did not have a safe and sustainable conservation strategy under controlled conditions.

Despite the number of regenerated plants obtained after a year of minimal growth conservation, based on the survival results, it can be recommended to reduce the storage time from 365 to 180 days in the case of Pebeco and Huerteño and to 270 days for Tacátzcuaro Especial. This strategy will guarantee higher numbers of regenerated plants.

Furthermore, endogenous contamination occurred in Pebeco, Tacátzcuaro Especial, and Huerteño varieties, so it is recommended to carry out additional disinfection treatments during the in vitro process and to isolate and identify garlic endophytic microorganisms to provide valuable information on its potential application. Regardless of contamination rates, regenerated aseptic plants are sufficient as starting material for a future micropropagation step. Results obtained during this study help to choose an adequate number of plants for medium-term conservation in future experiments with different varieties.

Even though this study was carried out with specific Mexican garlic varieties, the resulting conservation alternative can be transferred to other varieties generated from the same commercial variety "Taiwan"; therefore, its application can be considered global.

Author Contributions: Conceptualization, G.S.-C. and B.T.; resources, G.S.-C.; experimental procedures and writing—original draft preparation, B.T.; writing—review and editing, B.T., G.S.-C., V.M.G.-R., C.I.C.-C., L.X.Z.-M. and H.R.-V.; analysis of microorganisms found as contamination, L.X.Z.-M.; funding acquisition, G.S.-C. All authors have read and agreed to the published version of the manuscript.

Funding: This work was supported by the National Institute of Forestry, Agricultural and Livestock Research (INIFAP, Mexico) with project No. SIGI 1155634761 and partially supported by the National Council of Humanities Science and Technology (CONAHCYT, Mexico), Grant Number 747860 provided to B.T.

Data Availability Statement: Data are contained within the article.

Acknowledgments: We thank Luis Felipe Guzmán for his help and support during the achievement of this work and Elena Heredia García, INIFAP, Celaya, Guanajuato, Mexico, for the generous supply of germplasm.

Conflicts of Interest: The authors declare no conflict of interest.

References

1. Abdelrahman, M.; Hirata, S.; Mukae, T.; Yamada, T.; Sawada, Y.; El-Syaed, M.; Yamada, Y.; Sato, M.; Hirai, M.Y.; Shigyo, M. Comprehensive Metabolite Profiling in Genetic Resources of Garlic (*Allium sativum* L.) Collected from Different Geographical Regions. *Molecules* **2021**, *26*, 1415. [CrossRef] [PubMed]
2. Magryś, A.; Olender, A.; Tchórzewska, D. Antibacterial properties of *Allium sativum* L. against the most emerging multidrug-resistant bacteria and its synergy with antibiotics. *Arch. Microbiol.* **2021**, *203*, 2257–2268. [CrossRef]
3. Wang, Y.; Wei, K.; Han, X.; Zhao, D.; Zheng, Y.; Chao, J.; Gou, J.; Kong, F.; Zhang, C.-S. The Antifungal Effect of Garlic Essential Oil on *Phytophthora nicotianae* and the Inhibitory Component Involved. *Biomolecules* **2019**, *9*, 632. [CrossRef] [PubMed]
4. Botas, J.; Fernandes, Â.; Barros, L.; Alves, M.J.; Carvalho, A.M.; Ferreira, I.C.F.R. A comparative study of black and white *Allium sativum* L.: Nutritional composition and bioactive properties. *Molecules* **2019**, *24*, 11. [CrossRef]
5. UPOV. International Union for the Protection of New Varieties of Plants. 2021. Available online: https://www.upov.int/portal/index.html.es (accessed on 6 May 2023).
6. Mishra, R.K.; Jaiswal, R.K.; Kumar, D.; Saabale, P.R.; Singh, A. Management of major diseases and insect pests of onion and garlic: A comprehensive review. *J. Plant Breed. Crop. Sci.* **2014**, *6*, 160–170. [CrossRef]
7. Sharrock, S.; Hoft, R.; de Souza Dias, B.F. An overview of recent progress in the implementation of the Global Strategy for Plant Conservation—A global perspective. *Rodriguesia* **2018**, *69*, 1489–1511. [CrossRef]
8. Volk, G.M.; Bretting, P.K.; Byrne, P.F. Survey identifies essential plant genetic resources training program components. *Crop Sci.* **2019**, *59*, 2308–2316. [CrossRef]
9. FAO. Food and Agriculture Organization of the United Nations. 2023. Available online: https://www.fao.org (accessed on 6 May 2023).
10. Gulati, R. Strategies for sustaining plant germplasm evaluation and conservation a review. *Res. J. Life Sci. Bioinform. Pharm. Chem. Sci.* **2018**, *4*, 313–320. [CrossRef]
11. Oseni, O.M.; Pande, V.; Nailwal, T.K. A review on plant tissue culture, a technique for propagation and conservation of endangered plant species. *Int. J. Curr. Microbiol. App. Sci.* **2018**, *7*, 3778–3786. [CrossRef]
12. Panis, B.; Nagel, M.; den Houwe, I.V. Challenges and prospects for the conservation of crop genetic resources in field genebanks, in in vitro collections and/or in liquid nitrogen. *Plants* **2020**, *9*, 1634. [CrossRef]
13. De Lacerda, L.F.; Gomes, H.T.; Bartos, P.M.C.; Vasconcelos, J.M.; Vasconcelos-Filho, S.C.; Silva-Cardoso, I.M.A.; Scherwinski-Pereira, J.E. Growth, anatomy and histochemistry of fast growing species under in vitro conservation through mineral oil and low-temperature combination. *Plant Cell Tiss. Organ Cult.* **2021**, *144*, 143–156. [CrossRef]
14. Engelmann, F. Use of biotechnologies for the conservation of plant biodiversity. *In Vitro Cell. Dev. Biol. Plant.* **2011**, *47*, 5–16. [CrossRef]
15. Benke, A.; Kuldip; Shelke, P.; Singh, M. Slow growth in vitro conservation studies in garlic. *J. Allium Res.* **2018**, *1*, 98–101.
16. Pardo, A.; Rivero, S.; Alvarado, G. Conservación in vitro de microbulbos de ajo (*Allium sativum* L.). *Bioagro* **2014**, *25*, 115–122.

17. Hassan, N.A.; El-awady, A.A.; Gaber, A.; El-awady, M.; Khalaf, A. Slow-growth in vitro conservation of garlic cultivars grow in Egypt: Chemical characterization and molecular evaluation. *Glob. J. Mol. Sci.* **2007**, *2*, 67–75.
18. Ruta, C.; Lambardi, M.; Ozudogru, E.A. Biobanking of vegetable genetic resources by *in vitro* conservation and cryopreservation. *Biodivers. Conserv.* **2020**, *29*, 3495–3532. [CrossRef]
19. Murashige, T.; Skoog, F. A revised medium for rapid growth and bioassays with tobacco tissue cultures. *Physiol. Plant* **1962**, *15*, 473–497. [CrossRef]
20. Dunstan, D.I.; Short, K.C. Improved growth of tissue culture of the onion, *Allium cepa*. *Physiol. Plant* **1977**, *41*, 70–72. [CrossRef]
21. Khoury, C.; Laliberté, B.; Guarino, L. Trends in ex situ conservation of plant genetic resources: A review of global crop and regional conservation strategies. *Genet. Resour. Crop. Evol.* **2010**, *57*, 625–639. [CrossRef]
22. Tirado-Pérez, B.; Sandoval-Cancino, G. Cryopreservation of plant genetic resources: A legacy for humanity. *Afr. J. Biotechnol.* **2022**, *21*, 55–63.
23. Byrne, P.F.; Volk, G.M.; Gardner, C.; Gore, M.A.; Simon, P.W.; Smith, S. Sustaining the future of plant breeding: The critical role of the USDA-ARS national plant germplasm system. *Crop Sci.* **2018**, *58*, 451–468. [CrossRef]
24. Nievola, C.C.; Carvalho, C.P.; Carvalho, V.; Rodrigues, E. Rapid responses of plants to temperature changes. *Temperature* **2017**, *4*, 371–405. [CrossRef]
25. Li, S.L.; Li, Z.G.; Yang, L.T.; Li, Y.R.; He, Z.L. Differential effects of cold stress on chloroplasts structures and photosynthetic characteristics in cold-sensitive and cold-tolerant cultivars of sugarcane. *Sugar Tech.* **2018**, *20*, 11–20. [CrossRef]
26. Mawia, R.S.; Saleem, H.Z.; Fahed, B.; Ayman, S.O. Impact of sorbitol-induced osmotic stress on some biochemical traits of potato in vitro. *Iraqi J. Agric. Sci.* **2020**, *51*, 1038–1047. [CrossRef]
27. Sotiropoulos, T.E.; Molassiotis, A.N.; Mouhtaridou, G.I. Sucrose and sorbitol effects on shoot growth and proliferation *in vitro*, nutritional status and peroxidase and catalase isoenzymes of M 9 and MM 106 apple (*Malus domestica* Borkh.) rootstocks. *Eur. J. Hortic. Sci.* **2006**, *71*, 114–119.
28. Gelmesa, D.; Dechassa, N.; Mohammed, W.; Gebre, E.; Monneveux, P.; Bündig, C.; Winkelmann, T. In vitro screening of potato genotypes for osmotic stress tolerance. *Open Agric.* **2017**, *2*, 308–316. [CrossRef]
29. Alexopoulos, A.A.; Mavrommati, E.; Kartsonas, E.; Petropoulos, S.A. Effect of temperature and sucrose on in vitro seed germination and bulblet production of *Pancratium maritimum* L. *Agronomy* **2022**, *12*, 2786. [CrossRef]
30. Kwon, Y.H.; Jeong, J.H.; Lee, J.S.; Jeon, J.O.; Park, Y.U.; Min, J.H.; Chang, W.B.; Lee, S.Y.; Youn, C.K.; Kim, K.H. Multiple shoot induction and bulb mass proliferation system by in vitro immature spathe culture of elephant garlic (*Allium ampeloprasum* L.). *Korean J. Plant Res.* **2018**, *31*, 355–362. [CrossRef]
31. Youssef, N.M.; Shaaban, S.A.; Ghareeb, Z.F.; Taha, L.S. In vitro bulb formation of direct and indirect regeneration of *Lilium orientalis* cv. "Starfighter" plants. *BNRC* **2020**, *43*, 211. [CrossRef]
32. Niino, T.; Arizaga, M.V. Cryopreservation for preservation of potato genetic resources. *Breed. Sci.* **2015**, *65*, 41–52. [CrossRef]
33. De Souza Mendes, M.I.; Verde, D.D.S.V.; de Souza Ramos, A.P.; da Silva Gesteira, A.; dos Santos Soares Filho, W.; da Silva Souza, A. In vitro conservation of citrus rootstocks using paclobutrazol and analysis of plant viability and genetic stability. *Sci. Hortic.* **2021**, *286*, 110231. [CrossRef]
34. Cioloca, M.; Tican, A.; Bădărău, C.; Bărăscu, N.; Popa, M. In vitro medium term conservation of sweet potato genotypes using mannitol and sorbitol. *Rom. Agric. Res.* **2021**, *38*, 123–132. [CrossRef]
35. Dos Santos Silva, S.S.; de Souza, E.H.; Souza, F.V.D.; Nepomuceno, C.F.; de Carvalho Costa, M.A.P. Micropropagation and in vitro conservation of *Alcantarea nahoumii* (Bromeliaceae), an endemic and endangered species of the brazilian atlantic forest. *Acta Sci. Biol. Sci.* **2020**, *42*, e52940. [CrossRef]
36. Singh, C.R. Review on problems and its remedy in plant tissue culture. *Asian J. Biol. Sci.* **2018**, *11*, 1651672. [CrossRef]
37. Orlikowska, T.; Nowak, K.; Reed, B. Bacteria in the plant tissue culture environment. *Plant Cell Tissue Organ Cult.* **2017**, *128*, 487–508. [CrossRef]
38. Diriba-Shiferaw, G. Review of management strategies of constraints in garlic (*Allium sativum* L.) production. *J. Agric. Sci.* **2016**, *11*, 186–207. [CrossRef]
39. Pitekelabou, R.; Aïdam, A.V.; Kokou, K. Influence of various carbohydrates on the *in vitro* micropropagation of *Nauclea diderrichii* (De Wild T. Durand) Merrill, an endangered forest species in Togo. *Afr. J. Biotechnol.* **2015**, *14*, 1283–1289. [CrossRef]
40. Reid, S.J.; Abratt, V.R. Sucrose utilization in bacteria: Genetic organization and regulation. *Appl. Microbiol. Biotechnol.* **2005**, *67*, 312–321. [CrossRef] [PubMed]
41. Ray, S.S.; Ali, N. Biotic contamination and possible ways of sterilization: A review with reference to bamboo micropropagation. *Braz. Arch. Biol. Technol.* **2018**, *60*, e16160485. [CrossRef]
42. Wei, Z.Z.; Liu, M.Y.; Li, Z.; Zhou, Y.J. Research on the separation and bacteriostasis of garlic endophytes. *Heilongjiang Agric. Sci.* **2013**, *12*, 5–8.
43. Costa Júnior, P.S.P.; Cardoso, F.P.; Martins, A.D.; Buttrós, V.H.T.; Pasqual, M.; Dias, D.R.; Schwan, R.F.; Dória, J. Endophytic bacteria of garlic roots promote growth of micropropagated meristems. *Microbiol. Res.* **2020**, *241*, 126585. [CrossRef] [PubMed]
44. Sarropoulou, V.; Maloupa, E.; Grigoriadou, K. Cretan Dittany (*Origanum dictamnus* L.), a valuable local endemic plant: In vitro regeneration potential of different type of explants for conservation and sustainable exploitation. *Plants* **2023**, *12*, 182. [CrossRef] [PubMed]

45. Pérez-Molphe-Balch, E.; Pérez-Reyes, M.E.; De La Rosa-Carrillo, M.L. In vitro conservation of *Turbinicarpus* (Cactaceae) under slow growth conditions. *Haseltonia* **2012**, *17*, 51–57. [CrossRef]
46. Heredia, G.; Delgadillo, S. *El Ajo en Mexico: Origen, Mejoramiento Genético y Tecnología de Producción*; SAGAR, INIFAP, Campo Experimental Bajío: Celaya, Mexico, 2000; Volume 3, 102p.

plants

MDPI

Article

In Vitro Induction of Interspecific Hybrid and Polyploidy Derived from *Oryza officinalis* Wall

Meimei Tan [1,2,3,4,†], Ruoxin Chen [1,2,3,4,†], Xingran Chen [1,2,3,4], Muhammad Qasim Shahid [1,2,3,4], Xiangdong Liu [1,2,3,4] and Jinwen Wu [1,2,3,4,*]

1 State Key Laboratory for Conservation and Utilization of Subtropical Agro-Bioresources, South China Agricultural University, Guangzhou 510642, China; tanmeimei1115@163.com (M.T.); erdongre@163.com (R.C.); scau_cxr@163.com (X.C.); qasim@scau.edu.cn (M.Q.S.); xdliu@scau.edu.cn (X.L.)
2 Base Bank of Lingnan Rice Germplasm Resources, South China Agricultural University, Guangzhou 510642, China
3 Guangdong Provincial Key Laboratory of Plant Molecular Breeding, South China Agricultural University, Guangzhou 510642, China
4 College of Agriculture, South China Agricultural University, Guangzhou 510642, China
* Correspondence: jwwu@scau.edu.cn
† These authors contributed equally to this work.

Abstract: *Oryza officinalis* Wall is a potential genetic resource for rice breeding; however, its distant genome limits its crossing ability with cultivated rice. The interspecific hybridization of *O. officinalis* and cultivated rice, establishment of its tissue culture, and induction of polyploidy are ways to improve *O. officinalis*'s poor crossability. We developed an interspecific hybrid and studied its reproductive pollen development process in this work, and the results showed that abortive pollens (81.94%) and embryo sac abnormalities (91.04%) were the key causes of its high sterility. In order to induce callus formation in interspecific hybrid explants, two different culture media, namely Chu's N-6 medium (N6) and 1/2 Murashig and Skoog medium (1/2 MS), were employed. Additionally, two plant growth regulators (PGRs), namely 2,4-dichlorophenoxyacetic acid (2,4-D) and 6-benzylaminopurine (BA), along with L-proline (Pro) and acid hydrolyzed casein, were utilized in the experiment. The optimal N6 medium, supplemented with 2.0 mg·L^{-1} 2,4-D, produced the highest induction rate (80.56 ± 5.44)%. For callus differentiation and proliferation, the MS medium supplemented with 2.0 mg·L^{-1} BA + 0.2 mg·L^{-1} NAA produced the highest differentiation rate (75.00 ± 4.97)% and seedling emergence ratio (28.97 ± 4.67)%. The optimal combination for seedling rooting was the 1/2 MS medium supplemented with 2.0 mg L^{-1} NAA and 0.2 mg L^{-1} BA, which produced an average of 13.95 roots per plant. For polyploidy induction in the interspecific hybrid, the concentration of colchicine treatment at 400 mg·L^{-1} for three days is an ideal protocol. We devised tissue culture and interspecific hybrid polyploidy induction to overcome *O. officinalis*' poor crossability and introduce its favorable features into cultivated rice.

Keywords: rice; *Oryza officinalis* Wall; tissue culture; polyploidy

Citation: Tan, M.; Chen, R.; Chen, X.; Shahid, M.Q.; Liu, X.; Wu, J. In Vitro Induction of Interspecific Hybrid and Polyploidy Derived from *Oryza officinalis* Wall. *Plants* **2023**, *12*, 3001. https://doi.org/10.3390/plants12163001

Academic Editor: Iyyakkannu Sivanesan

Received: 7 July 2023
Revised: 17 August 2023
Accepted: 18 August 2023
Published: 20 August 2023

1. Introduction

Oryza officinalis Wall is a diploid wild rice species belonging to the CC genome, with 2n = 24 chromosomes [1]. Regarding the three existing wild rice species in China, *O. officinalis* is endemic to the Guang Dong, Hai Nan, Guang Xi, and Yun Nan provinces. *O. officinalis* has excellent disease resistance, insect resistance, and stress tolerance compared with cultivated rice [2–4]. *O. officinalis* plants have strong resistance to brown planthopper, white–back planthopper, and bacterial blight resistance [5,6]. *O. officinalis* also has the potential for fast growth and strong resistance; it has a potential role in germplasm improvement and novel variety breeding. Interspecific hybridization and backcross-selection represent prominent methodologies employed to effectively amalgamate the dominant genes of *O. officinalis* and cultivated rice, hence facilitating the successful breeding of novel germplasm [7,8]. To utilize the potential excellent genes of wild rice resources, conventional hybridization, embryo rescue, embryo culture, protoplast fusion, and heterologous additional lines are frequently used in wild rice [9,10]. The hybrids formed between *O. officinalis* and cultivated rice exhibit reproductive isolation, leading to significant male and female sterility. This reproductive barrier restricts the compatibility of remote hybridization and limits gene exchange and germplasm exploitation due to hybrid sterility [11]. Homologous chromosome pairing, abnormal megaspores, and abnormal microspores in the meiosis process could result in the limited utilization of *O. officinalis* [11]. Twenty-five monosomic alien addition lines (MAAL) were generated through the interspecific hybridization of *O. officinalis* and the successive backcrossing of hybrid progenies [12].

Plant tissue culture is the utilization of its cell totipotency. Under suitable conditions, plant organs, tissue cells, and protoplast can develop into a complete plant through differentiation and regeneration. Rice genotypes, explants type, basic media, PGRs, culture duration, temperature, photoperiod, and other factors can affect the rice callus formation and plant regeneration ability [13,14]. Among the different types of explants, the induction rate of rice seeds, young embryos, and young panicles is higher than that of anthers [15]. In addition, plant genotypes have a great influence on rice tissue culture. The cultivation of *japonica* rice is comparatively more feasible than that of *indica* rice, whereas cultivated rice, in general, presents greater ease of cultivation when compared to wild rice [16,17]. Leaf sheaths and buds were used to induce regenerated plants through the callus pathway and tissue culture seedlings in *O. officinalis* [18].

Polyploidy induction is an efficient strategy for improving agronomic traits and resistance in cultivated and wild plants [19–23]. Polyploidy induction methods could be divided into physical, chemical, and biological induction. The polyploidy induction of many crops, flowers, and fruits, such as *Catharanthus roseus* L., *Pyrus communis* L., and *Citrus sinensis*, has been frequently reported using colchicine [8,24,25]. Doubled haploid (DH) technology is also a very convenient approach for polyploidy induction in plant breeding, which can save time and corresponding costs [26,27]. A tetraploid fennel was induced by diploid fennel seeds with 0.05% (w/v) colchicine for 24 h [28]. Polyploidy was generated using lavender tissue culture seedlings into a subculture supplemented with 2% DMSO and 0.2–0.4% colchicine from 48 to 72 h.

O. officinalis accumulated the number of dominant genes that were absent or lost in cultivated rice during the long-term natural adaptation and selection process [2,3]. Interspecific hybridization is one of the main approaches for developing rice varieties by incorporating the dominant genes of *O. officinalis* into cultivated rice. In order to harness advantageous genetic traits found in *O. officinalis*, many techniques, like traditional hybridization, embryo rescue, embryo culture, and protoplast fusion, are commonly employed [9,10]. However, very little is known about the reproductive characteristics and tissue culture system of *O. officinalis*. In the present study, we constructed an interspecific hybrid of *O. officinalis* and cultivated rice, and observed its reproductive features using cytological analysis. Then, we developed an interspecific hybrid generated from *O. officinalis* using tissue culture and polyploidy induction. These results provide a theoretical basis for the further improvement of rice tissue culture technology, offering the possibility of transferring desirable traits into cultivated rice.

2. Results

2.1. Reproductive Characteristics of the Interspecific Hybrid Derived from O. officinalis and Cultivated Rice

This study used interspecific hybrid plants derived from *O. officinalis* and cultivated rice to investigate the reproductive characteristics of an interspecific rice hybrid constructed from *O. officinalis* (female parent, CC genome) and cultivated rice (male parent, AA genome). We detected the pollen and embryo sac fertility of interspecific hybrid plants, and found a high percentage of sterility in the interspecific hybrid (Figure 1). A total of 81.94% of abortive pollen grains was observed in interspecific hybrid plants, compared with its cultivated rice parent. Typical abortive pollens, stained abortive pollens, and spherical abortive pollens were frequently observed (Figure 1a–c). Additionally, we assessed the fertility of the embryo sac in interspecific hybrid plants. The sterility value of the embryo sac was found to be 91.04%. The main kind of embryo sac abortion seen in interspecific hybrid plants in this study was the degeneration of the megaspore mother cell (Figure 1d–f).

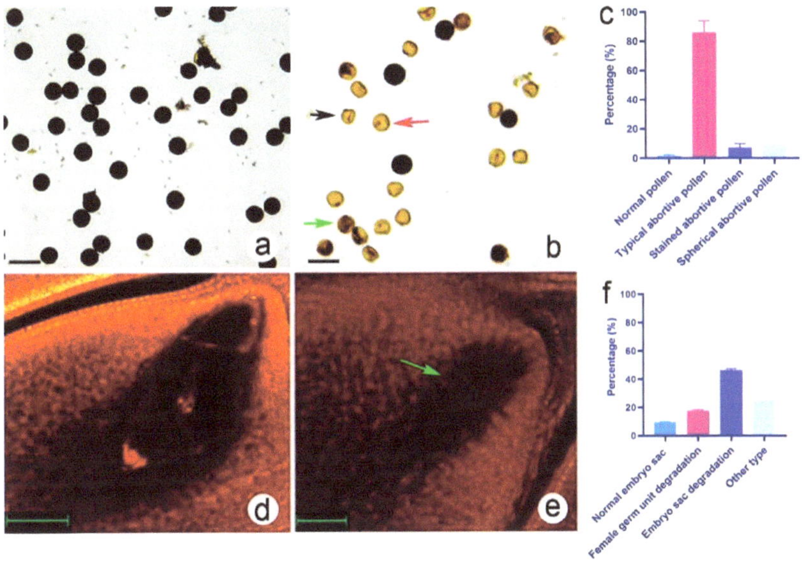

Figure 1. Comparison of the pollen and embryo sac fertility between the interspecific hybrid and its parent. (**a**) The pollen phenotype of cultivated rice; (**b**) the pollen phenotype of the interspecific hybrid (the black arrow indicates typical abortive pollens, the green arrow indicates stained abortive pollens, and the red arrow indicates spherical abortive pollens); (**c**) pollen fertility value of the interspecific hybrid; (**d**) embryo sac phenotype of cultivated rice; (**e**) abortive embryo sac of interspecific hybrid, the arrow indicates embryo sac degeneration; (**f**) embryo sac fertility value of the interspecific hybrid. Bar = 40 μm.

The cultivated rice parent displayed a normal pollen development process divided into eight stages, including the pre-meiotic stage, meiosis stage, early microspore stage, middle microspore stage, late microspore stage, early bicellular pollen stage, late bicellular pollen stage, and mature pollen stage (Figure S1). Compared with its parent, pollen development also experienced similar stage divisions. Several kinds of abnormalities in differential pollen development stages, such as abnormal pollen mother cells (PMCs), abnormal dyads, abnormal tetrads, abnormal microspores, and abnormal bicellular pollens, were observed in the interspecific hybrid plants during the pollen development process (Figure 2). Compared with its parent, several abnormal embryo sacs were also observed in the interspecific hybrid during the embryo sac development process (Figure 3). Megaspore mother cell degeneration, functional megaspore degeneration, and a small embryo sac (less than 2/3 of the normal embryo sac) also accounted for a certain proportion found in the hybrid plants (Figure 3). All of these anomalies resulted in substantial sterility rates in the hybrid plants.

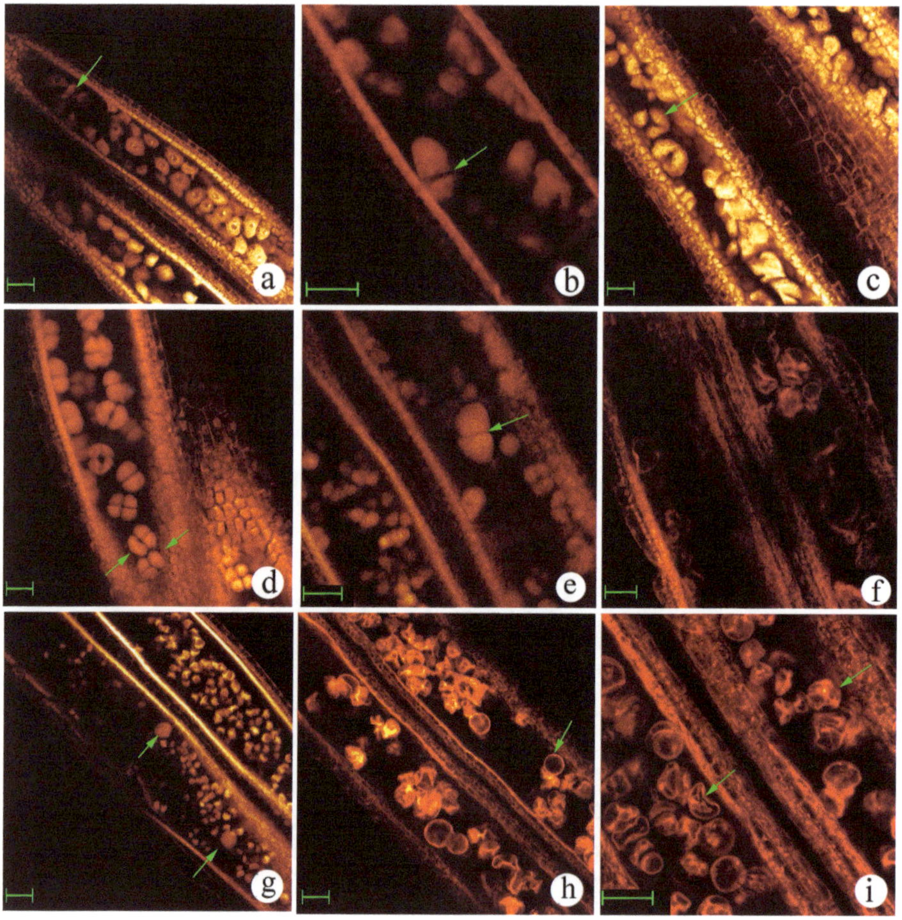

Figure 2. The pollen development process of interspecific hybrid. (**a**) Pre-meiotic stage (the arrow indicates the degraded pollen mother cell); (**b**) telophase I (the arrow indicates asynchronous pollen); (**c**) telophase II (the arrow indicates abnormal tetrad cell, triad); (**d**) tetrad stage (the arrows indicate abnormal tetrad cells, two triads); (**e**) tetrad stage (the arrow indicates the abnormal shape of the tetrad); (**f**) single microspore stage, degradation of microspore cells; (**g**) Middle single microspore stage (the arrows indicate the abnormal formation of microspores); (**h**) mature pollen stage (the arrow indicates spherical abortive pollen); (**i**) mature pollen stage (the arrows indicate typical abortive pollens). Bar = 40 μm.

2.2. Tissue Culture System of the Interspecific Hybrid Derived from O. officinalis and Cultivated Rice

To investigate the callus induction rate of the interspecific hybrid derived from *O. officinalis* and cultivated rice, four combinations of basic media, 2,4-D, Pro, and acid hydrolyzed casein, were used (Table 1). Differential medium combinations showed significant variations in callus induction (Figure 4). In this study, the rate of callus induction without Pro and acid-hydrolyzed casein showed the highest induction rate (80.56 ± 5.44)% (Table 1, Figure 4a–c). In comparison, the N6 medium supplemented with 2,4-D (2.0 mg·L^{-1}) + acid-hydrolyzed casein (0.3 g·L^{-1}) + Pro (3.0 g·L^{-1}) had the lowest induction rate ratio (7.78 ± 2.35)% (Table 1, Figure 4g–i). These results indicated that the N6 medium, supplemented with 2.0 mg·L^{-1} 2,4-D, without Pro and acid-hydrolyzed casein, was more suitable for the induction of the young panicle callus.

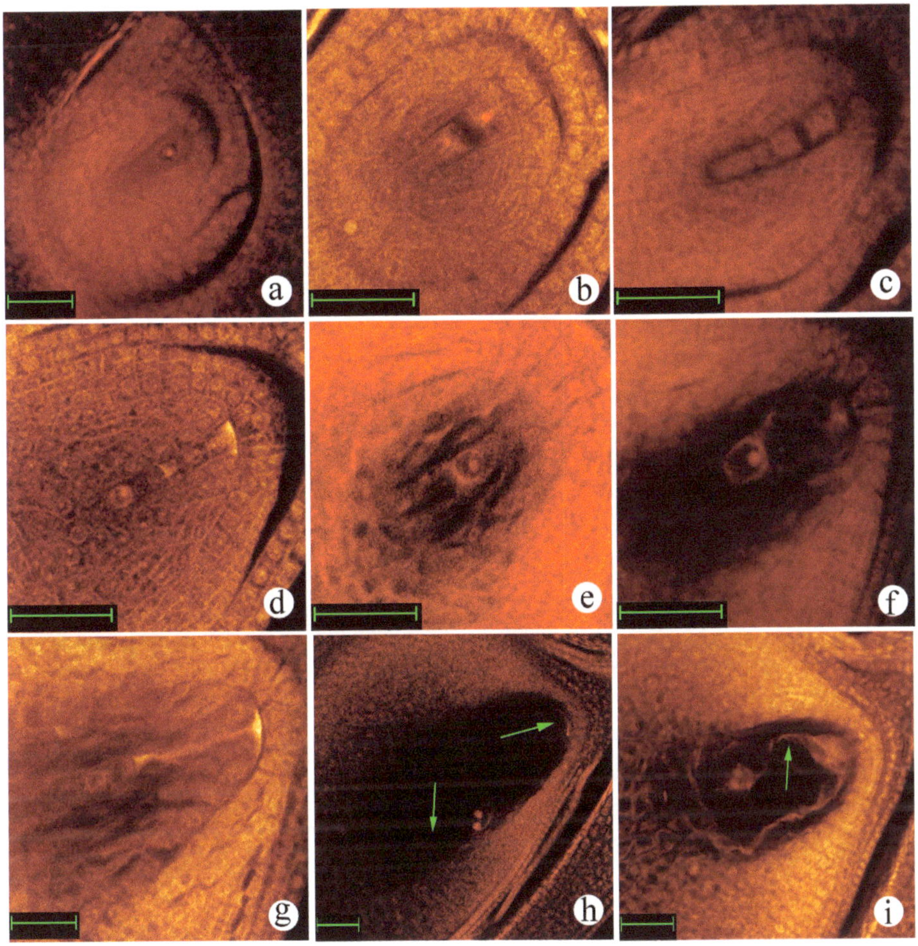

Figure 3. Embryo sac development of the interspecific hybrid. (**a**) Megasporocyte stage; (**b**) meiotic division stage, dyad stage; (**c**) meiotic division stage, tetrad stage; (**d**) functional megaspore stage; (**e**) mono-nucleate embryo sac stage; (**f**) two-nucleate embryo sac stage; (**g**) meiotic division stage, tetrad stage, abnormal degradation of megaspore cell; (**h**) mature embryo sac stage (the arrows indicate female germ unit and polar nuclei degradation); (**i**) mature embryo sac stage (the arrow indicates no female germ unit). Bar = 40 μm.

Table 1. Callus induction rate of different media and plant growth regulators.

Combination Name	Medium Name	2,4-D Concentration ($mg \cdot L^{-1}$)	Acid-Hydrolyzed Casein Concentration ($g \cdot L^{-1}$)	Pro Concentration ($g \cdot L^{-1}$)	Induction Rate (%)
Y1	N6	2.0	0.0	0.0	80.56 ± 5.44 a
Y2	N6	2.0	0.3	0.3	18.52 ± 3.17 bc
Y3	N6	2.0	0.3	3.0	7.78 ± 2.35 c
Y4	1/2MS	2.0	0.0	0.0	30.14 ± 5.18 b

Note: Data are shown as the mean and standard error (SE). Different letters represent significant differences at $p < 0.01$.

Due to the limitation of the callus, subculture and callus proliferation were conducted using the N6 medium supplemented with 2,4-D (2.0 $mg \cdot L^{-1}$) and different concentrations of BA (0.0, 0.2 and 0.5 mg·L^{-1}) (Figure 5, Table S1). These results demonstrated that the N6 medium supplemented with 2,4-D (2.0 mg·L^{-1}) + BA (0.2 mg·L^{-1}) for subculture had a significant effect on the callus proliferation rate, and increases the number of calli (Figure 5g–i, Table S1).

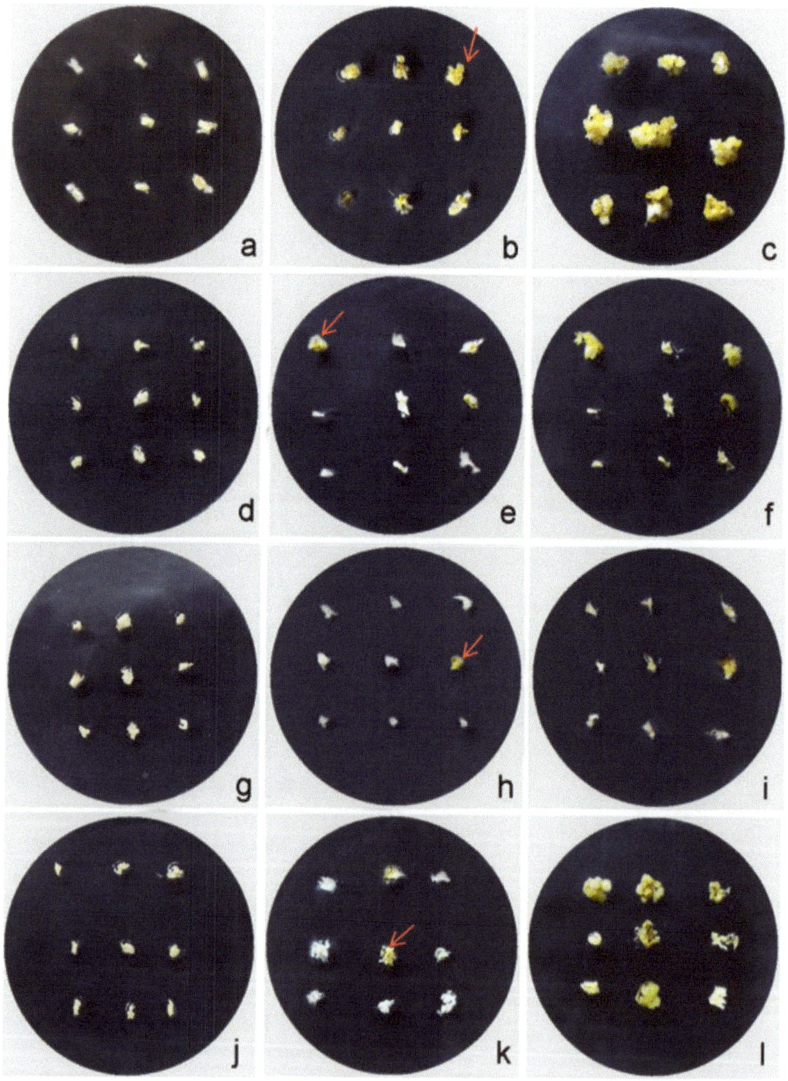

Figure 4. Callus induction of the interspecific hybrid in different media. (**a–c**) Callus. Induction in the Y1 medium at 0 d, 10 d, and 30 d, respectively. (**d–f**) Callus induction in the Y2 medium at 0 d, 10 d, and 30 d, respectively. (**g–i**) Callus induction in the Y3 medium at 0 d, 10 d and 30 d, respectively. (**j–l**) Callus induction in the Y4 medium at 0 d, 10 d and 30 d, respectively. The red arrow indicates the initiation of differentiation.

To improve the rate of callus differentiation and proliferation, the MS medium with three differential PGRs, including BA, kinetin (KIN), and α-naphthalene acetic acid (NAA), were used in this study (Table 2, Figure 6). BA combined with NAA showed better effects on callus bud differentiation. The F2 combination of the MS medium supplemented with BA (2.0 mg·L^{-1}) + NAA (0.2 mg·L^{-1}) demonstrated the highest differentiation and proliferation rate, with a 75% differentiation rate in the interspecific hybrid and a seedling emergence ratio of 28.97% (Figure 6d–f). These results suggest that the MS medium is more suitable for callus differentiation in the interspecific hybrid derived from *O. officinalis* and cultivated rice.

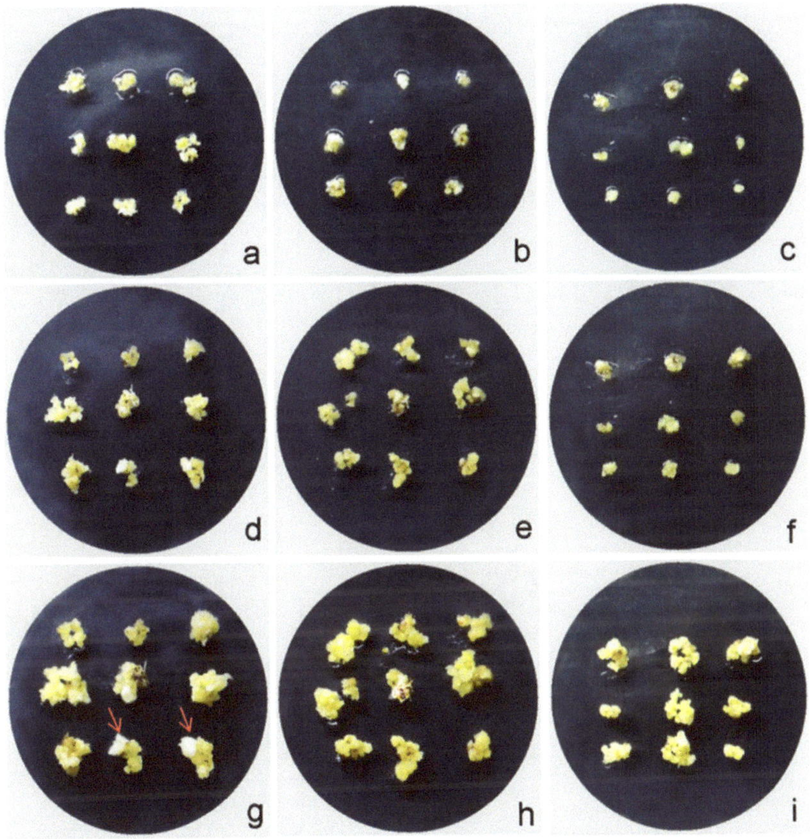

Figure 5. Effect of different BA concentrations on the callus proliferation of the interspecific hybrid. (**a**,**d**,**g**) Callus was sub-cultured in the J1 medium for 2 d, 10 d, and 25 d, respectively. (**b**,**e**,**h**) Callus was sub-cultured in the J2 medium for 2 d, 10 d, and 25 d, respectively. (**c**,**f**,**i**) Callus was sub-cultured in the J3 medium for 2 d, 10 d, and 25 d, respectively. The arrow indicates embryogenic degeneration of the callus with white color.

Table 2. Effects of exogenous hormone ratio on callus differentiation.

Combination Name	Differentiation Rate (%)	Seedling Emergence Rate (%)	Browning Rate (%)
F1	46.46 ± 7.61 b	12.12 ± 5.70 B	24.24 ± 4.93 B
F2	75.00 ± 4.97 a	28.97 ± 4.67 A	14.29 ± 3.38 B
F3	50.56 ± 6.25 b	9.72 ± 4.84 B	8.80 ± 3.71 B
F4	47.62 ± 8.11 b	5.56 ± 2.26 B	49.21 ± 8.54 A

Note: Data are shown as the mean and standard error (SE). Different capital letters in the same column represented significant differences at $p < 0.01$, and lowercase letters in the same column represented significant differences at $p < 0.05$.

Figure 6. Effect of the exogenous hormone ratio on callus differentiation. (**a–c**) Callus was differentiated and cultured in the F1 medium for 2 d, 9 d, and 40 d, respectively. (**d–f**) Callus was differentiated and cultured in the F2 medium for 2 d, 9 d, and 40 d, respectively. (**g–i**) Callus was differentiated and cultured in the F3 medium for 2 d, 9 d, and 40 d, respectively. (**j–l**) Callus was differentiated and cultured in the F4 medium for 2 d, 9 d, and 40 d differentiation of callus. The arrow points to the differentiated green dot.

2.3. *Influence of Exogenous Hormone Ratio on Shoot Rooting*

To evaluate the relationship between PGRs and rooting formation, three 1/2 MS basic media, combined with different concentrations of BA, NAA, 3-indole acetic acid (IAA), and activated charcoal (AC), were used (Table 3, Figures 7 and S2). In this study, the 1/2 MS basic medium combined with IAA, NAA, or BA promoted the rooting of seedlings.

The G1 medium + NAA (2.0 mg·L^{-1}) + BA (0.2 mg·L^{-1}) had the optimal effect on the root seedlings, with an average of 13.95 roots per plant (Table 3, Figure 7a,c).

Table 3. Influence of exogenous hormone ratio on the number of roots of the shoots.

Combination Name	Medium Name	IAA Concentration (mg·L^{-1})	NAA Concentration (mg·L^{-1})	BA Concentration (mg·L^{-1})	AC Concentration (g·L^{-1})	Root Number
G1	1/2MS	0.0	2.0	0.2	0.5	13.95 ± 1.04 a
G2	1/2MS	2.0	0.0	0.2	0.5	11.85 ± 1.72 a
G3	1/2MS	1.0	1.0	0.5	0.5	10.77 ± 1.25 a

Note: Data are shown as the mean and standard error (SE). Different lowercase letters represent significant differences at $p < 0.05$.

Figure 7. Tissue culture of strong rooting seedlings derived from the callus of the interspecific hybrid. (**a**,**c**) Root-strong seedlings cultured in the G1 medium for 1 d and 10 d, respectively. (**b**,**d**) Cultured in the G2 medium for 1 d and 10 d, respectively. The arrow indicates the rooting of the plants.

2.4. Induction of Interspecific Hybrid Derived from O. officinalis and Cultivated Rice by Colchicine

To induce polyploidy in interspecific hybrid plants, a colchicine-treated callus was utilized in this study. The callus survival rate and seedling emergence rate of colchicine co-cultured for 5 days at the same concentration were significantly lower than those for 3 days at the same concentration. We detected that 400–500 mg·L^{-1} of colchicine treatment showed a higher rate of seedling induction at 3 days compared with the 5 days. In comparison to other colchicine treatment concentrations, the 400 mg/L concentration demonstrated a better percentage of callus survival and seedling emergence (Table S2).

Higher concentrations of colchicine treatment led to a high percentage of callus browning (Figure 8). In this study, callus browning gradually appeared after culturing for 2 to 3 days. The death of the callus was easy to detect after culturing for 20 d. Among the four concentrations of colchicine treatment, the 400 mg·L^{-1} colchicine treatment also showed a low percentage of callus browning (Table S2). During the differentiation and re-differentiation process, we detected that the differentiation rate of the callus under colchicine treatment was slower than the no-treatment condition. Over a period of 2 to 20 days, the callus underwent a gradual process of browning and subsequent death Figures 8, 9 and S3). Differentiated seedlings treated with colchicine showed thicker and abnormal-shape leaves.

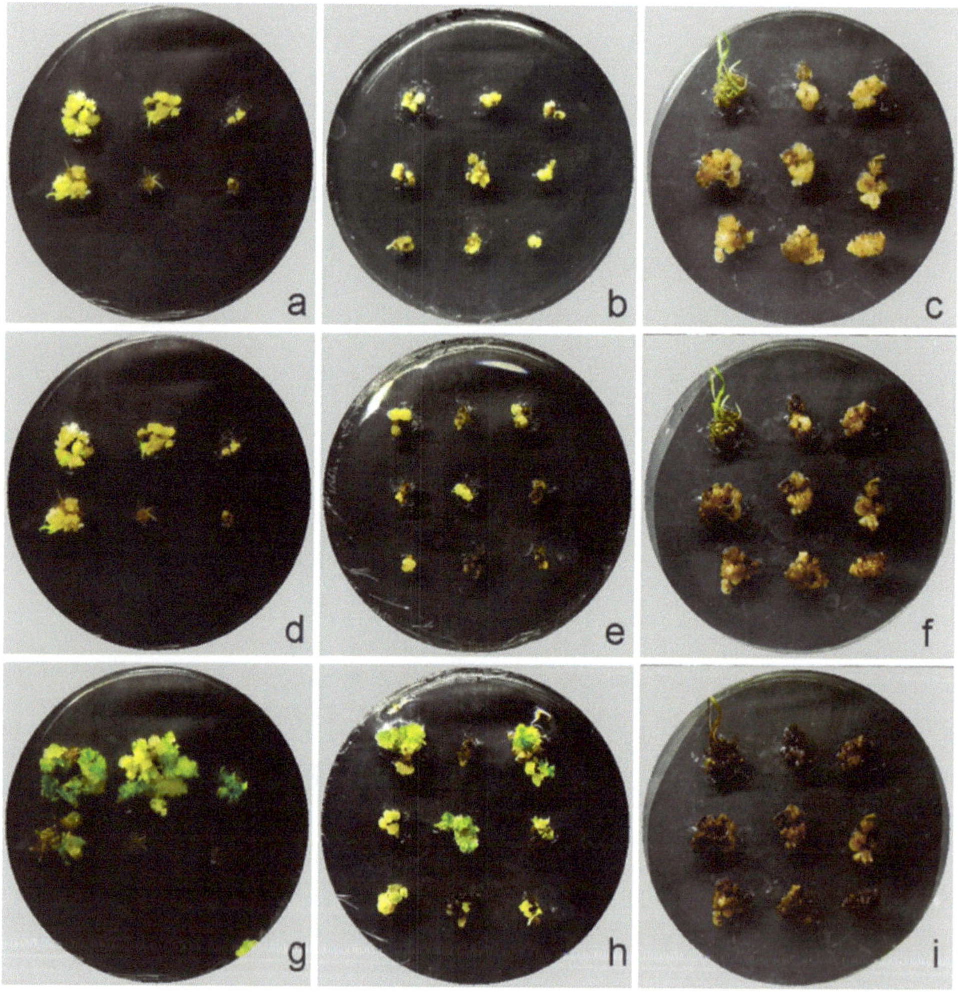

Figure 8. Callus status after co-culturing with different colchicine concentrations. (**a**,**d**,**g**) Callus was co-cultured with 400 mg·L^{-1} colchicine for 3 days and differentiated for 2 d, 10 d, and 20 d, respectively. (**b**,**e**,**h**) Callus was co-cultured with 500 mg·L^{-1} colchicine for 3 days and differentiated for 2 d, 10 d, and 20 d, respectively. (**c**,**f**,**i**) Callus was co-cultured with 600 mg·L^{-1} colchicine for 3 days and differentiated for 2 d, 10 d, and 20 d, respectively.

Figure 9. Growth status of mutant seedlings obtained through colchicine treatment. (**a–c**) The crouching and twisting leaves under colchicine treatment; (**d**) abnormal stout seedling under colchicine treatment.

2.5. Investigation of the Agronomic Traits of Colchicine-Induced Plants

The agronomic traits of the plants obtained from tissue culture after colchicine treatments were investigated. Compared with its original material, candidate materials of mixed-ploidy plants were obtained in this study (Figure 10). After treatment, the mixed-ploidy material showed no significant difference in average plant height, leaf length, leaf width, and stem diameter. Leaf width and plant thickness were consistent with the general morphological rule of polyploid plants (Table 4). However, its tiller number increased significantly, contrary to the tiller reduction characteristic in most polyploid rice genotypes.

Table 4. Comparison of agronomic traits between colchicine-induced plants and original materials.

Materials	Tiller Number	Plant Height (cm)	Leaf Length (cm)	Leaf Width (cm)	Stem Diameter (cm)
Control	8.67 ± 0.92	74.63 ± 1.21	48.88 ± 0.72	1.42 ± 0.04	0.44 ± 0.02
Mixed-ploidy plant	29.29 ± 3.06 **	70.99 ± 1.15	38.8 ± 1.54 **	1.91 ± 0.12 **	0.63 ± 0.05 *

Note: Control represents the original interspecific hybrid. * and ** represent statistically significant differences at $p < 0.05$ and $p < 0.01$, respectively.

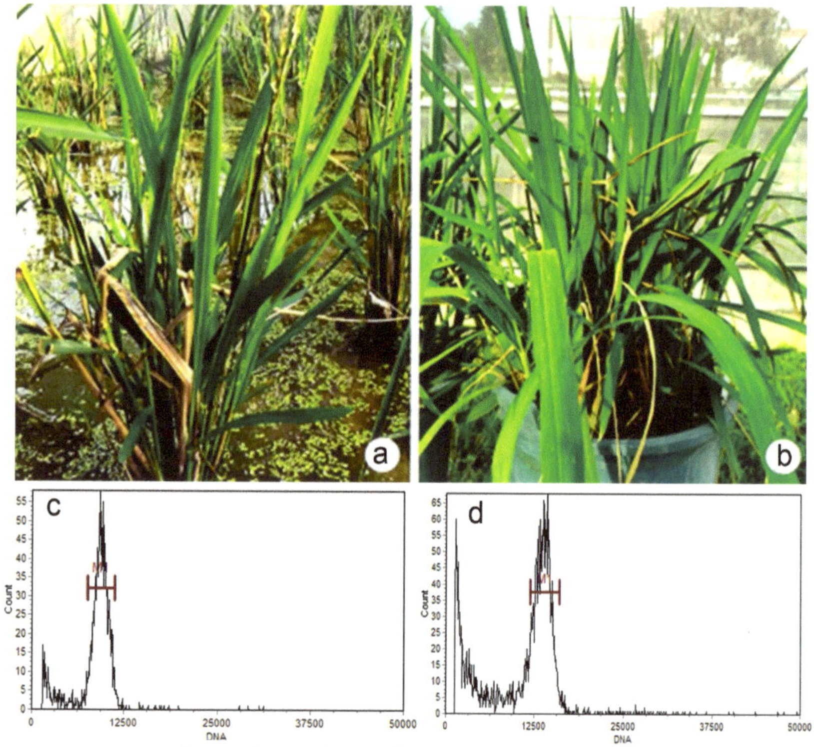

File name	Number of events	Genometric Mean	CV
Control.fcs	1,582	9,462.71	8.66

File name	Number of events	Genometric Mean	CV
Mix-poploidy.fcs	2,394	13,806.73	6.89

Figure 10. Comparison of mixed-ploidy plant and its original interspecific hybrid plant. (**a**) The phenotype of the interspecific hybrid plant; (**b**) the phenotype of the mixed-ploidy plant induced by colchicine treatment; (**c**) flow cytometric analysis of the interspecific hybrid plant; (**d**) flow cytometric analysis of induced candidate polyploid plants.

3. Discussion

3.1. Low Fertility Is the Major Limitation for the Utilization of the Hybrid Derived from O. officinalis and Cultivated Rice

Distant hybridization caused the recombination of highly heterologous genomes [13]. The incompatibility of chromosomes frequently led to the low fertility of distant hybrids, which is a significant barrier for their potential use. Pollen sterility and embryo sac abortion are important factors affecting distant rice hybrid fertility. Therefore, understanding the reproductive characteristics of interspecific hybrids is an important factor for using the potential germplasm of distant rice hybrid plants. Until now, very little has been known about the reproductive characteristics of interspecific hybrid plants of *O. officinalis*. From this study, we constructed hybrid plants of *O. officinalis* and cultivated rice to utilize the excellent gene resources of *O. officinalis*. Different types of abnormalities were detected in an interspecific hybrid of *O. officinalis* during rice pollen and embryo sac development. Abnormalities of the embryo sac, such as functional megaspore degradation, female reproductive unit degradation, and dysplasia, are frequently detected in *O. officinalis* × cultivated rice hybrids [29]. Our results proved the incompatibility of distant hybridization between *O. officinalis* and cultivated rice.

3.2. Factors Influencing Tissue Culture in the Interspecific Hybrid Derived from *O. officinalis* and Cultivated Rice

The type of explants, genotypes, exogenous PGRs, and other additives are important factors that affect the rice tissue culture. Rice anthers, young panicles, young embryos, and mature embryos were frequently used as explants in tissue culture [30]. From this study, high sterility was the major hindrance to the interspecific hybrid of *O. officinalis* and cultivated rice, leading to the difficulty in obtaining seed embryos. Therefore, we used young panicles as the explants, and preliminarily established the tissue culture of the interspecific hybrid of *O. officinalis* and cultivated rice. We screened the suitable culture formula of the interspecific hybrid, and the maximum emergence rate reached 28.97%. The emergence rate of this study was close to the other *O. officinalis* material in the Yunnan province [31].

Medium components have a great effect on interspecific hybrid callus induction. The nutrient broth (NB) basic medium produced the highest induction rate for the *Oryza alta* Swallen [23]. The N6 medium is suitable for interspecific hybrid callus induction, which was the same as the basic media for the callus induction of *indica* and *japonica* rice [32,33]. However, we detected that the MS medium was suitable for the callus induction culture of *O. officinalis*. This result differs from previous results and indicates that the suitable basic medium for induction probably has some relationship with rice genotypes.

PGRs and other additives are frequently used to improve the callus induction rate in interspecific hybrids. The type of exogenous PGRs added depends on the material. For example, 2,4-D can effectively promote callus formation in rice callus induction [34]. A 2.0–3.0 mg·L^{-1} 2,4-D concentration was suitable for callus induction in *indica* rice [35]. The combination of 6.0 mg·L^{-1} 2,4-D with NAA, IAA, and KIN was used to induce wild rice callus [36]. In the *O. officinalis*, the optimal medium for callus induction was 1.0 mg·L^{-1} 2,4-D combined with a small amount of cytokinin analogs [18]. From this study, 2.0 mg·L^{-1} 2,4-D was considered the most suitable concentration for callus induction in interspecific hybrid of *O. officinalis* and cultivated rice without adding other exogenous PGRs. The concentration of 2,4-D of interspecific hybrid was similar to the induction of *indica* rice and other wild rice material. All of these results indicated that 2,4-D played an important role in the induction of rice callus.

Cytokinins, such as BA or KIN, are essential and frequently added in the differentiation process. In the tissue culture of *O. alta*, 2.0 mg·L^{-1} KIN was used for callus differentiation, and the dosage was 10 times that of auxin analogue NAA [23]. In the *O. officinalis*, 2.0–4.0 mg·L^{-1} BA + 1.0–2.0 mg·L^{-1} KIN combined with 1.0 mg·L^{-1} IAA was the most suitable for differentiation and growth [31]. In this study, 2 mg·L^{-1} NAA combined with 0.2 mg·L^{-1} BA had the optimal effect in promoting differentiation. The ratio of exogenous PGRs was similar to Zhang et al. [23]. These results indicated that medium components, PGRs, and cytokinins play a critical role in the tissue culture of interspecific hybrids.

3.3. Polyploidy Induction of Interspecific Hybrids Derived from *O. officinalis* and Cultivated Rice

Colchicine-induced tetraploids have proven effective and were successfully employed in many plants [37,38]. A proper dosage range and treatment method of colchicine, as a frequently utilized polyploidy inducer substance, can successfully generate polyploidy plants [39–41]. However, colchicine treatment caused callus damage, resulting in the browning and death of the callus, and reduced differentiation ability. In the Gramineae plant, the induced polyploidy concentration of colchicine is always from 200 to 800 mg·L^{-1} [42]. This study found that 400–500 mg·L^{-1} of colchicine treatment showed a higher rate of 3-day seedling induction, which was comparable to earlier research on *O. alta* [42]. We speculated that the toxicity of colchicine in rice is similar within a specific range of concentration, and the toxicity has an enhanced effect when the concentration is higher than a certain limit.

Colchicine treatment led to the irreversible death of the callus cells [43]. High colchicine concentration always produces a large amount of toxins during the induction process. This study found that colchicine concentration was related to seedling emergence in interspecific hybrid plants. The toxic effect of colchicine is significantly increased when the colchicine concentration reaches 600 mg·L^{-1}. This result was similar to the polyploidy induction analysis of *O. alta*. The callus survival rate and seedling emergence rate of *O. alta* under the treatment of 600 mg·L^{-1} colchicine for 24 h was much higher than that of 200 to 400 mg·L^{-1} colchicine treatment [23]. In addition, we found that the colchicine treatment time also directly influenced the toxic effect in this study. Co-culturing for 5 days in colchicine treatment showed a stronger toxic effect than co-culturing for 3 days. This result is consistent with a previous study, which revealed that polyploidy induction of gerbera under colchicine treatment at 1% for 4 and 8 h showed a highly toxic effect compared to the treatment of 0.1% colchicine for 8 h [44]. All of these results indicated that colchicine concentration and treatment duration play a significant role in the induction of polyploidy in interspecific hybrids derived from *O. officinalis* and cultivated rice.

Polyploidy material exhibited non-significant differences in average plant height, leaf length, and stem diameter following treatment. Conforming to the general morphological norm of double-chromosome plants, leaf width and plant thickness were consistent. However, the number of tillers increased substantially compared to the tiller reduction typical of most polyploid rice genotypes [45].

4. Materials and Methods

4.1. Plant Material and Plant Growth Conditions

This investigation employed one interspecific hybrid created by *O. officinalis* and cultivated rice. The *O. officinalis* belongs to the CC genome and was utilized as the female parent, whereas cultivated rice belongs to the AA genome and was used as the male parent. All materials were grown at the farm (23°15′ N, 113°21′ E) of South China Agricultural University (SCAU), and management practices were according to the recommendations for the area. During the growing season, the average temperature was between 23 and 29 °C, with a relative humidity range of 74 to 88%. The young panicles of interspecific hybrid plants were collected and used for tissue culture, and the callus was employed in chromosome-doubling experiments by colchicine.

4.2. Cytological Observation of the Pollen and Embryo Sac Development Process

To verify pollen and embryo sac fertility in interspecific hybrid plants, whole-mount eosin B-staining confocal laser-scanning microscopy (WE-CLSM) was used. The young panicles were collected and fixed in an FAA solution for 48 h. Then, the samples were washed using 95% and 80% ethanol for 30 min, and kept in 70% ethanol at 4 °C until observation. Interspecific hybrid plant pollen and embryo sac fertility were detected with minor modifications to our previous research [46,47].

4.3. Tissue Culture of the Interspecific Hybrid Derived from *O. officinalis*

The young panicles of interspecific hybrid plants with a 0.5–5.0 cm range between their flag leaf cushion and the second to last leaf cushion were collected and stored in a refrigerator at 4 °C. The panicles were surface-sterilized with 75% ethanol for 30 s, and then disinfected in 0.1% $HgCl_2$ (mercury chloride) solution for 8–10 min. Then, samples were rinsed with sterile distilled water 4–6 times. Then, using five basic induction media with N6 or 1/2MS as the basic media and varying amounts of exogenous PGRs, immature panicles with 3.0 mm segments were treated, including 2,4-D (2.0 mg·L^{-1}), Pro (0, 0.3, and 3.0 g·L^{-1}) and acid-hydrolyzed casein (0, and 0.3 mg·L^{-1}).

To produce high-quality of calli, the N6 medium supplemented with 2,4-D (2.0 mg·L^{-1}) and three concentrations of BA (0, 0.2, and 0.5 mg·L^{-1}) were used for the subculture. All media were placed in the incubator at 28 °C and dark-cultured for 26 d. After the callus subculture, the basic medium of the MS and different concentrations of BA (0 and 2.0 mg·L^{-1}), KIN (0 and 2.0 mg·L^{-1}), and NAA (0.2, 0.4, and 1.0 mg·L^{-1}) were used for differentiation. The 1/2MS medium supplemented with 0.5 g·L^{-1} activated charcoal (AC) and different concentrations of IAA (0, 1.0, and 2.0 mg·L^{-1}), NAA (0, 1.0, and 2.0 mg·L^{-1}), and BA (0.2, and 0.5 mg·L^{-1}), was added to prepare the medium for promoting the rooting and vigorous seedlings.

4.4. Callus Treatment with Colchicine Solution

To induce an allopolyploid interspecific hybrid, the callus and cluster buds immersed in the liquid medium, which contained 1.5% dimethyl sulfoxide (DMSO) and five differential concentrations of colchicine (0, 300, 400, 500, and 600 mg·L^{-1}). The light yellow undifferentiated or partially green calli were immersed in a colchicine solution and cultured in a dark environment under a 150 rpm shaking table at 25 °C for 24 h. The treated callus was washed with sterile water and inoculated in the MS medium, including the different concentrations of BA (0, and 2.0 mg·L^{-1}), KIN (0, and 2.0 mg·L^{-1}), and NAA (0.2, 0.4, and 1.0 mg·L^{-1}) for differentiation. They were cultured under dark conditions at 28 °C for 2 days, and then the samples were transferred to a light-controlled environment to facilitate the development of seedlings.

4.5. Investigation of the Allopolyploid Hybrid and Statistical Analysis

To detect the variations in candidate polyploid plants, agronomic traits, including plant height, number of tillers, leaf shape, and stem diameter, were observed in this study. In accordance with the prior investigation [48], all characteristics were identified and subjected to statistical analysis. Excel 2016 and IBM SPSS Statistics 21.0 were used for data statistics and analysis in this study. The flow cytometric analysis was conducted based on our earlier research, with minor modifications [23].

5. Conclusions

The findings of this study suggest that the key factor contributing to the low fertility of the plant under investigation is the occurrence of pollen abortion and embryo sac abnormalities, as evidenced by the cytological analysis results. A tissue culture system has been generated for an interspecific hybrid derived from *O. officinalis* and cultivated rice. Moreover, the colchicine treatment at 400 mg·L^{-1} for 2 days is the optimal protocol to induce polyploidy in rice. The induction of polyploidy in *O. officinalis* and the tissue culture system described in this work may prove advantageous in addressing the issue of limited cross-compatibility in *O. officinalis*.

Supplementary Materials: The following supporting information can be downloaded at: https://www.mdpi.com/article/10.3390/plants12163001/s1, Figure S1: Normal pollen development of cultivated rice; Figure S2: Tissue culture of strong rooting seedlings cultured for 20 days. Figure S3: Callus differentiation at different durations in a 400 mg·L^{-1} colchicine co-culture. Table S1: Tissue culture for callus differentiation; Table S2: Effect of BA concentration on callus proliferation in the interspecific hybrid; Table S3: Effect of colchicine co-culture on callus.

Author Contributions: J.W. conceived and designed the experiments. M.T., R.C. and X.C. performed the experiments and analyzed the data. J.W., M.T., R.C., M.Q.S. and X.L. wrote and revised the paper. All authors have read and agreed to the published version of the manuscript.

Funding: This work was supported by the National Natural Science Foundation of China (32272154), the Guangdong Basic and Applied Basic Research Foundation (2021A1515010718, 2023A1515012672), and the Base Bank of Lingnan Rice Germplasm Resources Project (2023).

Institutional Review Board Statement: Not applicable.

Informed Consent Statement: Not applicable.

Data Availability Statement: Not applicable.

Acknowledgments: The authors thank Shuhong Yu, Zhixiong Chen, Lan Wan, and other lab members for assistance with the experiment.

Conflicts of Interest: The authors declare no conflict of interest.

References

1. Vaughan, D.A.; Morishima, H.; Kadowaki, K. Diversity in the Oryza genus. *Curr. Opin. Plant Biol.* **2003**, *6*, 139–146. [CrossRef]
2. Brar, D.S.; Khush, G.S. Alien introgression in rice. *Plant Mol. Biol.* **1997**, *35*, 35–47. [CrossRef]
3. Kitazumi, A.; Pabuayon, I.C.M.; Ohyanagi, H.; Fujita, M.; Osti, B.; Shenton, M.R. Potential of *Oryza officinalis* to augment the cold tolerance genetic mechanisms of *Oryza sativa* by network complementation. *Sci. Rep.* **2018**, *8*, 16346. [CrossRef]
4. Liu, R.; Zhang, H.; Chen, Z.; Shahid, M.Q.; Fu, X.; Liu, Y.; Liu, X.; Lu, Y. Development of drought-tolerant rice germplasm by screening and transforming TAC clones of *Oryza officinalis* Wall. *Sci. Agric. Sin.* **2014**, *47*, 1445–1457.
5. Zhang, W.L.; Dong, Y.; Yang, L.; Ma, B.J.; Ma, R.R.; Huang, F.D.; Wang, C.C.; Hu, H.T.; Li, C.S.; Yan, C.Q.; et al. Small brown planthopper resistance loci in wild rice (*Oryza officinalis*). *Mol. Genet. Genom.* **2014**, *289*, 373–382. [CrossRef]
6. Jiang, C.M.; Xiao, S.Q.; Li, D.Q.; Chen, L.; Zhang, Q.F.; Yin, F.Y.; Yu, T.Q.; Ke, X.; Zhang, D.Y. Identification and expression pattern analysis of bacterial blight ersistance genes in *Oryza officinalis* wall ex watt under Xanthomonas *Oryzae* Pv. *Oryzae Stress. Plant Mol. Biol. Rep.* **2019**, *37*, 436–449. [CrossRef]
7. Jena, K.K.; Khush, G.S. Introgression of genes from *Oryza officinalis* well ex watt to cultivated rice, *O. sativa* L. *Theor. Appl. Genet.* **1990**, *80*, 737–745. [CrossRef]
8. Huang, Z.; He, G.; Shu, L.; Zhang, Q. Identification and mapping of two brown planthopper resistance genes in rice. *Theor. Appl. Genet.* **2001**, *102*, 929–934. [CrossRef]
9. Liu, Y.G.; Nagaki, K.; Fujita, M.; Kawaura, K.; Uozumi, M.; Ogihara, Y. Development of an efficient maintenance and screening system for large-insert genomic DNA libraries of hexaploid wheat in a transformation-competent artificial chromosome (TAC) vector. *Plant J.* **2000**, *23*, 687–695. [CrossRef]
10. Song, J.Q.; Dong, F.G.; Lilly, J.W.; Stupar, R.M.; Jiang, J. Instability of bacterial artificial chromosome (BAC) clones containing tandemly repeated DNA sequences. *Genome* **2001**, *44*, 463–469. [CrossRef]
11. Fu, X.L.; Lu, Y.G.; Liu, X.D.; Li, J.Q.; Zhao, X.J. Cytological behavior of hybridization barriers between *Oryza sativa* and *Oryza officinalis*. *Agric. Sci. China* **2011**, *10*, 1489–1500. [CrossRef]
12. Tan, G.X.; Ren, X.; Weng, Q.M.; Shi, Z.Y.; Zhu, L.L.; He, G.C. Mapping of a new resistance gene to bacterial blight in rice line introgressed from *Oryza officinalis*. *J. Genet. Genom.* **2004**, *31*, 724–729.
13. McClintock, B. The significance of responses of the genome to challenge. *Science* **1984**, *226*, 792–801. [CrossRef]
14. Thengane, S.R.; Joshi, M.S.; Khuspe, S.S.; Mascarenhas, A.F. Anther culture in Helianthus annuus L.; influence of genotype and culture conditions on embryo induction and plant regeneration. *Plant Cell Rep.* **1994**, *13*, 222–226. [CrossRef]
15. Zheng, W.J.; Zhang, Y.Z.; Wang, C.H.; Chao, J.M.; Hu, J. Studies on the differences of tissue culture of different explants in rice and their progeny variation. *J. Anhui Agri Sci.* **2008**, *36*, 1368–1370.
16. Lin, Y.J.; Zhang, Q.F. Optimising the tissue culture conditions for high efficiency transformation of indica rice. *Plant Cell Rep.* **2005**, *23*, 540–547. [CrossRef]
17. Tie, W.W.; Zhou, F.; Wang, L.; Xie, W.B.; Chen, H.; Li, X.H.; Lin, Y.G. Reasons for lower transformation efficiency in indica rice using Agrobacterium tumefaciens-mediated transformation: Lessons from transformation assays and genome-wide expression profiling. *Plant Mol. Biol.* **2012**, *78*, 1–18. [CrossRef]
18. Wang, F.; Yuan, W.X.; Pao, G.L.; Qian, C.G.; Huang, Q.; Liu, J.; Yang, Y.; Chen, J.P. Rapid propagation of embryo rescue progeny of medicinal wild rice (*Oryza sativa* L.) by YF. *J Zhejiang Agri Sci.* **2017**, *58*, 642–643.
19. Shahid, M.Q.; Xu, H.; Lin, S.; Chen, Z.; Naeem, M.; Li, Y.; Liu, X. Genetic analysis and hybrid vigor study of grain yield and other quantitative traits in autotetraploid rice. *Pak. J. Bot.* **2012**, *44*, 237–246.
20. Ghaleb, M.A.A.; Li, C.; Shahid, M.Q.; Yu, H.; Liang, J.; Chen, R.; Wu, J.; Liu, X.D. Heterosis analysis and underlying molecular regulatory mechanism in a wide-compatible neo-tetraploid rice line with long panicles. *BMC Plant Biol.* **2020**, *20*, 83. [CrossRef]
21. Ghouri, F.; Shahid, M.J.; Liu, J.; Lai, M.; Sun, L.; Wu, J.; Liu, X.; Ali, S.; Shahid, M.Q. Polyploidy and zinc oxide nanoparticles alleviated Cd toxicity in rice by modulating oxidative stress and expression levels of sucrose and metal-transporter genes. *J. Hazard. Mater.* **2023**, *448*, 130991. [CrossRef] [PubMed]
22. Ghouri, F.; Shahid, M.J.; Liu, J.; Sun, L.; Riaz, M.; Imran, M.; Ali, S.; Liu, X.; Shahid, M.Q. The protective role of tetraploidy and nanoparticles in arsenic-stressed rice: Evidence from RNA sequencing, ultrastructural and physiological studies. *J. Hazard. Mater.* **2023**, *458*, 132019. [CrossRef]
23. Zhang, L.S.; Shivute, F.N.; Shahid, M.Q.; Kamara, N.; Wu, J.W.; Liu, X.D. In vitro induction of auto-allotetraploid in a newly developed wild rice line from *Oryza alta* Swallen. *Plant Cell Tiss. Organ. Cult.* **2019**, *139*, 577–587. [CrossRef]
24. Bouvier, L.; Guérif, P.; Djulbic, M.; Durel, C.E.; Chevreau, E.; Lespinasse, Y. Chromosome doubling of pear haploid plants and homozygosity assessment using isozyme and microsatellite markers. *Euphytica* **2002**, *123*, 255–262. [CrossRef]
25. Xing, S.H.; Guo, X.B.; Wang, Q.; Pan, Q.F.; Tian, Y.S.; Liu, P.; Zhao, J.Y.; Wang, G.F. Induction and flow cytometry Identification of tetraploids from seed-derived explants through colchicine treatments in *Catharanthus roseus* (L.) G. Don. *J. Biomed. Biotechnol.* **2011**, *2011*, 793198. [CrossRef]
26. Vahdati, K.; Sadat-Hosseini, M.; Martínez-Gómez, P.; Germanà, M.A. Production of haploid and doubled haploid lines in nut crops: Persian walnut, almond, and hazelnut. *Methods Mol. Biol.* **2021**, *2289*, 179–198. [CrossRef]
27. Hosseini Grouh, M.S.; Vahdati, K.; Lotfi, M.; Hassani, D.; Biranvand, N.P. Production of haploids in persian walnut through parthenogenesis induced by gamma-irradiated pollen. *J. Amer. Soc. Hort. Sci.* **2011**, *136*, 198–204. [CrossRef]

28. Sadat Noori, S.A.; Norouzi, M.; Karimzadeh, G.; Shirkool, K.; Niazian, M. Effect of colchicine-induced polyploidy on morphological characteristics and essential oil composition of ajowan (*Trachyspermum ammi* L.). *Plant Cell Tiss. Organ. Cult.* **2017**, *130*, 543–551. [CrossRef]
29. Fu, X.L.; Lu, Y.G.; Liu, X.D.; Li, J.Q.; Zhao, X.J. Comparatively embryological observations on interspecific hybridizations between *Oryza sativ* a with different ploidy levels and *O.officinalis*. *Chin. J. Rice Sci.* **2008**, *22*, 385–391. [CrossRef]
30. Guzmán, M.; Arias, F.J.Z. Increasing anther culture efficiency in rice (*Oryza sativa* L.) using anthers from ratooned plants. *Plant Sci.* **2000**, *151*, 107–114. [CrossRef]
31. Wang, L.X.; Chen, L.; Chen, Y.; Wang, B.; Fu, J.; Xiao, S.Q.; Ke, X.; Zhang, Q.F.; Lei, Y.T.; Yu, T.Q.; et al. Application of rapid propagation seedlings of yunnan medicinal *Oryza officinalis* Wall. ex wstt in breeding. *Liaoning Agric. Sci.* **2019**, *5*, 7–11.
32. Lee, K.; Jeon, H.; Kim, M. Optimization of a mature embryo-based in vitro culture system for high-frequency somatic embryogenic callus induction and plant regeneration from japonica rice cultivars. *Plant Cell Tissue Organ. Cult.* **2002**, *71*, 237–244. [CrossRef]
33. Liu, Y.F.; Liu, Y.Z.; He, H.; Wang, J.H.; Li, L.; Li, X.F. Factors Influening of callus Induction and plantlet regeneration from mature embryos in indica rice. *Plant Physiol. Commun.* **2004**, *40*, 319–322.
34. Chaturvedi, P.; Chowdhary, A. Enhancement of antioxidant compound in tylophora *indica* (*Asclepeadaceae*) callus. *Adv. Appl. Sci. Res.* **2013**, *4*, 325–330.
35. Ge, X.J.; Chu, Z.H.; Lin, Y.J.; Wang, S.P. A tissue culture system for different germplasms of indica rice. *Plant Cell Rep.* **2006**, *25*, 392–402. [CrossRef]
36. Pan, Y.H.; Liang, Y.T.; Chen, C.B.; Liang, S.C.; Tuan, J.Z.; Xu, J.P.; Huang, J.; Xu, Z.J. Study on differentiation and dedifferentiation of Oryza rufipogon Griff. *J. Hennan Agri Sci.* **2014**, *43*, 25–29. [CrossRef]
37. Aleza, P.; Juárez, J.; Cuenca, J.; Ollitrault, P.; Navarro, L. Recovery of citrus triploid hybrids by embryo rescue and flow cytometry from 2x × 2x sexual hybridisation and its application to extensive breeding programs. *Plant Cell Rep.* **2010**, *29*, 1023–1034. [CrossRef]
38. Urwin, N.A.R. Generation and characterisation of colchicine-induced polyploid Lavandula × intermedia. *Euphytica* **2014**, *197*, 331–339. [CrossRef]
39. Zhang, X.Y.; Liu, J.F.; Wang, L.Q. Polyploidy breeding and its application research progress of medicinal plants. *J. Jilin Norm. Univ. (Nat. Sci. Ed.)* **2009**, *30*, 128–131.
40. Dhooghe, E.; Van Laere, K.; Eeckhaut, T.; Leus, L.; Van Huylenbroeck, J. Mitotic chromosome doubling of plant tissues in vitro. *Plant Cell Tiss. Organ. Cult.* **2011**, *104*, 359–373. [CrossRef]
41. Xie, X.Q.; Agüero, C.B.; Wang, Y.J.; Walker, M.A. In vitro induction of tetraploids in Vitis × Muscadinia hybrids. *Plant Cell Tiss. Organ. Cult.* **2015**, *122*, 675–683. [CrossRef]
42. Luan, L.; Long, W.B.; Wang, X.; Chen, Y.; Qiu, L.; Tu, S.B.; Kong, F.L.; Xiao, X.Y. Research on creating autotetraploid rice using colchicum to induce the panicle callus of diploid rice. *J. Sichuan Uni. (Nat. Sci. Ed.)* **2009**, *46*, 829–837.
43. Xu, L.; Najeeb, U.; Naeem, M.S.; Naeem, M.S.; Daud, M.K.; Cao, J.S. Induction of tetraploidy in Juncus effusus by colchicine. *Biol. Plant* **2010**, *54*, 659–663. [CrossRef]
44. Gantait, S.; Mandal, N.; Bhattacharyya, S.; Das, P.K. Induction and identification of tetraploids using in vitro colchicine treatment of *Gerbera jamesonii* bolus cv. sciella. *Plant Cell Tiss. Organ. Cult.* **2011**, *106*, 485–493. [CrossRef]
45. Shahid, M.Q.; Li, Y.J.; Saleem, M.; Wei, C.M.; Naeem, M.; Liu, X.D. Yield and yield components in autotetraploid and diploid rice genotypes (*indica* and *japonica*) sown in early and late seasons. *Aust. J. Crop Sci.* **2013**, *7*, 632–641.
46. Ghouri, F.; Zhu, J.; Yu, H.; Wu, J.; Baloch, F.S.; Liu, X.D.; Shahid, M.Q. Deciphering global DNA variations and embryo sac fertility in autotetraploid rice line. *Turk. J. Agric.* **2019**, *43*, 554–568. [CrossRef]
47. Shahid, M.Q.; Liu, G.; Li, J.; Naeem, M.; Liu, X.D. Heterosis and gene action study of agronomic traits in diploid and autotetraploid rice. *Acta Agric. Scand. Sect. B Plant Soil. Sci.* **2011**, *61*, 23–32. [CrossRef]
48. Wu, J.W.; Hu, C.Y.; Shahid, M.Q.; Guo, H.B.; Zeng, Y.X.; Liu, X.D.; Lu, Y.G. Analysis on genetic diversification and heterosis in autotetraploid rice. *SpringerPlus* **2013**, *2*, 439. [CrossRef] [PubMed]

plants

MDPI

Review

The Cryopreservation of Medicinal and Ornamental Geophytes: Application and Challenges

Soumaya El Merzougui [1], Carla Benelli [2,*], Rachida El Boullani [1] and Mohammed Amine Serghini [1]

[1] Laboratory of Biotechnology and Valorization of Natural Resources, Department of Biology, Faculty of Sciences, Ibn Zohr University, Agadir 8106, Morocco; soumaya.merzougui@gmail.com (S.E.M.); rachidaelboullani@gmail.com (R.E.B.); m.serghini@uiz.ac.ma (M.A.S.)

[2] Institute of BioEconomy, National Research Council (CNR/IBE), Sesto Fiorentino, 50019 Florence, Italy

* Correspondence: carla.benelli@ibe.cnr.it

Abstract: Nowadays, plant genetic resources are often at risk of loss and destruction. Geophytes are herbaceous or perennial species that are annually renewed by bulbs, rhizomes, tuberous roots, or tubers. They are often subject to overexploitation, which, combined with other biotic and abiotic stresses, can make these plants more vulnerable to a decline in their diffusion. As a result, multiple endeavors have been undertaken to establish better conservation strategies. Plant cryopreservation at ultra-low temperatures in liquid nitrogen (−196 °C) has proven to be an effective, long-term, low-cost, and suitable conservation method for many plant species. Over the last two decades, major advances in cryobiology studies have enabled successful explants of multiple genera and types, including pollen, shoot tips, dormant buds, and zygotic and somatic embryos. This review provides an update on recent advances and developments in cryopreservation and its application to medicinal and ornamental geophytes. In addition, the review includes a brief summary of factors limiting the success of bulbous germplasm conservation. The critical analysis underpinning this review will benefit biologists and cryobiologists in their further studies on the optimization of geophyte cryopreservation protocols and will support a more complete and wider application of knowledge in this area.

Keywords: long-term conservation; plant germplasm; bulbous plant; pollen; shoot tip; buds

check for updates

Citation: El Merzougui, S.; Benelli, C.; El Boullani, R.; Serghini, M.A. The Cryopreservation of Medicinal and Ornamental Geophytes: Application and Challenges. *Plants* **2023**, *12*, 2143. https://doi.org/10.3390/plants12112143

Academic Editor: Andreas W. Ebert

Received: 14 April 2023
Revised: 12 May 2023
Accepted: 23 May 2023
Published: 29 May 2023

1. Introduction

The plant germplasm represents not only a vital genetic resource but also has great ecological, biological, therapeutic, social, economic, spiritual, and aesthetic significance and potential. Unfortunately, dramatic biodiversity loss has been recorded due to multiple factors. The main causes of this biodiversity loss are climatic changes, overexploitation, and the unreasonable collection of wild germplasm due to its enormous demand and value. Moreover, the degradation of natural habitats is caused by urban development along with the invasion of exotic species, pests, and diseases [1,2]. The loss of plant biodiversity has an impact not only on ecosystem performance but also on human livelihoods and food security. Over the last few years, great efforts and strategies have been employed for germplasm conservation.

The conservation of germplasm can be carried out by multiple strategies, including both in situ and ex situ conservation. The first approach is defined as the conservation of plants' germplasm in their natural habitat or environment [3]; it allows the preservation of ecosystems, species, and the co-evolution of plants with environmental changes. Nonetheless, this approach is often insufficient to achieve a complete safeguard for endangered species [1]. Ex situ conservation methods consist of germplasm conservation outside its natural ecosystem under controlled conditions. This conservation can be carried out via botanical gardens, seed banking, DNA storage, pollen storage, or field gene banks, and it is used to preserve threatened and vulnerable species. Seed storage is the most widely implicated pillar of ex situ preservation. Nevertheless, this technique is not practicable for some recalcitrant seeds or vegetatively propagated species, such as bulbous plants.

The application of biotechnology has helped the progress of plant germplasm ex situ conservation, starting with in vitro propagation. In addition to the large-scale production of disease-free plants, this technique allows medium-term conservation of about 1 to 5 years. The success of in vitro conservation relies on several factors, including the type of explant, the medium, and the culture conditions. However, the most critical factor depends on the acclimatization of plants to ex vitro conditions [1]. A more radical method for long-term, safe, and cost-effective storage is cryopreservation. This technique has proven to be a promising conservation method, and it has been applied successfully to many species [4,5] but it often requires an optimized in vitro culture protocol for the species tested [6,7]. Indeed, most of the explants used in cryopreservation derive from in vitro conditions, and their regrowth after liquid nitrogen depends on this propagation technique [8]. Bulbous geophytes are plants with an underground storage structure, such as a corm, rhizome, or tuber, which helps them to survive and sustain themselves in extreme conditions. Bulbous plants mainly belong to the *Amaryllidaceae*, *Alliaceae*, *Hyacinthaceae*, *Liliaceae*, and *Iridaceae* families. Geophyte plants are cultured naturally in many countries worldwide and are exploited for their enormous benefits for ornamental (flower bulb), economic, or medicinal purposes. The preservation of geophyte plants by the traditional method has many limitations, such as being costly, labor-intensive, and having the risk of natural disasters. Hence, it is important to use alternative methods, such as cryopreservation.

In this review, details about different approaches to the cryopreservation technique and its recent advances are summarized. Moreover, up-to-date information and studies on the application of cryopreservation to several ornamental and medicinal geophytes are highlighted. The large family of *Orchidaceae* has not been included in this review as it will be the single subject of a subsequent paper.

2. History, Principles, and Fundamentals of Cryopreservation

The application of the plant cryopreservation method has its origins in the use of cryobiology on mammalian species [9]. The first successful plant species storage at ultra-low temperatures was reported by Sakai [10] on mulberry twigs. Since then, cryopreservation has been successfully applied to a wide variety of species [9,11]. Cryopreservation is usually undertaken using two main approaches: (1) classical techniques, referred to as controlled freezing or slow cooling, and (2) vitrification-based procedures.

The first procedure is based on the osmotic regulation of cell contents and freeze-induced dehydration of plant material using a slow cooling system, before cryogenic storage in liquid nitrogen (LN) [12–14]. This technique involves controlled cooling rates alone or in combination with colligative cryoprotectants [15,16]. In fact, slow cooling is the determinant factor of cell survival after conservation. With a slow decrease in temperature, ice crystals are forced to form in the extracellular solution.

The second approach is the vitrification procedure. The basis of this technology is the transition that water undergoes from its liquid state to an amorphous 'glassy state' [12,16,17]. In this transition, crystalline ice structures are not formed; therefore, ice injuries are avoided. The vitrification state is achieved by the dehydration of cells, either by the treatment of plant tissue in a mixture of highly concentrated cryoprotective solutions (penetrating and non-penetrating agents) or by physical desiccation and subsequent very rapid cooling by direct immersion in LN. This procedure is simple, easy to apply, and does not need specific apparatus for controlled cooling; it has high reproducibility, and it has been applied successfully to a wide range of species using several explant types [16,18]. Therefore, it is amenable and more widely used compared to slow-controlled systems [14].

Different vitrification-based procedures can be mentioned: vitrification, encapsulation–dehydration, encapsulation–vitrification, droplet vitrification, and V- and D-cryoplate (Figure 1). Below is a short description of the procedures mentioned above.

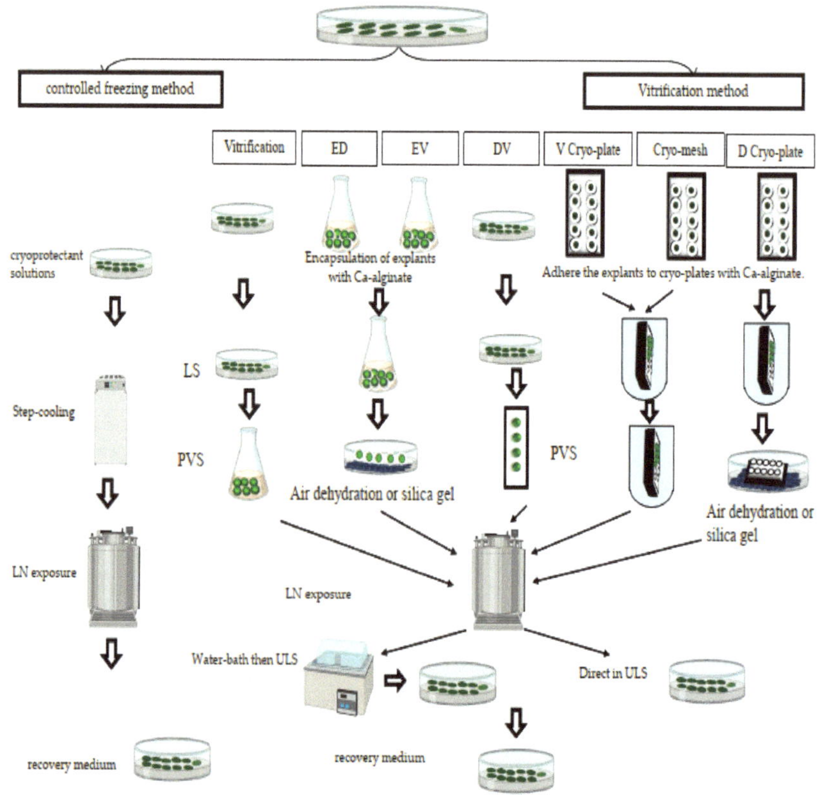

Figure 1. Cryopreservation techniques and the major steps in different procedures. ED: encapsulation–dehydration; EV: encapsulation–vitrification; DV: droplet vitrification; LS: loading solution; PVS: plant vitrification solution; LN: liquid nitrogen; ULS: unloading solution.

2.1. *Vitrification*

Vitrification involves the treatment of plant explants with a highly concentrated cryoprotective mixture of penetrating and non-penetrating agents. The additive penetrates the cell cumulatively. Consequently, it increases the solute concentration, while the non-penetrating agent acts synergistically by withdrawing water osmotically. The use of these additive solutions increases the viscosity of the cytoplasm cell, avoiding ice formation.

The complete procedure includes several steps in order to improve the survival of tissue/explants after their immersion in LN.

(i) Pretreatment of explants: this step is important because it contributes to the dehydration tolerance of plants. It consists, generally, of explants' treatment with a solution containing a high sucrose concentration for a short period (20–30 min): a loading solution (LS) [17].

(ii) Dehydration with complex mixtures of cryoprotectants, such as plant vitrification solutions (PVS2, PVS3). The most commonly employed mixture in cryopreservation is the PVS2 developed by Sakai et al. [19], which consists of 30% (*w*/*v*) glycerol, 15% (*w*/*v*) ethylene glycol, 15% (*w*/*v*) dimethyl sulfoxide (DMSO), and 0.4 M sucrose. Its exposure time to the explant can be a critical parameter in determining the possible toxic effect of the solution. The treatment period can vary (from 20 min to 200 min or more) depending on the type, size, and susceptibility of the explant. Moreover, most cryopreservation studies report that performing osmodehydration at 0 °C instead of 25 °C reduces the toxicity of PVS2 [7].

(iii) Rapid cooling by direct immersion in LN in order to promote the vitrification of internal solutes.

(iv) Rapid thawing of cryopreserved samples in a water bath at 40 °C for 1 min to 3 min. This step has to be completed very quickly to avoid devitrification, which would lead to ice crystal formation.

(v) Unloading treatment with a liquid medium containing 1.2 M sucrose or sorbitol (washing solution: ULS) is employed to remove the vitrification solution progressively and wash the cryopreserved explants.

2.2. Encapsulation–Dehydration and Encapsulation–Vitrification

The encapsulation–dehydration (ED) approach was developed by Fabre and Dereuddre [20]. This procedure is based on synthetic seed technology [21]. The dehydration process is promoted through water elimination by sterile airflow or silica gel [12,15,22]. In this method, the explant is first encapsulated in alginate-Ca beads, followed by treatment with a highly concentrated sucrose solution (0.7–1.5 M). The concentration is generally increased over several days in order to improve the intracellular solute concentration in the cells. The beads are then physically dehydrated under cabinet flow or with silica gel until an adequate moisture content is obtained (about 20–25%). Then, the beads are plunged into LN. The beads are thawed in a water bath at about 40 °C for 2–3 min. ED is laborious and time-consuming compared with the vitrification method, but it has the advantage of physical dehydration without cryoprotectant solution, which can be toxic for some explants, and overcomes the problems associated with the sensitivity of explants to PVS2 vitrification solution [22].

The encapsulation–vitrification (EV) technique is a combination of the two abovementioned techniques: encapsulation–dehydration and vitrification. The samples are encapsulated in alginate-Ca beads, and then the beads are treated with a vitrification solution to dehydrate before direct immersion in LN [7,14,23]. The time used for dehydration is greatly decreased in EV compared to that of ED.

2.3. Droplet Vitrification

In standard vitrification, the explant takes place in a cryotube before immersion in LN, while in the droplet vitrification (DV) procedure, the explants are placed on aluminum foil strips inside a drop of cryoprotectant solution. This procedure was developed with potatoes by Mix-Wagner et al. [24], using 10% DMSO solution as the drops with shoot tips on aluminum foil strips and rapid freezing by direct immersion in LN. The advantages of this method are the specific contact between the cryoprotectant and the explant inside the droplets and the use of aluminum foil, a metal with efficient thermal conductivity that can ensure homogenous cooling by uniform temperature dispersion. DV was applied for banana cryopreservation by Panis et al. [25], and it resulted in a significant increase in the regrowth rate compared to the classical vitrification protocol. The DV method is the most widely applied cryopreservation method for various species, including herbaceous and woody plants [11,26]. The positive findings reported could be due to the more direct contact of explants with LN, facilitating both their rapid freezing and thawing [23,27,28].

2.4. V- and D-Cryoplate

V-cryoplate (VC) combines the two techniques, droplet vitrification and encapsulation–vitrification, in which encapsulation plays an essential role [29]. The VC method was developed by Yamamoto et al. [28] in order to facilitate large-scale cryobanks and reduce the damage and loss of samples during the cryopreservation procedure. This technique uses aluminum cryoplates with 10–12 small circular wells, where the explants are placed and which are then filled with a drop of alginate solution containing 2–3% (w/v) sodium alginate in a calcium-free MS basal medium [23,27]. They are then left for 15 min to achieve total polymerization with the calcium chloride solution. The samples are adhered to the cryoplate during the whole procedure. The explants are physically dehydrated in the air or with silica gel before immersion in LN and thawing in an adequate solution at room temperature [27].

D-cryoplate (DC) also employs aluminum cryoplates, as in the VC procedure. The main difference between VC and DC is the dehydration step. In this case, after encapsulation, the explants are treated with a loading solution, dehydrated with a vitrification solution on cryoplates, frozen in LN, and thawed at room temperature [23,27,30]. This technique was developed by Niino et al. [30] to avoid the damage that could occur at different stages, including physical damage to the explant during the cryopreservation steps, excessive osmotic stress, and the chemical toxicity of PVS treatment [23].

These techniques are the latest advancements in cryopreservation, and the advantages of both methods are that they are easy and simple techniques to perform and that the samples become sealed to the cryoplates. Thus, the possibility of injuring or losing samples during the cryopreservation procedure is low [23]. The adhesion of samples to the cryoplates depends on the following factors: the size of wells and explants, the volume of alginate gel, sucrose, and/or glycerol in the alginate solution, and the tension of the surface of the cryoplate [27,31].

A significant advantage is the high rate of survival and growth of sensitive cryopreserved shoots; also, in this case, the result might be due to the fact of overcoming the damages associated with PVS toxicity and explants' sensitivity with respect to the classical vitrification procedure, where the explants are exposed, and to the high and rapid cooling and warming rates of treated materials, which improve recovery [23,30,32,33].

2.5. Cryo-Mesh

The cryo-mesh method was developed recently by Funnekotter et al. [34] for the successful cryopreservation of kangaroo paw (*Anigozanthos viridis*) shoot tips. The general procedure of this technique is similar to the VC method, and the main difference is the use of wire stainless steel mesh strips (cryo-mesh) instead of aluminum cryoplates.

3. Application of Cryopreservation to Some Medicinal and Ornamental Bulbous Plants

The choice of cryopreservation methods adopted, as well as the processes applied, including preconditioning, preculturing, cryoprotectant treatments, storage in LN, rewarming, and the recovery of samples, are the main factors to successful conservation (Figure 2). All of these steps have vital impacts on achieving a higher survival and growth rate of cryopreserved explants, but it should be emphasized that the major issues related to bulbous species propagation, such as poor generation of new buds, low growth rate, and contamination, should be overcome first with an optimized in vitro propagation protocol.

In this section, the main factors investigated and reviewed on the cryopreservation of different geophytes germplasm are emphasized and are summarized in Table 1.

3.1. Amaryllidaceae

Allium genus—Earlier taxonomic classifications placed the *Allium genus* in the *Liliaceae* family, but recently it has been indicated to be more closely related to the *Amaryllidaceae* family [35,36]. Garlic is reported for its wide range of therapeutic effects, including antihyperglycemic, antibacterial, antifungal, anticancer, and cardioprotective effects [37]. Presently, several papers have reported that cryopreservation is the most appropriate and reliable tool for space- and cost-efficiency and for providing an unlimited storage period for this species [38–40] The first report of cryopreservation was of *Allium wakegi*, with seven cultivars successfully cryopreserved by vitrification. Just after, Niwata [41] reported the conservation of *A. sativum* apical meristems using the same method. Over time, many studies tested cryopreservation procedures for *Allium*, including vitrification [42–47], encapsulation–dehydration [48,49], droplet vitrification [46,50–55], and cryoplate methods [23]. A variety of explants were examined in these different cryopreservation procedures, including shoot tips [23,42,50–53,56–58], unripe inflorescences and bulbil primordia [54,55], pollen [59], and embryonic callus [47].

Several factors have been highlighted to influence the success of allium cryopreservation, among them the effects of post-harvest storage and storage duration, which are linked to the dormancy of bulbs after harvest. The pre-acclimatization or cold storage of garlic bulbs and their storage duration have also proved to affect the survival and regeneration of cryopreserved garlic material [56]. Further, Lynch et al. [48] have also reported the importance of the physiological state and the duration of the storage of bulbs on the survival and regrowth of cryopreserved stem discs dissected from bulbs by encapsulation–dehydration. They obtained the highest survival of encapsulated stem discs (75%) with bulbs maintained for a period of 5 months at 10 °C, while the regrowth was 55% after 4 months. The most critical factors for the recovery and growth of cryopreserved garlic shoot tips were the vitrification solution and dehydration duration, followed by the material (source and size), preculture duration, sucrose concentration in the preculture medium, warming velocity, and unloading duration.

The type of PVS to apply in garlic cryopreservation was another issue discussed in several papers. According to Makowska et al. [45], the average survival of the three garlic accessions tested was high using PVS3 (76–83%) in comparison with the PVS2 treatment (0–37%). Further, Kim et al. [60] demonstrated that higher survival and regeneration rates of explants (90.6% and 83.2%, respectively) were obtained using PVS3 for 150 min under the DV protocol compared to the original PVS2 for 30 min (79% and 66%). Wang et al. [51,52], applying the DV procedure, observed that PVS3 was suitable for shoot tips of *A. cepa*, a small-bulb onion featuring short growth periods and resistance to diseases; the optimal regrowth of explants (58%) was obtained with 3 h of osmodehydration with PVS3. Moreover, in the same study, they detected, by differential scanning calorimetry (DSC), a 1.8% content of freezable water in the shoot tips dehydrated with PVS2 and no freezable water in those treated with PVS3. Further validation of this state was provided by histological observations; indeed, they reported that many cells in the apical dome survived after LN using PVS3 dehydration, while with PVS2 treatment, fewer cells survived. On the contrary, Volk et al. [53] used PVS2 for 30 min at 0 °C under the vitrification procedure, with a range of regrowth cryopreserved shoot tips from 25% to 75%, depending on the garlic accessions. Tanaka et al. [23] optimized the V-cryoplate procedure for *A. chinense* shoot tips, obtaining the maximum regrowth (100%) with a precultured treatment for 3 days in MS with 0.3 M sucrose and dehydration with PVS2 for 20–40 min. Survival and regrowth depend also on the size of the bulbs and the size of the explants used. Generally, the explants derived from large bulbs were better than the small ones, as reported by Keller [46]; moreover, the explants with sizes of 0.5–1 mm and lengths of 1–2 mm were preferred. Indeed, if the size is too large, the cryoprotective substances have difficulty entering all parts, and the thermal tensions can cause tissue ruptures during cooling or thawing; on the contrary, if the explants are too small, the ability to regrow can be lost [43,46]. The success of cryopreservation depends on the size of the final explants. Moreover, the type of explants used can also improve the success of cryopreservation in garlic. Wang et al. [52], applying the same DV protocol, observed different regrowth percentages using in vitro shoot tips (58%), adventitious buds (72%), and shoot tips from field bulbs (32%). Post-culture also has a significant impact on post-shoot tip regrowth. Keller [46] proved that a high regrowth rate of non-cryopreserved explants (95%) was obtained after preculture and post-culture on 0.3 M sucrose, and a 17.5% rate of cryopreserved explants was obtained with 0.44 M and 0.8 M sucrose preculture and post-culture treatments, respectively. Moreover, the author demonstrated the effect of explant types and cryopreservation procedures on the regrowth; the highest post-regrowth value was obtained from bulbils (72% and 80%) and in vitro plantlets (13% and 20%) using the vitrification and droplet vitrification techniques, respectively. Further, Wang et al. [51] also proved that more than 50% regrowth was obtained post-culture with 0.3 M sucrose for 2 days. However, post-regrowth of the shoot can depend on the genotypes/cultivars, as reported in some studies [50,52–54].

More recently, Tanaka et al. [58] reported the successful cryopreservation of *A. chinense* G. Don shoot tips by D-cryoplate for future applications in a gene bank. The authors embedded the explants on cryoplates with Ca-alginate. Thereafter, the cryoplates were immersed in an osmoprotection solution containing 2.0 M glycerol and 1.0 M sucrose and maintained for 30 min at 25 °C, followed by dehydration for 30 to 120 min at 25 °C. The cryoplate with explants was stored at −80 °C and −196 °C (LN); a high level of regrowth was recorded at both ultra-low temperatures. The average regrowth rates of the cryopreserved shoot tips recorded were 95.3% at −80 °C and −94.0% at 196 °C.

Cryopreservation is also a suitable method for the long-term preservation of *Allium* pollen. Ganeshan et al. [61,62] reported, in the first protocol for *A. cepa* pollen, that pollen viability and fertility were not influenced after 360 days of cryopreservation. In another study by Senula and Keller [59], different accessions within 78 *Allium* species showed a high mean germination of cryopreserved pollen, about 78%. In particular, the post-germination rate was cultivar-dependent, and the highest level, around 60%, was obtained with *A. cepa*, while only 25% was obtained with *A. obliquum*. The main papers on *Allium* cryopreservation are reported in Table 1.

Amaryllis genus—Amaryllis is a flowering geophyte that belongs to the *Amaryllidaceae* family, which comprises two species: *Amaryllis belladonna*, known as the 'belladonna lily', the only species reported in cryopreservation studies, and *Amaryllis paradisicola Snijma* [63]. The *Amaryllis* species have been used over time in folk medicine and as ornamental flowers. The application of cryopreservation is particularly important for this species as it is characterized by recalcitrant seeds with a short conservation period [64]. The earliest cryopreservation research was performed on embryonic axes by Sershen et al. [65]. In this study, they assessed fifteen species of *Amaryllidaceae*. The embryos were excised from the seeds of mature fruits, flash-dried until reaching a water content in the range of 0.4 to 0.1 g g^{-1}, and immersed in different cryoprotectant solutions (glycerol and sucrose alone or in combination). The embryo axes then underwent two cooling systems: (1) direct immersion of explants in liquid nitrogen or (2) slow cooling. The best post-thaw survival rate (~70%) was achieved in several accessions tested after the flash drying of embryos using glycerol as the cryoprotectant solution and rapid cooling, followed by thawing with direct immersion for 2 min in a preheated CaMg solution at 40 °C, and immersion of the cryopreserved embryos in a CaMg solution at 28 °C for 30 min for rehydration in dark conditions. The maximum survival rate (85%) was recorded in *A. belladonna*. This condition was also supported by CryoSEM analysis on zygotic explants [66], where glycerol allowed a less destructive densification of the zygotic tissues during desiccation, particularly in *A. belladonna*. In the same species, Berjak et al. [66] confirmed these results, highlighting that the major contributory factor in the post-viability and survival of embryos was the pre-conditioning, involving the cryoprotectant solution with glycerol. Indeed, the highest survival was observed with glycerol treatment, while no viability was observed after sucrose treatment [65–67]. In an additional specific study on *A. belladonna* zygotic embryos [66], the survival rate was improved to 100% by applying the same protocol as that of Sershen et al. [68]. The glycerol solution, in combination with partial drying, enhanced post-cryopreservation viability in recalcitrant zygotic embryos of *A. belladonna* by protecting the activities of some antioxidant enzymes during the steps of cryopreservation [68].

Galanthus genus—*Galanthus* spp., known as snowdrops, are herbaceous plants belonging to the *Amaryllidaceae* family and are grown naturally in Europe and the Middle East. This genus is well known for its alkaloid, flavonoid, and terpenoid content [69]. Cryobiology applications on *Galanthus* were first reported by Pawłowska [70] on somatic embryos by the encapsulation–dehydration technique. In this study, somatic embryos of *G. nivalis* L. and *G. elwesii* were encapsulated in MS medium enriched with 3% sodium alginate. After encapsulation, the explants were dehydrated by a quick method (in liquid media containing 0.75 M sucrose for 18 h) or by a gradual method (transfer into liquid media with increasing sucrose concentrations from 0.3 M to 1 M for 7 consecutive days). No post-regeneration was recorded after cryostorage, although the somatic embryos did not show browning. Later, Maślanka et al. [71] developed the droplet vitrification procedure, demonstrating successful cryopreservation of *G. elwesii* apical meristems. The explants were loaded into LS solution for 20 or 30 min at room temperature, then treated with PVS2 at 0 °C for 10, 20, and 30 min before plunging into LN. The thawing was performed with ULS 1.2 M sucrose in MS medium. In this work, the authors demonstrated that there is no significant effect of the duration of LS on the survival percentages, while the duration of PVS2 treatment has a strong effect on survival and regrowth. The high regrowth rates of 65.5% and 75% were obtained after 20 and 30 min of treatment, respectively, without statistical significance differences; the lowest regrowth rate (40%) was observed after 10 min of PVS2.

Narcissus genus—Narcissus is a perennial geophyte, widely spread in the Mediterranean basin and also in China and Japan. It is cultivated as a cut-flower and for its medicinal properties. Maślanka et al. [72] are the first and only authors to report on the PVS2-based droplet vitrification procedure for Narcissus L. 'Carlton' somatic embryos. The globular somatic embryos of different sizes, 1, 2, and 3 mm, were treated at room temperature for 20 min with LS. Thereafter, the explants were soaked in PVS2 for 10, 20, 30, 45, and 60 min at 0 °C. The somatic embryos were transferred in a droplet of PVS2 into a strip of sterile aluminum foil (2 cm × 0.5 cm), and the strips were plunged into LN. Later, rapid thawing was performed by quick soaking of the aluminum strips in ULS composed of 1.2 M sucrose in MS medium for 15 min at room temperature. The explants were placed into a semi-solid recovery medium composed of 1.2 M sucrose in MS medium for 1 day, then transferred to MS containing 30 g L^{-1} sucrose, 5 μM BAP, and 0.5 μM NAA in the dark. The main factors that may potentially influence the post-regrowth highlighted in this study were the effects of embryo size and PVS2 duration. The highest survival rates, 93.3% and 100%, were recorded using 2 mm and 3 mm somatic embryos, respectively, and applying the PVS2 treatment for more than 20 min. In the embryos with a size of 1 mm, the survival rate did not extend beyond 76.7%, but the growth rate (20%) was recorded only in these embryos after 60 min of PVS2, while no regeneration was achieved in large-size embryos.

3.2. *Aracaceae*

Colocasia genus—*Colocasia esculenta* (L.) Schott, known as taro, is an herbaceous plant belonging to the Aracaceae family. This plant is mainly cultivated in tropical and subtropical regions, and its corms are exported worldwide [73] for food purposes owing to their high starch, polysaccharide, and vitamin content and for medicinal value, including their antioxidant, antimetastatic, and anti-inflammatory properties, which occur as a result of secondary metabolites [74,75]. These nutritional, medicinal, and pharmaceutical properties make the taro corm a valuable plant genetic resource. However, a rapid decline of this plant in its natural habitat has been recorded due to overexploitation and its disease incidence caused by the pseudo-fungus *Phytophthora colocasiae*, leading to either stunting or failure to produce a corm. Thus, the development of long-term conservation is heavily needed [76]. During the last 20 years, several studies on cryobiology applications of the taro species have been reported. The first application was mentioned on embryogenic callus by Shimonishi et al. [77] using a slow freezing method. Later, other cryopreservation techniques were applied, including vitrification [78,79], droplet vitrification [80,81], and encapsulation–dehydration [76,82] on various explants, such as shoot tips [76,78,79,81,82], apical meristems [80], pollen [83], and axillary buds [78]. *C. esculenta* is a species characterized by its asynchronous flowering. To overcome this issue, an experiment was conducted to preserve the pollen in liquid nitrogen and then use the cryostored pollen for hybridization of taro [83]. The pollen was maintained in LN for different time intervals, ranging from 1 week to 2 months. After each conservation period, the viability and germination capacity were evaluated. No significant differences were recorded between fresh pollen and cryopreserved pollen from various taro accessions.

The main factors that influence the vitrification and encapsulation–dehydration procedures of taro shoot tips were discussed by Thinh and Takagi. [79]. In particular, they compared both procedures and found that vitrification was superior, showing higher levels of survival (75–100%; range of different taro cultivars), requiring a shorter time for the procedure, and not leading to callus formation from the cryopreserved shoot tip. Moreover, the small shoot tips were much more suitable for cryopreservation than the larger ones. Taro explants appear to have a low PVS2 sensitivity; this result was also confirmed by Thinh and Takagi [79], with the highest regeneration rate (94.7%) using the vitrification method rather than encapsulation/dehydration (85.5%). Further studies have noted more than 70% recovery recorded over a range of 20–40 min of PVS2 treatment [78,81].

The effect of sucrose duration and concentration on preculture treatment was observed in cryopreserved taro shoot tips by encapsulation–dehydration [76]. In this research, a survival rate of 63.9% was recorded in a medium containing 0.75 M sucrose for 2 days; the increase in concentration (more than 1 M) and duration of preculture (3–4 days) negatively influenced the recovery of the shoot tips. After sucrose preculture, the encapsulated explants were dehydrated to an 18–19% moisture content with silica gel or under laminar flow and were directly plunged into LN. Rapid thawing was performed in a 40 °C water bath for 3 min. Acedo et al. [82] also proved that a high sucrose concentration affects the viability of explants, with no sign of regrowth after LN treatment.

Taro apical meristems' cryopreservation by droplet vitrification was reported by Noor Camellia et al. [80]. The authors highlighted a suitable condition for a high survival rate, recorded at 77.8%, with LS (1.5 M glycerol + 5% DMSO + 0.4 M sucrose) for 20 min, followed by 10 min of osmodehydration with PVS3. The regeneration of cryopreserved meristems exposed to PVS3 for more than 10 min was stunted, and they often died after four weeks. The ULS, consisting of 1.2 M sucrose for 15 min, was also defined as a key factor for the post-regrowth of cryopreserved explants.

The application of droplet vitrification on eighteen taro cultivars from the Asia-Pacific region showed a high range of regeneration rates, from 75 to 100% [81], using PVS2 at 0 °C with an optimal exposure range (20–40 min), and was equally applicable among the different genotypes assessed.

Different findings for taro were reported with the application of various cryogenic techniques, but regardless of them, the survival or regrowth of different cultivars of the same species can be influenced by genotypic dependence. For example, Thinh and Takagi [79] had lower taro cryopreservation success with var. *esculenta* than var. *antiquorum*. Moreover, Sant et al. [81], applying the vitrification method to different taro cultivars, obtained a satisfactory rate of regeneration only in three out of the eight cultivars investigated; the other ones showed low regeneration in the range of 21–30%.

3.3. *Asparagaceae Family*

Asparagus genus—The *Asparagaceae* family comprises about 150 species dispersed throughout the tropical and subtropical regions and persisting up to 1500 m elevation. Several studies have reported long-term conservation of this genus by cryopreservation applications, but only on the *Asparagus officinalis* species. *A. officinalis* is a perennial herb cultivated in more than 60 countries worldwide; its cultivation is important due to its economic value and high nutritional, medicinal, and therapeutic value [84]. Kumu et al. [85] developed the first cryopreservation protocol using the controlled freezing technique for *A. officinalis* shoot tips with a high post-survival rate of up to 100%. Other cryopreservation approaches have been applied, including controlled freezing [86–88], vitrification [86,89,90], desiccation–vitrification [91,92], droplet vitrification [93], and encapsulation–dehydration [94]. In these protocols, a variety of explants were tested, comprising buds [86,91,92], embryogenic cells [89,90,95] shoot tips [85,87,93], and rhizome buds [94]. Overall, the survival and regrowth percentages of cryopreserved explants ranged from 63% to 90% and 70% to 80, respectively, using different techniques. In the literature, several studies clarify contributing factors to the post-survival and regrowth rates after freezing of *A. officinalis*. Firstly, the preculture step in a medium supplemented with a high sugar concentration (0.2–0.7 M) for 1 to 2 days is crucial. During the preculture, the amount of water content decreases and, meanwhile, the concentration of sugar and protein increases [95]. This step is effective for improving the freezing tolerance and survival rate of the shoot tips. Kumu et al. [85] reported that 2 days of preculture in medium with 4% DMSO, followed by treatment with 16% DMSO solution, were also effective for the viability of samples (100% survival). The dried buds of asparagus pretreated for 2 days at 25 °C in culture medium with 0.7 M sucrose significantly increased their survival after immersing in LN (7% untreated; 63% treated) [91]. Suzuki et al. [87] confirmed that the preculture treatment increases the freezing resistance by ultrastructural studies that revealed the heterogeneous composition in terms of the viability of cells in dome tissues and reported a relationship between histological changes induced by preculture on sugar-rich media and an increase in freezing resistance in asparagus shoot tips. This finding also concerned embryogenic cell suspensions, as reported by Jitsuyama et al. [95]. The preculture of embryogenic cells for 2 days in a medium enriched with 0.8 M sucrose increased the freezing tolerance remarkably. These results were confirmed by electrolyte leakage. By contrast, other studies are in contradiction to these, for example, the 90% post-regeneration of bud clusters without any preculture or cold acclimatization treatment reported by Kohmura et al. [86]. Moreover, Nishizawa et al. [90] found a high embryogenic cell survival rate (86%) without pretreatment using a PVS3 solution (50% (*w*/*v*) glycerol and 50% (*w*/*v*) sucrose in water) for 20 min at 0 °C. Mix-Wagner et al. [93] applied the DV protocol to eight genotypes of asparagus without cold hardening, and obtained a survival rate ranging from 48% to 90% and a regrowth rate from 31% to 71%.

The moisture content of asparagus explants is another important factor; therefore, chemical and physical dehydration have been applied. Physical desiccation of buds has been applied by Uragami et al. [91,92]: buds were subjected to dehydration at 25 °C in sealed Petri dishes containing 15 g of dry silica gel for 24 h. A total of 63% of the cryopreserved buds remained alive after this treatment, with a 70% regeneration rate. A better regrowth rate of 84% after cryopreservation was obtained by Carmona-Martín et al. [94] by applying the encapsulation–dehydration method on rhizome buds: the encapsulated buds were subjected to desiccation over 24 h in jars containing 100 g of silica gel until a 35% moisture content, after which the explants were immersed for one hour in crushed ice at 0 °C, then for one hour at −20 °C in LN, and plunged directly in liquid nitrogen. Moreover, the cryopreserved plants showed a good rooting percentage (43%) and a high rate of acclimatization (95%). *A. officinalis* buds were successfully osmodehydrated by PVS2 treatment for 120 min at 0 °C, with the average number of shoots produced from each segment at 3.5 [86]. Cultured callus and somatic embryos of asparagus were treated using the vitrification procedure before immersion in LN by PVS1 containing 22% glycerol, 15% ethylene glycol, 15% propylene glycol, and 7% DMS0 in MS medium with 0.5 M sorbitol [89] or by PVS3 consisting of 50% glycerol and 50% sucrose [90]. The latter study reported the survival of cryopreserved asparagus embryogenic cells by different cryogenic protocols: conventional slow freezing, simple freezing, and vitrification. Among the protocols tested, vitrification was the best in terms of the survival rate.

3.4. Asteraceae Family

Helianthus genus—The Jerusalem artichoke (*Helianthus tuberosus* L.) is a perennial plant that is originally native to Native America and is cultivated for the direct consumption of the tuber. It was introduced to Europe in the 19th century and later spread to Japan as a folk remedy for diabetes [96]. Recently, it received renewed interest due to its economic, pharmaceutic, and medicinal values. Jerusalem artichoke germplasm collection and accession are conserved in field collections or in national gene banks as tissue culture in different countries, including in the Chinese Academy of Agricultural Sciences and the United States Department of Agriculture-Agricultural Research Service (USDA-ARS) [97]. However, field collection is labor-demanding and very costly. Therefore, cryopreservation is the ideal solution and can ensure the storage of plants for an indefinite period. Cryopreservation of the

Jerusalem artichoke was reported by controlled freezing of the cell suspension, vitrification, and droplet vitrification on the shoot tip. Swan et al. [98] were the first to report *Helianthus tuberous* L cell suspension conservation using controlled freezing. Later, Harris et al. [99] indicated that the reduction in Tgase activity and a-tubulin tyrosination of the cryopreserved cell was related to a lack of post-thaw recovery when mannitol preculture treatment was applied. Volk and Richards [100] applied the vitrification method to cryopreserve shoot tips of nine Jerusalem artichoke cultivars, obtaining an average shoot regrowth of 34% after 15 min or 30 min of PVS2 treatment at 0 °C without significant differences in time exposure. More recently, Zhang et al. [97] developed an optimized droplet vitrification protocol for the shoot tips of four Jerusalem artichoke cultivars. In particular, they optimized protocols based on the following steps: preculture excised shoot tips in liquid MS medium with 0.4 M sucrose for 3 days, osmoprotection in loading solution for 30 min, treatment with PVS2 for 15 min at 0 °C, immersion in LN, and rapid thawing in MS liquid medium containing 1.2 M sucrose. Shoot tips of cv 'Shudi' showed the highest survival (93%) and regrowth (83%) rates. The survival and regrowth obtained from other cultivars tested an average of shoot tip survival and regrowth ranging from 44% to 72% and 37% to 53%, respectively.

Smallanthus genus—Yacon (*Smallanthus sonchifolius*) is a perennial plant of the Asteraceae family. It is native to the Andean region [101] and is currently widely cultivated in several countries for its highly nutritious properties and medicinal value. In particular, its tuberous roots have a high content of inulin-type fructo-oligosaccharides (FOS) [102]. Yacon plants are vegetatively propagated by tubers, thus their genetic variability is quite low [103]. This factor, combined with intensive agriculture practices and the broad selection of varieties for food purposes, suggests the need to adopt new conservation strategies in order to maintain the variability of this species [104]. The only known study of cryopreservation of the yacon is on the apical buds by the droplet vitrification method reported by Hammond et al. [104]. They used the apical buds excised from 2–3-week-old in vitro plantlets and precultured them in MS medium containing 0.3 M sucrose in dark conditions overnight, then treated them with LS for 20 min at room temperature. The study focused on the use of two plant vitrification solutions: PVS2 (15, 30, and 60 min) at 0 °C and PVS3 (30, 45, 60, and 75 min) at 22 °C using the DV procedure. Afterward, the cryopreserved buds were washed in an unloading liquid medium with 1.2 M sucrose for 15 min. This study reported no significant effect of LS in bud regrowth and no significant differences between PSV2 and PVS3 when they were used for 60 min. Moreover, they highlighted that hormone-free MS medium is optimal for the recovery of the post-cryopreserved buds. This protocol gave the maximum survival and regrowth at 87% and 90%, respectively. The optimal protocol developed was then applied to four clones of yacon, reporting a survival rate ranging from 79.7 to 94.1% and a regrowth rate of 66.3 to 75.4%.

3.5. Basellaceae Family

Ullucus genus—*Ullucus tuberosus* is a tuber crop, indigenous to the Andean region of South America. It is cultivated mainly for its edible tubers and economic value; this crop plays a significant role in the economies of the local communities of the Andean region [105]. However, a rapid decline in genetic diversity over the last decades has emerged as a result of the food selection of the Andean tuber variety, the intensive cultivation of the monocultural crop, and the marginalization of the species. Thus, advanced approaches for long-term conservation are heavily needed and should be complementary to traditional methods, such as field collections, as the risk of germplasm loss caused by climate events and unexpected diseases is highly possible [106]. Different procedures of cryopreservation, such as droplet vitrification [107,108], desiccation [108], V-cryoplate, and D-cryoplate [106], have been applied to the *Ullucus* genus, and, at present, the only explant type used is the shoot tip derived from in vitro culture.

The D-cryoplate method showed the highest shoot recovery rate with an average of 89.7% on the eleven lines of *Ullucus* tested [106], while the cryopreserved shoot tip regrowth using droplet vitrification was 52.5% [108]. Sánchez et al. [107], by applying the

DV procedure, emphasized the effect of preculture and cryoprotectant time on the recovery of the cryopreserved shoot tips. The study found that continuous preculture treatment at 18 °C had a positive effect on explant recovery, and PVS2 treatment for 60 min had the highest values (survival rate of 38.8% and regrowth rate of 34.6%). A longer exposure period decreased shoot tip survival. Furthermore, low temperatures on mother plants compared with culturing at a constant temperature of 18 °C had no significant effect on the regeneration of cryopreserved shoot tips. Zamecnikova et al. [108], with respect to the DV protocol, highlighted the significant effect of pretreatment of the shoots with 2 M sucrose and the application of PVS3 treatment for 1.5 h to achieve a high regeneration rate (52.5%) compared to the use of desiccation on silica gel with just a 10% shoot tip regrowth rate. The results by Arizaga et al. [106] showed superiority in terms of regrowth using the D-cryoplate procedure (76.7%) compared to V-cryoplate (43.3); moreover, for further improvement of the D-cryoplate procedure, they highlighted the positive effect of cold hardening for 3–4 weeks, the addition of 0.4 M sucrose in the polymerization solutions (sodium alginate and $CaCl_2$), a time exposure of LS for 30 min, and a dehydration time under an air laminar flow cabinet for 60 min at 25 °C. This optimal D-cryoplate procedure gave the maximum regrowth of 96.7% in three lines of *U. tuberosus*.

3.6. Berberidaceae Family

Podophyllum genus—Podophyllum is an herbaceous, rhizomatous plant with great medicinal value. Its medicinal properties are due to its high aryltetralin lignan content [109]. The only application of cryopreservation carried out on *Podophyllum hexandrum* seeds was reported by Kushwaha et al. [110]. In their study, 14 accessions of mature *P. hexandrum* seeds were extracted from ripe red berries and desiccated using a desiccator containing $CaCl_2$. Four moisture content levels were tested at 5, 10, 20, and 50% for the cryopreservation experiment. Thereafter, the seeds were directly immersed in LN. The highest survival rate was obtained with a 5% and 10% moisture content over a period of 24 months of conservation in LN, with a mean seed germination rate of 89%. While, with 20% moisture content, a good germination rate was recorded until 10 days of cryopreservation, beyond this period the germination stopped. No seed survival was observed with a 50% seed moisture content.

3.7. Colchicaceae Family

Gloriosa genus—*Gloriosa superba*, known as the flame lily, tiger lily, and glory lily, is a perennial plant belonging to the *Colchicaceae* family. It is grown naturally in Africa and Southeast Asia in tropical and subtropical regions. This plant is mainly exploited for its high medicinal value [111], in particular, for colchicine production. This species has been listed as endangered due to overharvesting, poor seed germination, and slow tuber multiplication. Thus, a well-developed protocol for its propagation, rapid multiplication, and effective conservation methods are becoming a priority; biotechnological approaches including in vitro culture and cryopreservation methods for the long term are now available.

However, the only cryopreservation application for this species is described by Rajasekharan et al. [112] for pollen conservation. In their study, the pollen was collected in the flowering phase in the early morning to ensure a high viability rate for explants, the pollen grains were desiccated to a moisture content of around 15–30% t before being plunged into liquid nitrogen. Cryostorage pollen was thawed at room temperature for 3–5 min, and viability was carried out with the hanging drop technique using Brewbaker and Kwack's medium enriched with 10% sucrose solution. The fertility capacity was also evaluated compared to fresh pollen and cryopreserved pollen germination. This procedure was effective for boosting field collections and hybridization programs.

3.8. Iridaceae Family

Crocus genus—The crocus genus includes sterile geophytes with vegetative-grown systems; their propagation relies entirely on corm multiplication. This genus is vulnerable to a large range of pathogens, environmental stresses, and diseases. Conventional conservation methods, such as bulb storage, are not often effective for this germplasm. Cryopreservation offers a cost-effective and efficient preservation method for this genus. MalekZadeh et al. [113] were the first to describe the successful vitrification procedure on the shoot tips of *Crocus sativus*. They indicated that the size of the corm is a crucial factor in explant survival after storage in LN. The highest survival rate of the shoot tips (85.8%) was observed but without any perceivable activity when they used an explant of 2 mm × 4 mm in size. Later, successfully cryopreserved embryogenic calluses (EC) of *Crocus hyemalis* and *Crocus moabiticus* were obtained by the vitrification technique [114]. The authors reported the importance of the type and duration of LS treatment combined with PVS2 incubation. After an application of LS for 20 min and a treatment with four stepwise increases in the concentration of PVS2 (20, 40, 60, and 100%) for 5 min at each concentration, and direct immersion in LN of the ECs, the survival and regrowth rates were of 66.7–50% and 50–44.4%, respectively, for *C. moabiticus* and *C. hyemalis*. The application of encapsulation–vitrification described by Baghdadi et al. [115] considerably improved the EC cryopreservation of both *Crocus* species. The maximum survival of cryopreserved encapsulated EC treated with PVS2 solution for 20 min at 25 °C, was 75.0% and 55.6%, respectively, for *C. hyemalis* and *C. moabiticus*, while the regrowth rate was 66.7% for both species.

Gladiolus genus—*Gladiolus* is cultivated for its ornamental and medicinal purposes; it has excellent potential in floriculture and international flower markets. The first application of cryopreservation to Gladiolus cultivars was reported by Rajasekharan et al. [116] using a controlled freezing approach. Later on, simple desiccation with preservation at −80 °C [117] and droplet vitrification techniques were used. The two main explants used were the pollen [116,118] and the shoot tips [119]. Pollen grains have a vital role in the reproductive process of flowering plants. Their conservation is a tool to maintain genetic resources and improve breeding programs' efficiency by overcoming physiological, seasonal, and geographical limitations [120]. The success of long-term conservation of pollen samples relies on many factors, such as the stage of pollen collection, the pretreatment of the samples, and the methodology applied [62]. Furthermore, post-thawing is yet another key factor that greatly influences the retrieval of stored pollen because it is related to pollen metabolism and the reactivation of post-cryopreserved metabolic processes [121]. *Gladiolus* pollen conservation was reported by Rajasekharan et al. [116]: in this study, five selected cultivars were tested, and the mean post-germination rate recorded was 52.8% compared to fresh pollen (without storage) at 58.1%. Of particular interest is that there was no decline in pollen viability between 1 and 10 years of cryogenic storage, as assessed in cv 'Jowagenaar'.

For shoot tips' long-term conservation, Joung et al. [119] was the first and only work that reported the successful cryopreservation of five *Gladiolus* genotypes. In this study, the shoot tips were excised from cormels and precultured in MS containing 2 mg/L kinetin and 3% sucrose in the dark at 4 °C for 16 h. The precultured shoot tips underwent LS treatment for 20 min, were transferred in drops on aluminum foil, and were treated with different PVS2 times before the immersion of the samples in LN. A rapid thawing was carried out in a liquid ULS (MS with 1.2 M sucrose) for 20 min at 25 °C. In this research, the key factors highlighted as influencing the post-regrowth of the vitrified shoot tips were the cormels' diameter and the PVS2 incubation time. Concerning the first parameter, the shoot tips excised from cormels that were less than 1.0 cm in diameter resulted in better shoot regrowth. In addition, the maximum post-regrowth rate of 54% was recorded after 120 min of PVS2 treatment in cv 'Peter Pan', while the regrowth rate was only 15% for explants of one tested *Gladiolus* breeding line with PVS treatment of 15 min. This suggested the need to optimize conditions for each genotype.

Iris genus—To date, only a few cryopreservation studies have been performed for the *Iris* genera. In two species cryopreservation was reported, in *Iris nigricans* the method used was encapsulation–dehydration of somatic embryos (SE) [122], and in *Iris pumila*, shoot tips were preserved by the vitrification procedure [123,124]. The different key factors discussed in these studies were the precultured and PVS2 timings. The preculturing of

Iris nigricans somatic embryos in a medium containing 0.75 M sucrose for 3 days at 22 °C, then at 30 °C for 1 day prior to freezing, was reported by Shibli. [122]. The loading solution treatment for 30 min at 25 °C for *Iris pumila* shoot tips was necessary for post-regeneration because no post-growth was observed when the loading solution was not applied [124]. In addition, the study of *Iris pumila* cryopreservation emphasized other factors that could influence the post-survival and regrowth of Iris, such as sample size. Shibli [122] indicated that sizes of 2–4 mm in somatic embryos had the maximum survival and regrowth rates (54% and 60%, respectively), higher than smaller (1–2 mm) or larger (4–6 mm) ones. In addition, for shoot tips, as cited by Jevremović et al. [124], a size of 2 mm showed important outcomes for successful regeneration. Moreover, when cold hardening is applied to shoot tips, these explants can survive after being exposed to PVS2 solution both at 0 and 25 °C. However, the highest tolerance to PVS2 was achieved at 0 °C with an osmodehydration time of 5–20 min; these conditions helped to increase tolerance to ultra-rapid storage in LN, showing high survival (65% for 15 min) and regrowth (55% for 20 min) rates [124]. The rooting of cryopreserved *I. pumila* shoots was achieved with success (90%) on hormone-free MS medium, and the clonal fidelity of the cryopreserved plants and control was also recognized [124].

3.9. Liliaceae Family

Chlorophytum genus—*Chlorophytum* is a large genus belonging to the Liliaceae family and is distributed in tropical and subtropical regions. This genus includes various perennial species that are well known for their horticultural, ornamental, and medicinal values. Recently, this genus has received more attention due to the recent discovery of its pharmacology characteristics, such as anticancer and immunomodulatory activities [125]. This increased attention has led to overexploitation and habitat destruction. Moreover, the overexploitation along with the low germination rate of these species are the main reasons for the plants' declining distribution. Various genus members have recently been listed as critically endangered species, such as *Chlorophytum borivilianum*, a tropical species. Thus, cryopreservation could be considered the most promising method for protection. The first and only cryopreservation method for this species was developed on meristems of *Chlorophytum borivilianum* by the vitrification procedure by Chauhan et al. [126]. In this study, the in vitro shoots were pre-acclimated for two months in MS media with 12% sucrose: the meristem explants were excised and precultured in MS enriched with 12% sucrose and 50 mg/L ABA for 48 h at 25 °C. Good survival and regeneration levels were obtained at 66% and 33%, respectively, with 20 min in LS (13.7% sucrose and 18.4% glycerol) at 25 °C and 30 min of PVS2 treatment at 0 °C. The factors highlighted were the optimal duration of PVS2 treatment and the effect of the preculture on the medium with ABA. Indeed, the preculture of explants in MS supplemented with 50 mg/L ABA and 12% sucrose had a significant role in the post-regrowth of the meristems, with better results than the preculture in a medium with 12% sucrose and 0.5 M glycerol. The authors asserted that the ABA effect could be associated with the freezing tolerance of protein synthesis.

Fritillaria genus—*Fritillaria* is a genus that includes about 100 to 140 species of bulbous plants distributed in the northern hemisphere and temperate regions. Fritillaria bulbs are known for their substantial medicinal and horticultural values [127]. The high demand and wide uses of the bulbs exert a strong pressure on the species and a consequential genetic erosion risk; hence, their conservation is required. To date, two cryobiology approaches have been used for the Fritillaria genus: the controlled freezing of *F. thunbergii* callus and pollen and the vitrification of *F. anhuiensis* shoot tips and *F. cirrhosa* callus. The key factors highlighted for the successful cryopreservation of *Fritillaria* are the explant age and desiccation level. Indeed, Su [128] reported that the highest viability (56.4%) was obtained using pollen from a flowering day with a 20% water content. Moreover, other experiments proved that the precultured period, the concentration of the cryoprotectants, and the duration of osmodehydration by PVS2 treatment need to be controlled. Zhu et al. [129] were the first to report the use of the vitrification technique for *F. anhuiensis* shoot tip cryopreservation. In their study, a high survival rate of 79.9% was reported after a 3-day preculture of the shoot tips in MS medium enriched with 0.4 M sucrose, then 20 min of 60% PSV2 treatment at 25 °C, and a step in ice-cooled vitrification solution for 60 min before rapid cooling in LN. Further, according to Wang et al. [130], an 80% survival rate of *F. cirrhosa* callus was obtained after 6 days in MS enriched with 30 g/L sucrose plus 5% DMSO, followed by a treatment solution composed of 60% PVS2 for 25 min, then osmodehydration in 100% PVS2 for 60 min at 2 °C. A longer pretreatment duration of 9 days in a medium supplemented with 10% DMSO containing 0.5 mol/L sorbitol was instead required to obtain a high survival rate (87.4%) of *F. cirrhosa* callus [131].

Lilium genus—Lilium is one of the most economically significant genera in many countries. It includes a wide number of ornamental species, which are required commercially for their attractive flowers, food properties, and medicinal functions [132,133]. Recently, several research papers uncovered its various medicinal and pharmacological properties, including antioxidant, antidepressant, and anti-inflammatory activity [134]. Several species, including *L. maculatum* var. bukosanense, *L. polyphyllum*, and *L. tsingtauense*, have been classified as endangered species [135].

Therefore, it is critical to preserve the *Lilium* genus, and cryopreservation can be the best choice for the conservation of lily germplasm resources as it is considered a cost-efficient preservation method for the long term. Moreover, Lilium genetic improvement programs are dependent on the provision of genetic resources, and Lilium cryobank germplasm can provide vegetal material useful for breeding programs. A detailed review reports on cryobiotechnologies applied to *Lilium* species [135].

The first application on shoot tips of *L. speciosum* was reported by Bouman and De Klerk [136] using a two-step freezing method, but they only obtained an 8% survival rate. Since then, different lilium explant types have been preserved with cryopreservation methods, including seeds [137–142], shoot tips [143–151], meristems [144,145,152], embryonic axes [137,138], and pollen [153]. Different cryopreservation procedures have been applied, such as droplet vitrification [147–149,151,152], encapsulation–dehydration [137,139,142], encapsulation–vitrification [138], vitrification [145,152,153], and desiccation [140]. Several factors have been highlighted in previous researches that had crucial effects on the post-survival and regeneration levels of the cryopreserved explants. The cold hardening of the shoot tips was identified as affecting the shoots post-cryopreservation in some studies. Indeed, the cold hardening of scale segments from 7 to 30 days gave good shoot formation (72%) from excised apical meristems with the vitrification procedure [145]. A similar result was also obtained after a cold hardening at 4 °C for 7 days of scale segments in droplet vitrification on five lily accessions with wider survival (57.7–89.5%) and regeneration (52.7–87.5%) ranges [149]. Urbaniec-Kiepura and Bach [154] compared the effect of two storage temperatures (5 °C or 20 °C) for *L. martagon* bulblets on the growth of their meristems after vitrification; the material stored at 20 °C in a medium containing 3% sucrose showed a survival rate of 65% and a regeneration rate of 87%. In contrast, in the study carried out by Yin et al. [147] using DV, the cryopreservation of shoot tips without any cold hardening was reported: six lily accessions were able to survive and regrow into shoots following DV, although this capacity varied among the genotypes tested. With the same protocol, Yin Z. et al. [148], from cryopreserved shoot tips of Lilium oriental hybrid 'Siberia', obtained embryogenic callus (with a 70% frequency) and shoot regrowth (90%).

The LS composition and treatment duration were optimized in several works using droplet vitrification. Yi [149] demonstrated that the high survival and regeneration rates, 89.5% and 87.5%, respectively, in Lilium species tested were obtained with an LS called LD1 (MS medium added of 35% PVS2) for 60 min at 23 °C, followed by PVS3 treatment for 240 min. Using this protocol, approximately 160 accessions of lily germplasm were preserved [143], with survival rates ranging from 58.3% to 66.4% and regeneration rates ranging from 54.3% to 58.5%. A recent study reported that pretreatment for 40 min with LS containing 35% of PVS3 was an efficient step in DV on the regrowth of cryopreserved adventitious buds of lily [155]. A further investigation reported by Yin et al. [151] obtained the highest regeneration frequency of shoots (more than 90%) with LS, containing 0.4 M sucrose, for 20 min at room temperature.

PVS2 duration was also a factor investigated in several works. A range of 3–7 h at 0 °C was applied by Wang et al. [148], with maximum survival and regrowth rates of 84% and 72%, respectively, obtained after 7 h of PVS2 treatment. Yin et al. [147,151] confirmed that the higher shoot regrowth rates (more than 90%) were recorded after 3–4 h of PVS2 treatment. Further, Yi et al. [143] obtained good survival (63.3%) and regrowth (56.7%) rates after 4–6 h of PVS2 treatment. In the case of using PVS3, Yi et al. [149] reported that the highest shoot survival and regeneration rates (89.5% and 87.5%) occurred after PVS3 treatment for 240 min at 23 °C for all of the cryopreserved lily accessions assessed. Bi et al. [146] evaluated the effect of the size of the shoot tips on the survival frequency of shoots after cryopreservation; the highest percentage (92.5%) was obtained using shoot tips of 1 mm, while with shoot tips of 2 or 3 mm, a slight survival decrease was observed but without significant effects. Xu et al. [150] instead showed that the rapid cooling of shoot tips can be performed in the vitrification technique by using 200 μL Eppendorf tubes instead of 1 ml cryotubes: the reduction in the volume of the vitrification solution can help to improve the heat exchange rate and to enhance cell survival after cryopreservation. Indeed, the size of the tube influenced the survival rate, which was recorded at 95.8% with an Eppendorf tube (200 μL) and 75.1% using a cryotube (1 mL).

As reported in Table 1, several cryopreservation procedures were applied to various explants of Lilium species. The different results obtained can be caused by any of several factors, including donor plant variety, status, and culture condition, as well as by different cultivars, even when using the same method. Chen et al. [152] compared cryopreservation by droplet vitrification and by vitrification procedures. They reported that DV is more efficient, leading to an increase in the survival and regeneration percentages of some lily cultivars compared to vitrification. The main difference between the two procedures is the freezing rate: the cooling rate in droplet vitrification is faster than that in vitrification.

For the lily seed cryostorage, Kaviani et al. [138] found a high germination rate of 75% after the pretreatment of the seed with sucrose and dehydration. A combination of cold hardening and pretreatment of the seed affected its germination capacity, as reported by Urbaniec-Kiepura and Bach [141]. In their study, a 100% germination rate was found for seeds that were stored at 15 °C for twenty-six weeks, treated in 0.75 M sucrose, followed by air desiccation (moisture content seed: 13.1%), and direct immersion in LN. The encapsulation of the seed could also reduce seed-freezing injuries [137,142]. The good germination rate of the seeds (50%) was obtained after encapsulation–dehydration and LN storage, while the non-encapsulated seeds did not survive [137]. In another study, the encapsulation–vitrification procedure determined only 10% of regrowth in cryopreserved seeds and embryogenic axes [137]. The main factor for post-germination was the moisture content of the seed, with the best range between 10% and 20% [137,139,141,142]. Interesting findings were obtained when different cryopreservation procedures were compared on seeds of *Lilium ledebourii* stored at 2–4 °C for 6 weeks [156]. Applying vitrification, encapsulation–dehydration, pretreatment with glycerol, and desiccation, the authors reported good germination after seed cryostorage without significant differences among the procedures tested, although the glycerol treatment and vitrification showed the highest seed germination after LN (97.5% and 97.4%, respectively). Moreover, the study highlighted the desiccation procedure (94.8%) as the best treatment because it does not need any chemical compounds.

As an explant for Lilium cryopreservation, pollen was also investigated [143] due to its importance in the breeding program. Pollen grains were desiccated at 4 °C for 2 h on silica gel (with a moisture content of about 7.3–7.7%), maintained at 20 °C for 20 min, and then plunged into LN. The germination percentage recorded after thawing was 51% for 'Sorbonne' lily accession and 48% for 'Siberia' accession after 420 days in LN. Further, Xu et al. [153] applied rapid cooling and vitrification procedures to the Lilium oriental hybrid, 'Siberia', reporting, respectively, 58.8% and 70.3% pollen viability. For vitrification, they used suspension pollen in an LS for 20 min at 25 °C and PVS2 treatment for 50 min on ice. The loading treatment with the dehydration step was critical for the post-survival of pollen; omitting one of those steps during the vitrification procedure reduced pollen viability significantly.

Tulip genus—Few studies on cryopreservation applications have been reported for the *Tulipa* L. genus [157,158]. The droplet vitrification procedure on apical meristems is the unique protocol that has been applied to this species. This approach could be a new prospect for establishing gene banks of ornamental bulbous plants, which are mainly propagated vegetatively and thus cannot be stored in seed banks. This study identified two factors that can impact the effectiveness of cryopreservation and bulb apical meristem recovery. The cold hardening of the bulbs and the PVS2 treatment time. The pre-storage of bulbs at 5 °C for 10 weeks before cryopreservation significantly improved their recovery rate after liquid nitrogen. Therefore, prior cold treatment of the bulb was fundamental for a better cryopreserved survival rate. Furthermore, the maximum survival and regrowth rates of cryopreserved apical meristems (100%) were recorded under 30 min or 60 min of PVS2 treatment [158].

3.10. *Primulaceae Family*

Cyclamen genus—The Cyclamen genus, classified in the *Primulaceae* family, includes 22 perennial species originating from the Mediterranean region [159]. Several species were cultivated for horticultural/ornamental interest, such as *C. purpurascens* Mill, and others for medicinal purposes due to their interesting biological properties [160]. To date, only two cryopreservation procedures have been successfully performed in cyclamen: controlled freezing and vitrification [161]. Winkelmann et al. [162], for the first time, reported the use of embryogenic suspension cultures of *C. persicum* for cryopreservation by the controlled–freezing method. They investigated the influence of the type and concentration of the cryoprotective solutions employed in the preculturing and pretreatment, in addition to the time of the pretreatment, on the cryopreserved explants' regrowth. After testing the different concentrations of cryoprotectants (0.09, 0.2, 0.4, or 0.6 M sucrose or 0.4 M sorbitol) during the preculture step, the most successful combination was 0.6 M sucrose for preculture followed by 10% DMSO for pretreatment. These steps significantly improved the regrowth rate (75%) of embryonic callus culture. Furthermore, the optimal pretreatment time was 2–4 days. This tested protocol could be a useful tool for embryogenic cell lines of other cyclamen genotypes for storage in LN [163]. Later on, another study conducted by Izgu et al. [164] reported effective cryopreservation protocols for embryogenic callus in several cyclamen species (*C. cilicium*, *C. mirabile*, *C. parviflorum*, and *C. pseudibericum*) by the vitrification procedure. In this research, the authors, in addition to checking the genetic stability after cryopreservation, focused on factors influencing the post-recovery and regrowth of the cryopreserved explants, including the type of osmotic reagents used as cryoprotectants and the duration and temperature of the PVS2 solution. The highest regrowth rate of embryogenic callus without somaclonal variation was obtained using a pretreatment of 48 h at 4 °C in a medium with 0.5 M sucrose and a PVS2 treatment for 60 min at 0 °C (Izgu T; personal communication).

3.11. *Ranunculaceae Family*

Aconitum genus—The genus *Aconitum* belongs to the Ranunculaceae and is composed of approximately 400 species. Several plants in this genus are well known for their medicinal value due to their high secondary metabolite content, especially those in the alkaloid group. This genus has gained attention for the properties described above, which has led to its worldwide demand. *Aconitum heterophyllum*, known as 'Ativisha', is grown naturally in sub-alpine and alpine zones of the Himalayas. Its nontoxic tuberous roots are widely used in homeopathy and traditional Indian and Chinese medicine [165] Due to its high value and demand, it has become increasingly threatened by illegal collection and marketing [166]. Therefore, adequate measures must be taken to conserve it. To our knowledge, the only application of cryopreservation was demonstrated by Kushwaha et al. [110] on *A. heterophyllum* seeds. In this study, the seeds were removed manually from dry follicles, then dried using a desiccant containing $CaCl_2$ until a 5% or 10% of moisture content, followed by their direct immersion in LN. The germination rate of the desiccated, non-cryopreserved seeds recorded with both moisture contents tested (5% and 10%) was 90% and 89%, respectively. After one month of cryopreservation, the seeds of accession 'Chamba' with a 5% moisture content had a slight but significant decline in germination, while after 24 months, the germination rate remained constant (88%) without significant differences from that observed at one month.

3.12. *Zingeberaceae Family*

Curcuma genus—The genus Curcuma is well known for its multivarious uses as a spice, medicine, cosmetic, dye, flavoring, and ornament. Its name, Turmeric, is derived from the Latin word '*terra merita*' meaning meritorious land, which refers to the color of ground turmeric, which resembles a mineral pigment. The species that belong to the genus are currently threatened due to high anthropogenic interference and habitat destruction. The most widespread is *Curcuma domestica* Val. Syn. *Curcuma longa* is well known for its significant commercial and medicinal value. However, many species of the Curcuma genus have recently received more attention and generated worldwide commercial demand as ornamental plants, for example, *C. alismatifolia*, *C. amada*, *C. angustifolia*, *C. aromatica*, and *C. zedoaria* [167].

Curcuma species are characterized by low genetic variation as a result of sexual incompatibility due to their triploid nature. For this reason, a constant loss of genetic variability causes serious threats to extinction. Appropriate methods for medium- and long-term conservation are required to safeguard these species for future generations. The only cryopreservation study for these species is on *C. longa* by Islam [168], carried out with the DV procedure using in vitro axillary buds. In this study, the buds were treated with LS for 20 min, then incubated in three osmoprotection solutions, PVS [89], PVS2 [19], and Steponkus solution [169], for different time periods. The strength of the PVS2 solution compared to PVS2 0.60x and PVS2 0.80X was found to be more effective on the buds' survival (53.3%), with an incubation time of 20 min. Moreover, the use of medium-sized buds (3–4 mm, a pretreatment on MS medium with 0.3 M sucrose) and thawing in a 1.2 M sucrose solution for 10 min led to the highest survival rate (80%) after cryopreservation.

Kaempferia genus—*Kaempferia galanga* L. is an endangered medicinal species cultivated in tropical Asia for its aromatic rhizome, which has a wide range of medicinal applications. Several uses of aromatic ginger for health benefits, food, and nutritional purposes are reported [170]. The rhizomes contain a high-value volatile oil, several alkaloids, starch, protein, aminoacids, minerals, and fatty matter, and the leaves and flowers contain flavonoids. Conventional propagation is by rhizomes, which remain dormant during drought periods and sprout in the spring. During maintenance in the field, the plants are under environmental pressures, such as diseases, pests, and extreme weather conditions. To avoid a loss of germplasm resources, in vitro propagation [171–173] and cryopreservation [174,175] as ex situ conservation techniques have been applied. Preetha et al. [174] reported the first cryopreservation study on *K. galanga*, assessing the effect of sucrose concentration during preculture, treatment with PVS2, and recovery medium on the cryopreservation of shoot tips. The optimized protocol included an overnight preculture treatment with 0.4 M sucrose, followed by dehydration with PVS2 for 20 min at 0 °C, immersion in LN, and thawing at 40 °C. The cryopreserved shoot tips were transferred on suitable medium recovery (MS + BA + GA3), showing survival and regrowth rates of 66.6% and 46.6%, respectively; the cryopreserved shoots also showed rooting after 30 days. In particular, the 20 min PVS2 treatment at 0 °C led to high shoot tip survival and regrowth rates. No significant difference was found after genetic analysis (RAPD). More recently, Preetha et al. [175] have applied encapsulation–dehydration to the shoot tips of *K. galanga*. The authors evaluated the effect of sucrose concentration and duration on the preculture treatment of encapsulated explants.

The best result in terms of shoot recovery (around 50%) was obtained by preculturing with 0.3 M sucrose for 3 days together with 4 h of dehydration of beads under a laminar airflow cabinet, resulting in a bead moisture content of about 20–30%.

Zingiber genus—*Zingiber officinale* has been cultivated from time immemorial in India and China. It is used as an ingredient in many spice mixes in food preparation, and it is important for traditional Indian, Chinese, and Japanese medicine. Over 800 accessions of ginger germplasm are available in the National Conservatory for Ginger [176] at the Indian Institute of Spices Research (IISR). The main limitation involved in the conservation of the ginger germplasm relates to soil-borne diseases (*Pythium* spp.; *Pseudomonas solanacearum*). In addition, leaf fleck virus infection also raises several conservation issues. Considering that diseases are extremely difficult to control under field conditions, the ginger germplasm at IISR is actually preserved in specially made cement tubs under 50% shade as a nucleus gene bank to safeguard the purity of the germplasm. For the above reasons, it is important to apply a complementary strategy to protect genetic resources from diseases or other natural disasters. The first communication of the cryopreservation of ginger was reported by Geetha [177], using encapsulation–dehydration with a success rate of 40–50%. Later, Yamuna et al. [178] developed an efficient cryopreservation protocol for in vitro shoot buds of *Zingiber officinale* by comparing encapsulation–dehydration (ED), encapsulation–vitrification (EV), and vitrification (Vitr) procedures. The most effective cryopreservation procedure for ginger was obtained by vitrification, with an 80% shoot regrowth rate after cryopreservation. In this study, after preculturing the shoot buds in a medium with 0.3 M sucrose for 72 h, five different cryoprotectant mixtures and a PVS2 solution were applied. The latter was the most suitable cryoprotectant for shoot regrowth (80%) after liquid nitrogen and thawing in a water bath for 1 min. Moreover, the research highlighted the dehydration steps as the most fundamental factor for the successful application of ED and EV procedures. In ED, the regrowth of the cryopreserved ginger explants was greatly influenced by the moisture content of the precultured beads. Indeed, preculturing by increasing the sucrose concentration of the medium and dehydrating for 6 h under an air cabinet flow until a 21% bead moisture content, gave a recovery rate of 41% of the encapsulated ginger shoot buds. While, in EV, a treatment with 2 M glycerol and 1.6 M sucrose for 3 h at 25 àC showed a higher cryopreserved shoot regrowth rate (66%) when compared to the sucrose solution alone (24%).

Table 1. Cryopreservation of medicinal and ornamental geophytes. Best results are reported for each species (terminology and values are the same as mentioned by the authors).

Species	Explant Type	Cryopreservation Techniques	Survival/Regrowth (%)	References
		Amaryllidaceae		
Allium cepa	Pollen	Des	R: 60	[59]
Allium cepa	Shoot tips	DV	R: 58	[52]
	Adventitious buds		R: 72	
	Shoot tips from bulbs		R: 32	
Allium chinense	Shoot tips	DV	R: 100	[23]
Allium sativum	Shoot apices from bulbs	Vitr	R: 90	[43]
Allium sativum	Shoot tips from bulbs	Vitr	R: 75	[53]
Allium sativum	Shoot tips	Vitr	R: 70	[44]
Allium sativum	Apices bulbs	DV	R: 100	[179]
Allium sativum	Immature involucres (bulbil primordia and floral buds)	DV	S: 90 R: 83	[60]

Table 1. *Cont.*

Species	Explant Type	Cryopreservation Techniques	Survival/Regrowth (%)	References
Allium sativum	Unripe inflorescences	DV	S: 79.9 R: 78.2	[54]
Allium sativum	Stem discs	ED	S: 75 R: 55	[48]
Allium sativum	Shoot apices	DV	S: 82.6 R: 75.9	[50]
Allium sativum	Shoot tips	D-cryoplate	R: 94	[58]
Amaryllis belladonna	Embryo axes	Des and cryoprotection	S: 85	[65]
Amaryllis belladonna	Zygotic embryos	Des and cryoprotection	S: 100	[66]
Galanthus elwesii	Apical meristems	DV	S: 96.7 R: 75.5	[71]
Narcissus L.	Somatic embryos	DV	S: 100	[72]
		Aracaceae		
Colocasia esculenta	Embryogenic callus	CF	S: 75	[77]
Colocasia esculenta	Shoot tips	Vitr	S: ~80	[78]
Colocasia esculenta	Shoot tips	Vitr	S: 100	[79]
		ED	S: 85.5	
Colocasia esculenta	Embryogenic callus	EV	S: 60	[76]
Colocasia esculenta	Shoot tips	Vitr	S: 30	[180]
Colocasia esculenta	Shoot tips	DV	S: 100	[81]
Colocasia esculenta	Shoot tips	ED	S: 65	[76]
Colocasia esculenta	Apical meristems	DV	S: 77.8	[80]
Colocasia esculenta	Shoot tips	ED	S: 33	[82]
Colocasia esculenta	Pollen	Directly in LN	V: 86 * G: 16 *	[83]
		Asparagaceae		
Asparagus officinalis	Shoot tips	CF	S: ~100	[85]
Asparagus officinalis	Somatic embryos	Vitr	S: 65	[89]
Asparagus officinalis	Cultured cells	Vitr	S: 48	[89]
Asparagus officinalis	Single node segments with axillary bud	Des	S: 63 R: 70	[91]
Asparagus officinalis	Segment bud clusters	Vitr	R: ~90	[86]
Asparagus officinalis	Embryogenic suspension cells	Vitr	S: 86	[90]
Asparagus officinalis	Shoot tips	DV	S: 90 R: 71	[93]
Asparagus officinalis	Rhizome buds	ED	R: 84	[94]

Table 1. *Cont.*

Species	Explant Type	Cryopreservation Techniques	Survival/Regrowth (%)	References
		Asteraceae		
Helianthus tuberosus	Shoot tips	Vitr	R: 34	[100]
Helianthus tuberosus	Shoot tips	DV	S: 93 R: 83	[97]
Smallanthus sonchifolius	Apical buds	DV	S: 90 R: 87	[104]
		Berberidaceae		
Podophyllum hexandrum	Mature seeds	Des	G: 89	[110]
		Basellaceae		
Ullucus tuberosus	Shoot tips	DV	S: 38.8 R: 34.6	[107]
Ullucus tuberosus	Shoot tips	DV	R: 52.5	[108]
		Desiccation	R: 10	
Ullucus tuberosus	Shoots tips	V-cryoplate	R: 43.3	[106]
		D-cryoplate	R: 96.7	
		Colchicaceae		
Gloriosa superba	Pollen	Des	N.R.	[112]
		Iridaceae		
Crocus sativus	Shoot tips	Vitr	S: 85.8	[113]
Crocus hyemalis	Embryogenic callus	Vitr	S: 66.7 R: 50.0	[114]
		EV	S: 75 R: 66.7	[115]
Crocus moabiticus	Embryogenic callus	Vitr	S: 50 R: 44.4	[114]
		EV	S: 55.6 R: 66.7	[115]
Gladiolus spp	Shoot tips	DV	R: 54.0	[119]
Gladiolus	Pollen	CF	G: 52.8	[116]
Iris pumila	Shoot tips	Vitr	S: 63 R: 55	[124]
Iris nigricans	Somatic embryos	ED	S: 60 R: 54	[122]
		Liliaceae		
Chlorophytum borivilianum	Meristems	Vitr	S: 66 R: 33	[126]
Fritillaria anhuiensis	Shoot tips	Vitr	S: 79.9 R: 52.3	[129]
Fritillaria cirrhosa	Callus	Vitr	S: 80	[130]
Fritillaria cirrhosa	Callus	Vitr	S: 87.4	[131]

Table 1. *Cont.*

Species	Explant Type	Cryopreservation Techniques	Survival/Regrowth (%)	References
Fritillaria thunbergii	Pollen	CF	V: 56.4	[128]
Fritillaria thunbergii	Callus	CF	N.R.	[181]
Lilium spp	Apical meristems	Vitr	R: 85	[145]
Lilium japonicum	Scale segments with adventitious buds	ED	R: 90	[182]
Lilium spp	Shoot tips	DV	S: 89.5 R: 87.5	[149]
Lilium spp	Shoot tips	DV	S: 95 R: 87.5	[151]
Lilium oriental hybrid 'Siberia'	Shoot tips	DV	R: 90	[148]
Lilium spp	Shoot meristems	Vitr	S: 90	[144]
Lilium oriental hybrid 'Siberia	Shoot tips	DV	S: 92.5	[146]
Lilium lancifolium	Shoot tips	Vitr	S: 95	[150]
Lilium oriental hybrid 'Siberia	Small leaf squares with adventitious bud	Vitr	S: 85 R: 72	[183]
Lilium spp	Apical meristems	DV	S: 83.8 R: 67.6	[152]
Lilium martagon	Meristem	DV	S: 65 R: 87	[154]
Lilium spp	Adventitious bulbs	DV	R: 65.7	[155]
Lilium ledebourii	Seeds	Des	G: 94.8	[156]
		Vitr	G: 97.4	
		Glycerol pretreatment	G: 97.5	
		ED	G: 69.4	
Lilium ledebourii	Seeds	Des	G: 100	[141]
Lilium ledebourii	Seeds	ED	G: 50	[139]
Lilium ledebourii	Seeds	Pre-growth dehydration	G: 75	[140]
Lilium ledebourii	Seed	EV	R: 10	[138]
	Embryonic axes		R: 10	
Lilium oriental hybrids	Pollen	Des	G: 51	[184]
Lilium oriental hybrid 'Siberia'	Pollen	Rapid cooling	V: 58.8	[153]
		Vitr	V: 70.3	
Tulipa tarda	Apical meristems	DV	S: 90 R: 40	[157]
Tulipa tarda	Apical meristems	DV	R: 100	[158]

Table 1. *Cont.*

Species	Explant Type	Cryopreservation Techniques	Survival/Regrowth (%)	References
		Primulaceae		
Cyclamen persicum	Embryogenic suspension cultures	CF	R: 75	[162]
Cyclamen persicum	Embryogenic callus	Vitr	R: 90	[164]
C. cilicium			R: 78	
C. mirabile			R: 80	
C. parviflorum			R: 70	
C. pseudibericum			R:75	
		Ranunculaceae		
Aconitum heterophyllum	Mature seeds	Des	G: 88	[110]
		Zingeberaceae		
Curcuma longa	Axillary buds	DV	S: 80	[168]
Kaempferia galanga	Shoot tips	Vitr	S: 66.7 R: 46.7	[174]
Kaempferia galanga	Shoot tips	ED	S: 56.7 R: 33.3	[174]
Zingiber officinale	Shoot buds	Vitr	R: 80	[178]

DV: droplet vitrification; Vitr: vitrification; EV: encapsulation–vitrification; ED: encapsulation–dehydration; CF: controlled freezing; Des: desiccation; LN: liquid nitrogen; G: germinability; V: viability; S: survival; R: regrowth; N.R.: not reported. * value after 72.

4. Conclusions

The World Health Organization (WHO) has reported that about 75% of the world's population utilizes plant-derived medicines for their health and that around 21,000 plants are noted to have medicinal potential [185]. Despite the potential of new synthetic molecules, plants continue to be used as a major source of medicine throughout the world.

Many geophyte species are known for their pharmacological and therapeutic attributes because of their high content of natural bioactive compounds, such as carbohydrates, proteins, and mineral potentials, as well as for their ornamental and economic value. Given their potential value and due to their overexploitation combined with disease diffusion, environmental pollution, habitat fragmentation, climate change, urban expansion, and tourism, many plants from this group have been included in the list of Rare, Endangered, and Threatened plants. Therefore, the conservation and sustainable exploitation of plant genetic resources must be considered a priority if the future requirements of generations to come are to be met.

Furthermore, several geophytes are essential to the agricultural economies of many countries. As a source of carbohydrates for many people, they are a valuable basic food and can improve the livelihoods of the small farmers who cultivate them. Although geophytes have developed specialized organs to tolerate adverse environmental conditions, modified climatic conditions can deeply damage agricultural production, thereby affecting the lives and financial security of many people.

Biotechnological approaches to selecting, propagating, and preserving endangered genotypes are important tools in conservation activities. Cryopreservation can be a complementary method to conventional conservation to ensure that there is a back-up for the long-term storage of geophyte genetic resources. Some countries worldwide have developed cryobanks to preserve the main species of cultivated geophytes, such as cassava, potato, sweet potato, yam, and two of the species that are included in this review, garlic and taro. Currently, the practical application of the cryopreservation method is limited to the main economically important crops [186]. To encourage the further development and dissemination of the use of cryopreservation, it would appear advisable that simple and well-described protocols continue to evolve and adequate facilities and trained personnel be made available [187]. All these conditions will allow the development of new cryobanks in the future.

This review reported on the latest advances in cryopreservation and highlighted the key factors impacting the survival and regrowth of cryopreserved geophyte explants. We believe that this work will help researchers and the next generation of cryobiologists in their ongoing study of cryobiology.

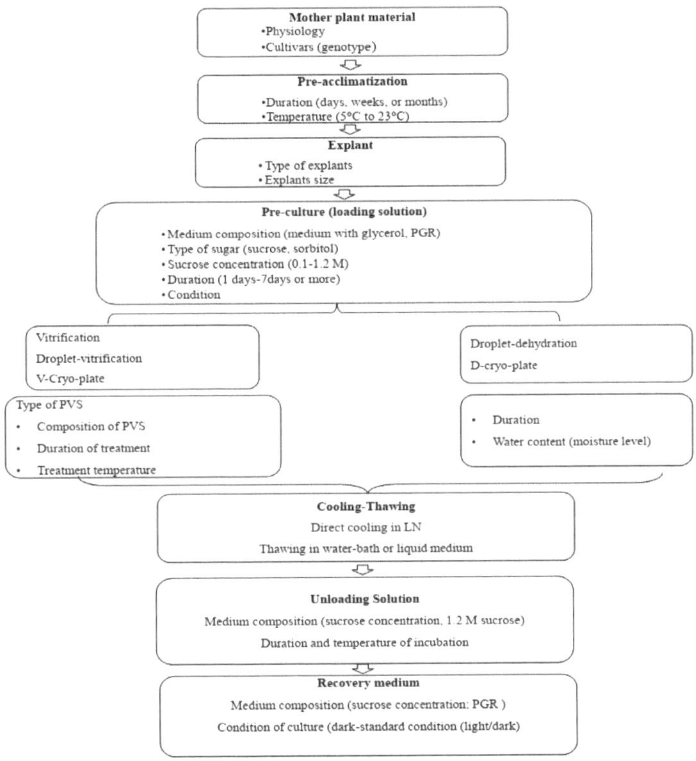

Figure 2. The major factors influencing the explants' survival and regrowth in cryopreservation procedures.

Author Contributions: Conceptualization, S.E.M., M.A.S. and C.B.; methodology, S.E.M. and C.B.; writing—original draft preparation, S.E.M. and C.B.; writing—review and editing, S.E.M., R.E.B. and C.B.; review and editing, S.E.M., M.A.S. and C.B.; supervision, M.A.S. and C.B. All authors have read and agreed to the published version of the manuscript.

Funding: This research received no external funding.

Data Availability Statement: The data presented are available in all the publications cited in this review.

Acknowledgments: The authors express their sincere gratitude to the Hassan II Academy of Sciences and Technology (SafranVal project), the National Centre for Scientific and Technical Research (PPR/2015/33 project), and the University of Ibn Zohr for providing financial assistance. The National Council Research of Italy and Ibn Zohr University Morocco (Bilateral Project 2022–2023: 'Improving cultivation, quality and conservation of saffron accessions from Morocco and Italy') are acknowledged for financial support.

Conflicts of Interest: The authors declare no conflict of interest.

References

1. Reed, B.M.; Sarasan, V.; Kane, M.; Bunn, E.; Pence, V.C. Biodiversity conservation and conservation biotechnology tools. *Vitr. Cell. Dev. Biol.-Plant* **2011**, *47*, 1–4. [CrossRef]
2. Coelho, N.; Gonçalves, S.; Romano, A. Endemic Plant Species Conservation: Biotechnological Approaches. *Plants* **2020**, *9*, 345. [CrossRef] [PubMed]
3. Frankel, O. Natural variation and its conservation. In *Genetic Diversity in Plants*; Springer: Berlin/Heidelberg, Germany, 1977; pp. 21–44.
4. Gonzalez-Arnao, M.T.; Martinez-Montero, M.E.; Cruz-Cruz, C.A.; Engelmann, F. Advances in cryogenic techniques for the long-term preservation of plant biodiversity. *Biotechnol. Biodivers* **2014**, *4*, 129–170.
5. Jenderek, M.M.; Reed, B.M. Cryopreserved storage of clonal germplasm in the USDA National Plant Germplasm System. *Vitr. Cell. Dev. Biol.-Plant* **2017**, *53*, 299–308. [CrossRef]
6. Engelmann, F. Use of biotechnologies for the conservation of plant biodiversity. *Vitr. Cell. Dev. Biol.-Plant* **2011**, *47*, 5–16. [CrossRef]
7. Sakai, A.; Engelmann, F. Vitrification, encapsulation-vitrification and droplet-vitrification: A review. *CryoLetters* **2007**, *28*, 151–172.
8. Popova, E.; Kulichenko, I.; Kim, H.-H. Critical Role of Regrowth Conditions in Post-Cryopreservation of In Vitro Plant Germplasm. *Biology* **2023**, *12*, 542. [CrossRef]
9. Engelmann, F. In vitro conservation methods. *Biotechnol. Agric. Ser* **1997**, *19*, 119–162.
10. Sakai, A. Survival of the Twig of Woody Plants at −196 °C. *Nature* **1960**, *185*, 393–394. [CrossRef]
11. Wang, M.-R.; Lambardi, M.; Engelmann, F.; Pathirana, R.; Panis, B.; Volk, G.M.; Wang, Q.-C. Advances in cryopreservation of in vitro-derived propagules: Technologies and explant sources. *Plant Cell Tissue Organ Cult. (PCTOC)* **2021**, *144*, 7–20. [CrossRef]
12. Benito, M.G.; Clavero-Ramírez, I.; López-Aranda, J. The use of cryopreservation for germplasm conservation of vegetatively propagated crops. *Span. J. Agric. Res* **2004**, *2*, 341–351. [CrossRef]
13. Reed, B.M.; Uchendu, E. Controlled rate cooling. In *Plant Cryopreservation: A Practical Guide*; Springer: New York, NY, USA, 2008; pp. 77–92.
14. Kaczmarczyk, A.; Funnekotter, B.; Menon, A.; Phang, P.Y.; Al-Hanbali, A.; Bunn, E.; Mancera, R. Current issues in plant cryopreservation. In *Current Frontiers in Cryobiology*; IntechOpen: London, UK, 2012; Volume 6, pp. 417–438.
15. Reed, B.M.; Kovalchuk, I.; Kushnarenko, S.; Meier-Dinkel, A.; Schoenweiss, K.; Pluta, S.; Straczynska, K.; Benson, E.E. Evaluation of critical points in technology transfer of cryopreservation protocols to international plant conservation laboratories. *CryoLetters* **2004**, *25*, 341–352. [PubMed]
16. Day, J.G.; Harding, K.C.; Nadarajan, J.; Benson, E.E. Cryopreservation. In *Molecular Biomethods Handbook*; Springer: Berlin/Heidelberg, Germany, 2008; pp. 917–947.
17. Benson, E.E. Cryopreservation. In *Plant Conservation Biotechnology*; CRC Press: Boca Raton, FL, USA, 1999; pp. 109–122.
18. Reinhoud, P.J.; Van Iren, F.; Kijne, J.W. Cryopreservation of undifferentiated plant cells. In *Cryopreservation of Tropical Plant Germplasm: Current Research Progress and Application*; Engelmann, F., Takagi, H., Eds.; JIRCAS, IPGRI: Rome, Italy, 2000; pp. 91–102.
19. Sakai, A.; Kobayashi, S.; Oiyama, I. Cryopreservation of nucellar cells of navel orange (*Citrus sinensis* Osb. var. brasiliensis Tanaka) by vitrification. *Plant Cell Rep.* **1990**, *9*, 30–33. [CrossRef] [PubMed]
20. Fabre, J.; Dereuddre, J. Encapsulation-dehydration: A new approach to cryopreservation of *Solanum* shoot-tips. *CryoLetters* **1990**, *11*, 413–426.
21. Sakai, A.; Hirai, D.; Takao, N. Development of PVS-based vitrification and encapsulation–vitrification protocols. In *Plant Cryopreservation: A Practical Guide*; Springer: New York, NY, USA, 2008; pp. 33–57.
22. Engelmann, F.; Arnao, M.-T.G.; Wu, Y.; Escobar, R. Development of encapsulation dehydration. In *Plant Cryopreservation: A Practical Guide*; Springer: New York, NY, USA, 2008; pp. 59–75.
23. Tanaka, D.; Sakuma, Y.; Yamamoto, S.; Matsumoto, T.; Niino, T. Development of the V cryo-plate method for cryopreservation of in vitro rakkyo (*Allium chinense* G. Don). In Proceedings of the Acta Hortic 1234, Bangkok, Thailand, 26–28 March 2018; pp. 287–292.
24. Mix-Wagner, G.; Schumacher, H.; Cross, R. Recovery of potato apices after several years of storage in liquid nitrogen. *CryoLetters* **2003**, *24*, 33–41.
25. Panis, B.; Piette, B.; Swennen, R. Droplet vitrification of apical meristems: A cryopreservation protocol applicable to all Musaceae. *Plant Sci.* **2005**, *168*, 45–55. [CrossRef]
26. Panis, B.; Piette, B.; Andre, E.; Van den Houwe, I.; Swennen, R. Droplet vitrification: The first generic cryopreservation protocol for organized plant tissues? In Proceedings of the I International Symposium on Cryopreservation in Horticultural Species, Leuven, Belgium, 5–8 April 2011; pp. 157–162.
27. Niino, T.; Yamamoto, S.; Matsumoto, T.; Engelmann, F.; Valle Arizaga, M.; Tanaka, D. Development of V and D cryo-plate methods as effective protocols for cryobanking. In Proceedings of the III International Symposium on Plant Cryopreservation 1234, Bangkok, Thailand, 26–28 March 2018; pp. 249–262.
28. Yamamoto, S.-I.; Rafique, T.; Priyantha, W.S.; Fukui, K.; Matsumoto, T.; Niino, T. Development of a cryopreservation procedure using aluminium cryo-plates. *CryoLetters* **2011**, *32*, 256–265.
29. da Silva, J.A.T.; Kim, H., Engelmann, F. Chrysanthemum low-temperature storage and cryopreservation: A review. *Plant Cell Tissue Organ Cult. (PCTOC)* **2015**, *120*, 423–440. [CrossRef]
30. Niino, T.; Yamamoto, S.-i.; Fukui, K.; Martínez, C.R.C.; Arizaga, M.V.; Matsumoto, T.; Engelmann, F. Dehydration improves cryopreservation of mat rush (*Juncus decipiens* Nakai) basal stem buds on cryo-plates. *CryoLetters* **2013**, *34*, 549–560.

31. Niino, T.; Watanabe, K.; Nohara, N.; Rafique, T.; Yamamoto, S.-I.; Fukui, K.; Arizaga, M.V.; Martinez, C.R.C.; Matsumoto, T.; Engelmann, F. Cryopreservation of mat rush lateral buds by air dehydration using aluminum cryo-plate. *Plant Biotechnol.* **2014**, *31*, 281–287. [CrossRef]
32. Niino, T.; Gupta, S.; Meena, D. 4. Cryopreservation of Shoot Apices Using V Cryo-Plate and D Cryo-Plate Method. In *Laboratory Manual for Eighth International Training Course on In Vitro and Cryopreservation Approaches for Conservation of Plant Genetic Resources;* Agrawal, A., Gowthami, R., Srivastava, V., Malhotra, E.V., Pandey, R., Sharma, N., Gupta, S., Bansal, S., Chaudhury, R., Rana, J.C., et al., Eds.; Malhotra Publishing House: New Delhi, India, 2019; pp. 17–23.
33. Matsumoto, T.; Yamamoto, S.-I.; Fukui, K.; Rafique, T.; Engelmann, F.; Niino, T. Cryopreservation of persimmon shoot tips from dormant buds using the D cryo-plate technique. *Hortic. J.* **2015**, *84*, 106–110. [CrossRef]
34. Funnekotter, B.; Bunn, E.; Mancera, R.L. Cryo-mesh: A simple alternative cryopreservation protocol. *CryoLetters* **2017**, *38*, 155–159. [PubMed]
35. Stavělíková, H. Morphological characteristics of garlic (*Allium sativum L.*) genetic resources collection–Information. *Inf. Hortic. Sci.* **2008**, *35*, 130–135. [CrossRef]
36. Garden, K.R.B. Available online: https://www.kew.org/science/collections-and-resources/data-and-digital/names-and-taxonomy (accessed on 24 May 2023).
37. Kumar, R.; Kumar, D. Comprehensive metabolomics and antioxidant activity of *Allium* species viz *Allium semenovii, A. sativum* and A. cepa: An important spice. *Food Res. Int.* **2023**, *166*, 112584. [CrossRef] [PubMed]
38. Kim, H.-H.; Popova, E.; Shin, D.-J.; Yi, J.-Y.; Kim, C.H.; Lee, J.-S.; Yoon, M.-K.; Engelmann, F. Cryobanking of Korean Allium germplasm collections: Results from a 10 year experience. *CryoLetters* **2012**, *33*, 45–57.
39. Zanke, C.; Zamecnik, J.; Kotlińska, T.; Olas, M.; Keller, E. Cryopreservation of garlic for the establishment of a European core collection. In Proceedings of the I International Symposium on Cryopreservation in Horticultural Species 908, Leuven, Belgium, 5–8 April 2009; pp. 431–438.
40. Keller, E.R.J.; Senula, A.; Zanke, C. Alliaceae in cryopreservation, achievements and constraints. In Proceedings of the I International Symposium on Cryopreservation in Horticultural Species 908, Leuven, Belgium, 5–8 April 2009; pp. 495–508.
41. Niwata, E. Cryopreservation of apical meristems of garlic (*Allium sativum L.*) and high subsequent plant regeneration. *Cryo-Lett* **1995**, *16*, 102–107.
42. Subbarayan, K.; Rolletschek, H.; Senula, A.; Ulagappan, K.; Hajirezaei, M.-R.; Keller, E.J. Influence of oxygen deficiency and the role of specific amino acids in cryopreservation of garlic shoot tips. *BMC Biotechnol.* **2015**, *15*, 40. [CrossRef] [PubMed]
43. Baek, H.-J.; Kim, H.-H.; Cho, E.-G.; Chae, Y.-A.; Engelmann, F. Importance of explant size and origin and of preconditioning treatments for cryopreservation of garlic shoot apices by vitrification. *CryoLetters* **2003**, *24*, 381–388.
44. Keller, E. Improvement of cryopreservation results in garlic using low temperature preculture and high-quality in vitro plantlets. *CryoLetters* **2005**, *26*, 357–366.
45. Makowska, Z.; Keller, J.; Engelmann, F. Cryopreservation of apices isolated from garlic (*Allium sativum* L.) bulbils and cloves. *CryoLetters* **1999**, *20*, 175–182.
46. Keller, E. Cryopreservation of *Allium sativum* L.(garlic). In *Cryopreservation of Plant Germplasm II*; Springer: Berlin/Heidelberg, Germany, 2002; pp. 37–47.
47. Sudarmonowati, E. Cryopreservation of garlic (*Allium sativum*) cv. Lumbu Hijau using vitrification technique. *Ann. Bogor.* **2001**, *8*, 39–47.
48. Lynch, P.T.; Souch, G.R.; Harding, K. Effects of post-harvest storage of *Allium sativum* bulbs on the cryopreservation of stem-discs by encapsulation/dehydration. *J. Hortic. Sci. Biotechnol* **2012**, *87*, 588–592. [CrossRef]
49. Lynch, P.T.; Souch, G.R.; Zámeník, J.; Harding, K. Optimization of water content for the cryopreservation of *Allium sativum* in vitro cultures by encapsulation-dehydration. *CryoLetters* **2016**, *37*, 308–317. [PubMed]
50. Liu, X.-X.; Wen, Y.-B.; Cheng, Z.-H.; Mou, S.-W. Establishment of a garlic cryopreservation protocol for shoot apices from adventitious buds in vitro. *Sci. Hortic* **2017**, *226*, 10–18. [CrossRef]
51. Wang, M.-R.; Zhang, Z.; Haugslien, S.; Sivertsen, A.; Rasmussen, M.; Wang, Q.-C.; Blystad, D.-R. Cryopreservation of shallot (*Allium cepa* var. aggregatum) shoot tips by droplet-vitrification. In Proceedings of the III International Symposium on Plant Cryopreservation 1234, Bangkok, Thailand, 26–28 March 2018; pp. 241–248.
52. Wang, M.-R.; Zhang, Z.; Zámečník, J.; Bilavčík, A.; Blystad, D.-R.; Haugslien, S.; Wang, Q.-C. Droplet-vitrification for shoot tip cryopreservation of shallot (*Allium cepa* var. aggregatum): Effects of PVS3 and PVS2 on shoot regrowth. *Plant Cell Tissue Organ Cult. (PCTOC)* **2020**, *140*, 185–195. [CrossRef]
53. Volk, G.M.; Maness, N.; Rotindo, K. Cryopreservation of garlic (*Allium sativum* L.) using plant vitrification solution 2. *CryoLetters* **2004**, *25*, 219–226.
54. Kim, H.-H.; Lee, J.-K.; Hwang, H.-S.; Engelmann, F. Cryopreservation of garlic germplasm collections using the droplet-vitrification technique. *CryoLetters* **2007**, *28*, 471–482.
55. Taesan, K.; Joung-Kwan, L.; Jae-Jun, J.; Sang-Sik, N.; Engelmann, F.; Haeng-Hoon, K.; Yun-Keol, L. Implementation of cryopreservation for garlic genetic resources by the droplet vitrification procedure. In Proceedings of the XXVII International Horticultural Congress-IHC2006: II International Symposium on Plant Genetic Resources of Horticultural 760, Seoul, Republic of Korea, 14–18 August 2006; pp. 209–215.

56. Kim, H.; Cho, E.; Park, S. Analysis of factors affecting the cryopreservation of garlic shoot tips. *J. Biomed. Nanotechnol* **2006**, *2*, 129–132. [CrossRef]
57. Mahajan, R. In vitro and cryopreservation techniques for conservation of snow mountain garlic. In *Protocols for In Vitro Cultures and Secondary Metabolite Analysis of Aromatic and Medicinal Plants*, 2nd ed.; Springer: Berlin/Heidelberg, Germany, 2016; pp. 335–346.
58. Tanaka, D.; Sakuma, Y.; Yamamoto, S.-I.; Valle Arizaga, M.; Niino, T.; Matsumoto, T. Development of–80° C storage for Allium shoot tips using D cryo-plate method. *Plant Cell Tissue Organ Cult. (PCTOC)* **2021**, *144*, 115–122. [CrossRef]
59. Senula, A.; Keller, E. Pollen cryopreservation to support maintenance of a wild species collection of the genus Allium. In Proceedings of the II International Symposium on Plant Cryopreservation 1039, Fort Collins, CO, USA, 11–14 August 2013; pp. 289–296.
60. Kim, H.-H.; Lee, J.-K.; Yoon, J.-W.; Ji, J.-J.; Nam, S.-S.; Hwang, H.-S.; Cho, E.-G.; Engelmann, F. Cryopreservation of garlic bulbil primordia by the droplet-vitrification procedure. *CryoLetters* **2006**, *27*, 143–153.
61. Ganeshan, S. Viability and fertilizing capacity of onion pollen (*Allium cepa* L.) stored in liquid nitrogen (−196 °C). *Trop Agric (Trinidad)* **1986**, *63*, 46–48.
62. Ganeshan, S.; Rajasekharan, P.; Shashikumar, S.; Decruze, W. Cryopreservation of pollen. In *Plant cryopreservation: A Practical Guide*; Springer: New York, NY, USA, 2008; pp. 443–464.
63. Dennehy, Z.; Bilsborrow, J.; Culham, A.; David, J.; Könyves, K. The complete plastome of the South African species, *Amaryllis belladonna* L.(Amaryllidaceae). *Mitochondrial DNA Part B* **2021**, *6*, 3393–3395. [CrossRef] [PubMed]
64. Sershen, N.; Berjak, P.; Pammenter, N.W. Desiccation sensitivity of excised embryonic axes of selected amaryllid species. *Seed Sci. Res.* **2008**, *18*, 1–11.
65. Sershen, N.; Pammenter, N.; Berjak, P.; James, W.-S. Cryopreservation of embryonic axes of selected amaryllid species. *CryoLetters* **2007**, *28*, 387–399.
66. Sershen, N.; Berjak, P.; Pammenter, N.; Wesley-Smith, J. The effects of various parameters during processing for cryopreservation on the ultrastructure and viability of recalcitrant zygotic embryos of *Amaryllis belladonna*. *Protoplasma* **2012**, *249*, 155–169. [CrossRef] [PubMed]
67. Ballesteros, D.; Varghese, B.; Berjak, P.; Pammenter, N.W. Uneven drying of zygotic embryos and embryonic axes of recalcitrant seeds: Challenges and considerations for cryopreservation. *Cryobiology* **2014**, *69*, 100–109. [CrossRef]
68. Sershen, N.; Varghese, B.; Pammenter, N.W.; Berjak, P. Cryo-tolerance of zygotic embryos from recalcitrant seeds in relation to oxidative stress—A case study on two amaryllid species. *J. Plant Physiol.* **2012**, *169*, 999–1011. [CrossRef]
69. Berkov, S.; Bastida, J.; Sidjimova, B.; Viladomat, F.; Codina, C. Alkaloid diversity in *Galanthus elwesii* and *Galanthus nivalis*. *Chem. Biodivers.* **2011**, *8*, 115–130. [CrossRef]
70. Pawłowska, B. Employment of encapsulation-dehydration method for liquid nitrogen cryopreservation of ornamental plant explants propagated. *Folia Hortic* **2008**, *20*, 61–71. [CrossRef]
71. Maślanka, M.; Panis, B.; Bach, A. Cryopreservation of *Galanthus elwesii* Hook. apical meristems by droplet vitrification. *CryoLetters* **2013**, *34*, 1–9.
72. Maślanka, M.; Panis, B.; Malik, M. Cryopreservation of Narcissus L.'Carlton'somatic embryos by droplet vitrification. *Propag. Ornam. Plants* **2016**, *16*, 28–35.
73. Macharia, M.W.; Runo, S.M.; Muchugi, A.N.; Palapala, V. Genetic structure and diversity of East African taro [Colocasia esculenta (L.) Schott]. *Afr. J. Biotechnol* **2014**, *13*, 2950–2955.
74. Baltazar-Bernal, O.; Spinoso-Castillo, J.L.; Mancilla-Álvarez, E.; Bello-Bello, J.J. arbuscular mycorrhizal fungi induce tolerance to salinity stress in Taro plantlets (*Colocasia esculenta* L. Schott) during acclimatization. *Plants* **2022**, *11*, 1780. [CrossRef] [PubMed]
75. Rashmi, D.; Raghu, N.; Gopenath, T.; Palanisamy, P.; Bakthavatchalam, P.; Karthikeyan, M.; Gnanasekaran, A.; Ranjith, M.; Chandrashekrappa, G.; Basalingappa, K. Taro (*Colocasia esculenta*): An overview. *J. Med. Plants Stud* **2018**, *6*, 156–161.
76. Seng.-Ron, H.; Ming.-Hua, Y. A simple cryopreservation protocol for in vitro-grown shoot tips of Chinese genuine red bud taro (*Colocasia esculenta* L. Schott Var. *Cormosus* CV. Hongyayu) by encapsulation-dehydration. *Sci. Hortic.* **2013**, *162*, 226–233. [CrossRef]
77. Shimonishi, K.; Karube, M.; Ishikawa, M. Cryopreservation of taro (*Colocasia esculenta*) embryogenic callus by slow prefreezing. *Jpn. J. Breed* **1993**, *43*, 187.
78. Takagi, H.; Thinh, N.T.; Islam, O.; Senboku, T.; Sakai, A. Cryopreservation of in vitro-grown shoot tips of taro (*Colocasia esculenta* (L.) Schott) by vitrification. 1. Investigation of basic conditions of the vitrification procedure. *Plant Cell Rep.* **1997**, *16*, 594–599. [CrossRef]
79. Thinh, N.T.; Takagi, H. Cryopreservation of *Colocasia esculenta* L. Schott (Taro). In *Biotechnology in Agriculture and Forestry*; Towill, L.E., Bajaj, Y.P.S., Eds.; Springer: Berlin/Heidelberg, Germany, 2002; Volume 50, pp. 96–118.
80. Noor Camellia, N.; Abdul Muhaimin, A.; Umikalsum, M.; Rozlaily, Z. Cryopreservation of In vitro Grown Taro (*Colocasia esculenta*) Apical Meristem using a Droplet-vitrification PVS3 Technique. *Tran. Malays. Soc. Plant Physiol* **2021**, *28*, 240–245.
81. Sant, R.; Panis, B.; Taylor, M.; Tyagi, A. Cryopreservation of shoot-tips by droplet vitrification applicable to all taro (*Colocasia esculenta* var. esculenta) accessions. *Plant Cell Tiss. Org. Cult.* **2007**, *92*, 107–111. [CrossRef]
82. Acedo, V.; Damasco, O.; Laurena, A.; Cruz, P.S.; Namuco, L.; Lalusin, A. In vitro Shoot Tip Response to Encapsulation Dehydration Procedure for Cryopreservation of Philippine Taro [*Colocasia esculenta* (L.) Schott]. *Int. J. Res. Agric. Sci* **2018**, *5*, 2348–3997.

83. Mukherjee, A.; George, J.; Pillai, R.; Chakrabarti, S.K.; Naskar, S.K.; Patro, R.; Nayak, S.; Lebot, V. Development of taro (*Colocasia esculenta* (L.) Schott) hybrids overcoming its asynchrony in flowering using cryostored pollen. *Euphytica* **2016**, *212*, 29–36. [CrossRef]
84. Guo, Q.; Wang, N.; Liu, H.; Li, Z.; Lu, L.; Wang, C. The bioactive compounds and biological functions of *Asparagus officinalis* L.–A review. *J. Funct. Foods* **2020**, *65*, 103727. [CrossRef]
85. Kumu, Y.; Harada, T.; Yakuwa, T. Development of a Whole Plant from a Shoot Tip of *Asparagus officinalis* L. Frozen Down to −196 °C. *J. Fac. Agric. Hokkaido Univ* **1983**, *61*, 285–294.
86. Kohmura, H.; Sakai, A.; Chokyu, S.; Yakuwa, T. Cryopreservation of in vitro-cultured multiple bud clusters of asparagus (*Asparagus officinalis* L. cv Hiroshimagreen (2n = 30) by the techniques of vitrification. *Plant Cell Rep.* **1992**, *11*, 433–437. [CrossRef]
87. Suzuki, T.; Kaneko, M.; Harada, T. Increase in Freezing Resistance of Excised Shoot Tips of *Asparagus officinalis* L. by Preculture on Sugar-Rich Media. *Cryobiology* **1997**, *34*, 264–275. [CrossRef]
88. Suzuki, T.; Kaneko, M.; Harada, T.; Yakuwa, T. Enhanced Formation of Roots and Subsequent Promotion of Growth of Shoots on Cryopreserved Nodal Segments of *Asparagus officinalis* L. *Cryobiology* **1998**, *36*, 194–205. [CrossRef] [PubMed]
89. Uragami, A.; Sakai, A.; Nagai, M.; Takahashi, T. Survival of cultured cells and somatic embryos of *Asparagus officinalis* cryopreserved by vitrification. *Plant Cell Rep.* **1989**, *8*, 418–421. [CrossRef] [PubMed]
90. Nishizawa, S.; Sakai, A.; Amano, Y.; Matsuzawa, T. Cryopreservation of asparagus (*Asparagus officinalis L.*) embryogenic suspension cells and subsequent plant regeneration by vitrification. *Plant Sci.* **1993**, *91*, 67–73. [CrossRef]
91. Uragami, A.; Sakai, A.; Nagai, M. Cryopreservation of dried axillary buds from plantlets of *Asparagus officinalis* L. grown in vitro. *Plant Cell Rep.* **1990**, *9*, 328–331. [CrossRef]
92. Uragami, A. Cryopreservation of Asparagus. In *Cryopreservation of Plant Germplasm I*; Springer: Berlin/Heidelberg, Germany, 1995; pp. 332–343.
93. Mix-Wagner, G.; Conner, A.; Cross, R. Survival and recovery of asparagus shoot tips after cryopreservation using the "droplet" method. *N. Z. J. Crop Hortic. Sci.* **2000**, *28*, 283–287. [CrossRef]
94. Carmona-Martín, E.; Regalado, J.; Perán-Quesada, R.; Encina, C. Cryopreservation of rhizome buds of *Asparagus officinalis* L.(cv. Morado de Huétor) and evaluation of their genetic stability. *Plant Cell Tissue Organ Cult. (PCTOC)* **2018**, *133*, 395–403. [CrossRef]
95. Jitsuyama, Y.; Suzuki, T.; Harada, T.; Fujikawa, S. Sucrose incubation increases freezing tolerance of asparagus (*Asparagus officinalis* L.) embryogenic cell suspensions. *CryoLetters* **2002**, *23*, 103–112.
96. Bakku, R.K.; Gupta, R.; Min, C.-W.; Kim, S.-T.; Takahashi, G.; Shibato, J.; Shioda, S.; Takenoya, F.; Agrawal, G.K.; Rakwal, R. Unravelling the *Helianthus tuberosus* L.(Jerusalem artichoke, Kiku-Imo) tuber proteome by label-free quantitative proteomics. *Molecules* **2022**, *27*, 1111. [CrossRef] [PubMed]
97. Zhang, J.-M.; Han, L.; Lu, X.-X.; Volk, G.M.; Xin, X.; Yin, G.-K.; He, J.-J.; Wang, L.; Chen, X.-L. Cryopreservation of Jerusalem artichoke cultivars using an improved droplet-vitrification method. *Plant Cell Tissue Organ Cult. (PCTOC)* **2017**, *128*, 577–587. [CrossRef]
98. Swan, T.; O'Hare, D.; Gill, R.; Lynch, P. Influence of preculture conditions on the post-thaw recovery of suspension cultures of Jerusalem artichoke (*Helianthus tuberosus* L.). *CryoLetters* **1999**, *20*, 325–336.
99. Harris, W.; Lynch, P.; Hargreaves, A.; Bonner, P. Cryopreservation of *Helianthus tuberosus* cell suspension cultures: The effect of preculture treatments on cytoskeletal proteins and transglutaminase activity. *CryoLetters* **2004**, *25*, 213–217.
100. Volk, G.; Richards, K. Preservation methods for Jerusalem artichoke cultivars. *HortScience* **2006**, *41*, 80–83. [CrossRef]
101. Chessum, K.; Chen, T.; Kam, R.; Yan, M. A Comprehensive Chemical and Nutritional Analysis of New Zealand Yacon Concentrate. *Foods* **2022**, *12*, 74. [CrossRef] [PubMed]
102. Campos, D.; Betalleluz-Pallardel, I.; Chirinos, R.; Aguilar-Galvez, A.; Noratto, G.; Pedreschi, R. Prebiotic effects of yacon (*Smallanthus sonchifolius* Poepp. & Endl), a source of fructooligosaccharides and phenolic compounds with antioxidant activity. *Food Chem.* **2012**, *135*, 1592–1599.
103. Žiarovská, J.; Padilla-González, G.F.; Viehmannová, I.; Fernández, E. Genetic and chemical diversity among yacon [*Smallanthus sonchifolius* (Poepp. et Endl.) H. Robinson] accessions based on iPBS markers and metabolomic fingerprinting. *Plant Physiol. Biochem.* **2019**, *141*, 183–192. [CrossRef]
104. Hammond, S.D.H.; Viehmannova, I.; Zamecnik, J.; Panis, B.; Faltus, M. Droplet-vitrification methods for apical bud cryopreservation of yacon [*Smallanthus sonchifolius* (Poepp. and Endl.) H. Rob.]. *Plant Cell Tissue Organ Cult. (PCTOC)* **2021**, *147*, 197–208. [CrossRef]
105. Hammond Hammond, S.D.; Viehmannova, I.; Zamecnik, J.; Panis, B.; Hlasna Cepkova, P. Efficient slow-growth conservation and assessment of clonal fidelity of *Ullucus tuberosus* Caldas microshoots. *Plant Cell Tissue Organ Cult. (PCTOC)* **2019**, *138*, 559–570. [CrossRef]
106. Arizaga, M.V.; Yamamoto, S.-I.; Tanaka, D.; Fukui, K.; Nohara, N.; Nishikawa, T.; Watanabe, K.; Niino, T. Cryopreservation of in vitro shoot tips of ulluco (*Ullucus tuberosus* Cal.) using D cryo-plate method. *CryoLetters* **2017**, *38*, 419–427.
107. Sánchez, D.; Panta, A.; Tay, D.; Roca, W. Cryopreservation of ulluco (*Ullucus tuberosus* CAL.) and oca (*Oxalis tuberosa* Mol.) shoot tips using the PVS2 droplet-vitrification method. In Proceedings of the I International Symposium on Cryopreservation in Horticultural Species 908, Leuven, Belgium, 5–8 April 2009; pp. 339–346.

108. Zamecnikova, J.; Fernandez, E.; Viehmannova, I.; Zamecnik, J.; Faltus, M. Preparation of shoot tips by sucrose and dehydration pre-treatment of *Ullucus tuberosus* Cal. for cryopreservation. In Proceedings of the I International Symposium on Cryopreservation in Horticultural Species 908, Leuven, Belgium, 5–8 April 2009; pp. 331–338.
109. Lata, H.; Moraes, R.M.; Bertoni, B.; Pereira, A.M. In vitro germplasm conservation of *Podophyllum peltatum* L. under slow growth conditions. *Vitr. Cell. Dev. Biol.-Plant* **2010**, *46*, 22–27. [CrossRef]
110. Kushwaha, R.; Chanda, S.; Ogra, R.; Bhattacharya, A. Cryopreservation criteria of *Podophyllum hexandrum* and *Aconitum heterophyllum* seeds based on storage behavior. *Seed Technol* **2010**, *32*, 117–127.
111. Haldhar, R.; Prasad, D.; Bahadur, I.; Dagdag, O.; Berisha, A. Evaluation of *Gloriosa superba* seeds extract as corrosion inhibition for low carbon steel in sulfuric acidic medium: A combined experimental and computational studies. *J. Mol. Liq* **2021**, *323*, 114958. [CrossRef]
112. Rajasekharan, P.; Rohini, M.; Harsha, R.; Anilkumar, G. *Pollen Cryopreservation Protocol for Gloriosa superba L. In.Pollen Cryopreserv. Protoc.*; Springer: Berlin/Heidelberg, Germany, 2023; pp. 383–391.
113. MalekZadeh, S.; Khosrowshahli, M.; Taeb, M. Cryopreservation of the axial meristem of *Crocus sativus* L. *Cryobiology* **2009**, *59*, 412. [CrossRef]
114. Baghdadi, S.; Makhadmeh, I.; Syouf, M.; Arabiat, A.; Shibli, R.; Shatnawai, M. Cryopreservation by vitrification of embryogenic callus of wild crocus (*Crocus hyemalis* and *Crocus moabiticus*). In Proceedings of the I International Symposium on Cryopreservation in Horticultural Species 908, Leuven, Belgium, 5–8 April 2009; pp. 239–246.
115. Baghdadi, S.; Shibli, R.; Syouf, M.; Shatnawai, M.; Arabiat, A.; Makhadmeh, I. Cryopreservation by encapsulationvitrification of embryogenic callus of wild Crocus (*Crocus hyemalis* and *Crocus moabiticus*). *Jordan J. Agric. Sci.* **2010**, *6*, 436–443.
116. Rajasekharan, P.; Rao, T.; Janakiram, T.; Ganeshan, S. Freeze preservation of gladiolus pollen. *Euphytica* **1994**, *80*, 105–109. [CrossRef]
117. Geng, X.; Huang, B.; Luo, F.; Wu, Z.; Qiu, J. A study on cryopreservation of Gladiolus pollens. *J. Nanjing For. Univ. (Nat. Sci. Ed.)* **2011**, *35*, 7–12.
118. Rajasekharan, P.; Rao, T. Pollen Cryopreservation Protocol for Gladiolus. In *Pollen Cryopreservation Protocols*; Springer: New York, NY, USA, 2023; pp. 373–382.
119. Joung, H.Y.; Cantor, M.; Kamo, K.; Ellis, D. Cryopreservation of Gladiolus cultivars. In Proceedings of the XXVII International Horticultural Congress-IHC2006: II International Symposium on Plant Genetic Resources of Horticultural 760, Seoul, Republic of Korea, 14–18 August 2006; pp. 225–231.
120. Rajasekharan, P.; Ganeshan, S. Current perspectives on pollen cryopreservation in horticultural species. In Proceedings of the III International Symposium on Plant Cryopreservation 1234, Bangkok, Thailand, 26–28 March 2018; pp. 47–56.
121. Rajasekharan, P.; Rohini, M. (Eds.) Pollen Cryopreservation: Advances and Prospects. In *Pollen Cryopreservation Protocols*; Springer Protocols Handbooks: Humana, NY, USA, 2023; pp. 1–18.
122. Shibli, R.A. Cryopreservation of black iris (*Iris nigricans*) somatic embryos by encapsulation-dehydration. *CryoLetters* **2000**, *21*, 39–46. [PubMed]
123. Jevremović, S.; Bohanec, B.; Lambardi, M.; De Carlo, A.; Benelli, C.; Subotiü, A. Clonal fidelity of *Iris pumila* plants regenerated after cryopreservation by vitrification. *Cryopreserv. Crop Species Eur* **2011**, *871*, 40.
124. Jevremović, S.; Subotić, A.; Benelli, C.; De Carlo, A.; Lambardi, M. Cryopreservation of *Iris pumila* shoot tips by vitrification. *Acta Hortic* **2011**, *908*, 355–359. [CrossRef]
125. Munyao, J.N.; Dong, X.; Yang, J.-X.; Mbandi, E.M.; Wanga, V.O.; Oulo, M.A.; Saina, J.K.; Musili, P.M.; Hu, G.-W. Complete chloroplast genomes of *Chlorophytum comosum* and *Chlorophytum gallabatense*: Genome structures, comparative and phylogenetic analysis. *Plants* **2020**, *9*, 296. [CrossRef] [PubMed]
126. Chauhan, R.; Singh, V.; Keshavkant, S.; Quraishi, A. Vitrification-based cryopreservation of in vitro-grown apical meristems of *Chlorophytum borivilianum* Sant et Fernand: A critically endangered species. *Proc. Natl. Acad. Sci. USA India Sect. B Biol. Sci* **2021**, *91*, 471–476. [CrossRef]
127. Day, P.D.; Berger, M.; Hill, L.; Fay, M.F.; Leitch, A.R.; Leitch, I.J.; Kelly, L.J. Evolutionary relationships in the medicinally important genus Fritillaria L.(Liliaceae). *Mol. Phylogen. Evol.* **2014**, *80*, 11–19. [CrossRef]
128. Su, X. Methods of cryopreservation of pollen in *Fritillaria thunbergii* Miq. *China J. Chin. Mater. Med.* **1991**, *16*, 399–401, 447.
129. Zhu, Y.; Chen, F.; Xue, J.; Zhang, A.; Sheng, W.; Song, Y. Study on preservation of *Fritillaria anhuiensis* by vitrification in vitro. *China J. Chin. Mater. Med.* **2011**, *36*, 2462–2464.
130. Wang, Y.; Lin, K.; Liu, Y.; Ma, L.; Dai, Y.; Xu, S.; Wang, X. Research on Cryopreservation of Cultured *Fritillaria Cirrhosa* D. Don Tissues by Vitrification. *J. Chengdu Univ. (Nat. Sci. Ed.)* **2012**, *31*, 11–13.
131. Yue-hua, W. The Key Cryopreservation Techniques of *Fritillaria cirrhosa* D. Don Callus. *J. Anhui Agric. Sci.* **2011**, *23*, 14023–14026.
132. Bacchetta, L.; Remotti, P.C.; Bernardini, C.; Saccardo, F. Adventitious shoot regeneration from leaf explants and stem nodes of Lilium. *Plant Cell Tiss. Org. Cult.* **2003**, *74*, 37–44. [CrossRef]
133. Nhut, D.T.; Van Le, B.; Tran Thanh Van, K. Manipulation of the morphogenetic pathways of *Lilium longiflorum* transverse thin cell layer explants by auxin and cytokinin. *Vitr. Cell. Dev. Biol.-Plant* **2001**, *37*, 44–49. [CrossRef]
134. Wang, P.; Li, J.; Attia, F.A.K.; Kang, W.; Wei, J.; Liu, Z.; Li, C. A critical review on chemical constituents and pharmacological effects of *Lilium*. *Food Sci. Hum. Wellness* **2019**, *8*, 330–336. [CrossRef]

135. Li, J.-W.; Zhang, X.-C.; Wang, M.-R.; Bi, W.-L.; Faisal, M.; da Silva, J.A.T.; Volk, G.M.; Wang, Q.-C. Development, progress and future prospects in cryobiotechnology of *Lilium* spp. *Plant Methods* **2019**, *15*, 125. [CrossRef]
136. Bouman, H.; De Klerk, G.J. Cryopreservation of lily meristems. In Proceedings of the V International Symposium on Flower Bulbs 266, Seattle, WA, USA, 10–14 July 1989; pp. 331–338.
137. Kaviani, B. Cryopreservation by Encapsulation-dehydration for Long-term Storage of Some Important Germplasm: Seed of Lily ['*Lilium ledebourii*'(Baker) Bioss.], Embryonic Axe of Persian Lilac ('*Melia azedarach*' L.), and Tea ('*Camellia sinensis*' L.). *Plant Omics* **2010**, *3*, 177.
138. Kaviani, B.; Dehkaei, M.P.; Hashemabadi, D.; Darabi, A. Cryopreservation of *Lilium ledebourii* (Baker) Bioss by encapsulation-vitrification and in vivo media for planting of germplasms. *Am.-Eurasian J. Agric. Environ. Sci.* **2010**, *8*, 556–560.
139. Kaviani, B.; Safari-Motlagh, M.; Padasht-Dehkaei, M.; Darabi, A.H.; Rafizadeh, A. Cryopreservation of lily [*Lilium ledebourii* (Baker) Bioss.] germplasm by encapsulation-dehydration. *Int. J. Bot* **2008**, *4*, 491–493. [CrossRef]
140. Kaviani, B.; Abadi, D.; Torkashvand, A.; Hoor, S. Cryopreservation of seeds of lily [*Lilium Iedebourii* (Baker) Bioss]: Use of sucrose and dehydration. *Afr. J. Biotechnol.* **2009**, *8*, 3809–3810. [CrossRef]
141. Urbaniec-Kiepura, M.; Bach, A. Effect of pre-storage on *Lilium martagon* L. seed longevity following cryopreservation. *CryoLetters* **2014**, *35*, 462–472.
142. Kaviani, B. Cryopreservation of lily seeds by encapsulation-dehydration. In Proceedings of the I International Symposium on Cryopreservation in Horticultural Species 908, Leuven, Belgium, 5–8 April 2009; pp. 253–256.
143. Yi, J.Y.; Lee, G.A.; Lee, S.Y.; Chung, J.W. Development of cryo-banking system of Lilium species. *Flower Res J.* **2014**, *22*, 185–189. [CrossRef]
144. Bouman, H.; Tiekstra, A.; Petutschnig, E.; Homan, M.; Schreurs, R. Cryopreservation of *Lilium* species and cultivars. In Proceedings of the 21st International Eucarpia Symposium on Classical versus Molecular Breeding of Ornamentals. Acta Hort 612, München, Germany, 25–29 August 2003; pp. 147–154.
145. Matsumoto, T.; Sakai, A.; Yamada, K. Cryopreservation of in vitro-grown apical meristems of lily by vitrification. *Plant Cell Tiss. Org. Cult.* **1995**, *41*, 237–241. [CrossRef]
146. Bi, W.L.; Yin, Z.F.; Guo, L.; Chen, L.; Pan, C.; Wang, Q.C. Plant regeneration from shoot regrowth and de novo embryo-like structures from cryopreserved shoot tips of *Lilium* spp. *Vitr. Cell. Dev. Biol.-Plant* **2015**, *51*, 390–398. [CrossRef]
147. Yin, Z.-F.; Bi, W.-L.; Chen, L.; Wang, Q.-C. Plant regeneration and cryopreservation of *Lilium*. In Proceedings of the III International Symposium on the Genus Lilium Acta Hortic.1027, Zhangzhou, China, 1–3 April 2014; pp. 185–196.
148. Wang, Q.C.; Volk, G.M.; Yin, Z.; Chen, L.; Bi, W. Somatic embryogenesis and organogenesis from cryopreserved shoot tips of *Lilium* oriental hybrid'Siberia'. In Proceedings of the II International Symposium on Plant Cryopreservation, Fort Collins, CO, USA, 11–14 August 2013; pp. 193–200.
149. Yi, J.Y.; Lee, G.A.; Chung, J.W.; Lee, S.Y.; Lim, K.B. Efficient cryopreservation of *Lilium* spp. shoot tips using droplet-vitrification. *Plant Breed. Biotechnol* **2013**, *1*, 131–136. [CrossRef]
150. Xu, J.; Liu, Y.; Li, B.L.; Liu, Q.; Shi, Y.; Jia, M.X.; Shen, C. Protective effect of reducing cryotube volume on *Lilium lancifolium* and Euonymus fortunei shoot tips survival after cryopreservation by vitrification. *Acta Hortic* **2015**, *1104*, 255–258.
151. Yin, Z.-F.; Bi, W.-L.; Chen, L.; Zhao, B.; Volk, G.M.; Wang, Q.-C. An efficient, widely applicable cryopreservation of Lilium shoot tips by droplet vitrification. *Acta Physiol. Plant* **2014**, *36*, 1683–1692. [CrossRef]
152. Chen, X.-L.; Li, J.-H.; Xin, X.; Zhang, Z.-E.; Xin, P.-P.; Lu, X.-X. Cryopreservation of in vitro-grown apical meristems of *Lilium* by droplet-vitrification. *South Afr. J. Bot* **2011**, *77*, 397–403. [CrossRef]
153. Jin, X.; Qian, L.; Mengxue, J.; Yan, L.; Bingling, L.; Yin, S. Generation of reactive oxygen species during cryopreservation may improve Lilium× siberia pollen viability. *Vitr. Cell. Dev. Biol.-Plant* **2014**, *50*, 369–375.
154. Urbaniec-Kiepura, M.; Bach, A. Cryopreservation of *Lilium martagon* L. meristems by droplet-vitrification and evaluation of their physiological stability. *CryoLetters* **2017**, *38*, 78–89.
155. Song, J.Y.; Lee, Y.Y.; Yi, J.Y.; Lee, J.R.; Yoon, M.S. Preconditioning for Cryopreservation of in vitro Grown Bulblets of Lily using Droplet-Vitrification. *Korean J. Plant Resour* **2020**, *33*, 689–695.
156. Mohajeri, M.R.; Zare, A.G.; Shahab, M.A.N.; Kalate, S.J. Seed germination of *Lilium ledebourii* (Baker) Boiss. after cryopreservation. *J. Rangel. Sci* **2014**, *4*, 279–286.
157. Maślanka, M.; Bach, A. Cryopreservation of Tulipa tarda Stapf. apical meristems by droplet vitrification. *Acta Biol. Crac.* **2013**, *55*, 58. [CrossRef]
158. Maślanka, M.; Szewczyk, A. Droplet-vitrification cryopreservation of Tulipa tarda Stapf. apical meristems. *Plant Cell Tissue Organ Cult. (PCTOC)* **2020**, *144*, 91–95. [CrossRef]
159. Zengin, G.; Mahomoodally, M.F.; Sinan, K.I.; Picot-Allain, M.C.N.; Yildiztugay, E.; Cziáky, Z.; Jekő, J.; Saleem, H.; Ahemad, N. Chemical characterization, antioxidant, enzyme inhibitory and cytotoxic properties of two geophytes: *Crocus pallasii* and *Cyclamen* cilicium. *Food Res. Int.* **2020**, *133*, 109129. [CrossRef]
160. Mahomoodally, M.F.; Picot-Allain, M.C.N.; Zengin, G.; Llorent-Martínez, E.J.; Stefanucci, A.; Ak, G.; Senkardes, I.; Tomczyk, M.; Mollica, A. Chemical profiles and biological potential of tuber extracts from *Cyclamen coum* Mill. *Biocatal. Agric. Biotechnol* **2021**, *33*, 102008. [CrossRef]
161. Tagipur, M.E.; Seker, G.; Teixeira da Silva, J.A.; Mendi, Y.Y. Somatic embryogenesis, cryopreservation, and in vitro mutagenesis in Cyclamen. In *Somatic Embryogenesis in Ornamentals and Its Applications*; Springer: New Delhi, India, 2016; pp. 155–167.

162. Winkelmann, T.; Mußmann, V.; Serek, M. Cryopreservation of embryogenic suspension cultures of *Cyclamen persicum* Mill. *Plant Cell Rep.* **2004**, *23*, 1–8. [CrossRef]
163. Mubmann, V.; Serek, M.; Winkelmann, T. Cryopreservation of Embryogenic Suspension Cultures of *Cyclamen persicum Mill.* In Proceedings of the V International Symposium on In Vitro Culture and Horticultural Breeding 725, Debrecen, Hungary, 12–17 September 2004; pp. 391–396.
164. Izgu, T.; Sevindik, B.; Simsek, O.; Curuk, P.; Kacar, Y.A.; Mendi, Y.Y. Analysis of Genetic Stability in Some *Cyclamen* Species Endemic to Turkey Following Cryopreservation By Vitrification. In Proceedings of the Plant and Animal Genome Conference, XXIV, San Diego, CA, USA, 9–13 January 2015.
165. Malhotra, N.; Shivani, S. Aconitum heterophyllum. In *Himalayan Medicinal Plants*; Malhotra, N., Singh, M., Eds.; Academic Press: Cambridge, MA, USA, 2021; pp. 5–25.
166. Wani, T.A.; Kaloo, Z.A.; Dangroo, N.A. *Aconitum heterophyllum* Wall. ex Royle: A critically endangered medicinal herb with rich potential for use in medicine. *J. Integr. Med* **2022**, *20*, 104–113. [CrossRef]
167. Nair, K.P. *Turmeric (Curcuma Longa L.) and Ginger (Zingiber Officinale Rosc.)-World's Invaluable Medicinal Spices: The Agronomy and Economy of Turmeric and Ginger*; Springer: Berlin/Heidelberg, Germany, 2019.
168. Islam, M. Genetic Diversity of the Genus Curcuma in Bangladesh and Further Biotechnological Approaches for In Vitro Regeneration and Long-Term Conservation of *C. longa* Germplasm. Ph.D. Thesis, University of Hannover, Hannover, Germany, 2004.
169. Langis, R.; Steponkus, P.L. Cryopreservation of rye protoplasts by vitrification. *Plant Physiol.* **1990**, *92*, 666–671. [CrossRef]
170. Preetha, T.S.; Hemanthakumar, A.S.; Krishnan, P.N. A comprehensive review of *Kaempferia galanga* L.(Zingiberaceae): A high sought medicinal plant in Tropical Asia. *J. Med. Plants Stud* **2016**, *4*, 270–276.
171. Haque, S.M.; Ghosh, B. Micropropagation of *Kaempferia angustifolia* Roscoe-an aromatic, essential oil yielding, underutilized medicinal plant of Zingiberaceae family. *J. Crop Sci. Biotechnol* **2018**, *21*, 147–153. [CrossRef]
172. Resmi, J.; Bindu, M.; Suja, G. In vitro Multiplication of *Kaempferia Galanga* L.-An Important Medicinal Plant. *J. Krishi Vigyan* **2022**, *11*, 223–228. [CrossRef]
173. Hemanthakumar, A.; Preetha, T. Macro-and Micropropagation of Plants for Income Generation. In *Conservation and Sustainable Utilization of Bioresources*; Springer: Berlin/Heidelberg, Germany, 2023; pp. 409–450.
174. Preetha, T.; Hemanthakumar, A.; Krishnan, P. Shoot tip cryopreservation by vitrification in *Kaempferia galanga* L. an endangered, overexploited medicinal plant in Tropical Asia. *IOSR J. Pharm. Biol. Sci* **2013**, *8*, 19–23. [CrossRef]
175. Preetha, T.S.; Hemanthakumar, A.S.; Krishnan, P.N. Cryopreservation of aromatic ginger *Kaempferia galanga* L. by encapsulation-dehydration. *Not. Sci. Biol.* **2021**, *13*, 11024. [CrossRef]
176. Ravindran, P.N.; Nirmal Babu, K. (Eds.) *Ginger: The Genus Zingiber*, 1st ed.; CRC Press: Washington, DC, USA, 2005; pp. 15–87.
177. Geetha, S.P. In Vitro Technology for Genetic Conservation of Some Genera of Zingiberaceae. Ph.D. Thesis, University of Calicut, Thenhipalam, India, 2002.
178. Yamuna, G.; Sumathi, V.; Geetha, S.; Praveen, K.; Swapna, N.; Nirmal Babu, K. Cryopreservation of in vitro grown shoots of ginger (*Zingiber officinale* Rosc.). *CryoLetters* **2007**, *28*, 241–252. [PubMed]
179. Ellis, D.; Skogerboe, D.; Andre, C.; Hellier, B.; Volk, G. Implementation of garlic cryopreservation techniques in the national plant germplasm system. *CryoLetters* **2006**, *27*, 99–106. [PubMed]
180. Sant, R.; Taylor, M.; Tyagi, A. Cryopreservation of in vitro-grown shoot-tips of tropical taro (*Colocasia esculenta* var. esculenta) by vitrification. *CryoLetters* **2006**, *27*, 133–142.
181. Xin, S. Preservation of the Callus Tissues under the Condition of Superlower Temperature (−196 °C) in Fritillaria thunbergii Miq. *Mod. Appl. Pharm.* **1990**, *4*, 18–19. [CrossRef]
182. Matsumoto, T.; Sakai, A.; Yamada, K. Cryopreservation of in vitro-grown meristems of wasabi (Wasabia japonica) by encapsulation-vitrification method. *CryoLetters* **1995**, *16*, 189–196.
183. Pan, C.; Liu, J.; Bi, W.L.; Chen, H.; Engelmann, F.; Wang, Q.C. Cryopreservation of small leaf squares-bearing adventitious buds of Lilium Oriental hybrid 'Siberia'by vitrification. *Plant Cell Tissue Organ Cult. (PCTOC)* **2018**, *133*, 159–164. [CrossRef]
184. Shi, T.T.Z.G.L.; Wen, S.M.; Lu, J.L. Cryopreservation of Lilium pollens. *North Hortic* **2013**, *13*, 64–66.
185. Pan, S.-Y.; Gerhard, L.; Gao, S.-H.; Zhou, S.-F.; Yu, Z.-L.; Chen, H.-Q.; Zhang, S.-F.; Tang, M.-K.; Sun, J.-N.; Ko, K.-M. Historical perspective of traditional indigenous medical practices: The current renaissance and conservation of herbal resources. *Evid.-Based Complement. Altern. Med.* **2014**, *2014*, 525340. [CrossRef]
186. Ruta, C.; Lambardi, M.; Ozudogru, E.A. Biobanking of vegetable genetic resources by in vitro conservation and cryopreservation. *Biodivers. Conserv.* **2020**, *29*, 3495–3532. [CrossRef]
187. Benelli, C. Plant cryopreservation: A look at the present and the future. *Plants* **2021**, *10*, 2744. [CrossRef] [PubMed]

plants

MDPI

Review

Development, Management and Utilization of a Kiwifruit (*Actinidia* spp.) In Vitro Collection: A New Zealand Perspective

Jayanthi Nadarajan *, Azadeh Esfandiari, Liya Mathew, Jasmine Divinagracia, Claudia Wiedow and Ed Morgan

Food Industry Science Centre, The New Zealand Institute for Plant and Food Research Limited, Palmerston North 4410, New Zealand
* Correspondence: jayanthi.nadarajan@plantandfood.co.nz

Abstract: The New Zealand Institute for Plant and Food Research Limited (PFR) supports a large kiwifruit breeding program that includes more than twenty *Actinidia* species. Almost all the kiwifruit accessions are held as field collections across a range of locations, though not all plants are at multiple locations. An in vitro collection of kiwifruit in New Zealand was established upon the arrival of *Pseudomonas syringae* pv. *Actinadiae*-biovar 3 in 2010. The value of an in vitro collection has been emphasized by restrictions on importation of new plants into New Zealand and increasing awareness of the array of biotic and abiotic threats to field collections. The PFR in vitro collection currently holds about 450 genotypes from various species, mostly *A. chinensis* var. *chinensis* and *A. chinensis* var. *deliciosa*. These collections and the in vitro facilities are used for germplasm conservation, identification of disease-free plants, reference collections and making plants available to users. Management of such a diverse collection requires appropriate protocols, excellent documentation, training, sample tracking and databasing and true-to-type testing, as well as specialized facilities and resources. This review also discusses the New Zealand biosecurity and compliance regime governing kiwifruit plant movement, and how protocols employed by the facility aid the movement of pathogen-free plants within and from New Zealand.

Keywords: *Actinidia* spp.; *Pseudomonas syringae actinidiae*; tissue culture; medium term in vitro storage; germplasm conservation; cryopreservation; biosecurity and compliance

Citation: Nadarajan, J.; Esfandiari, A.; Mathew, L.; Divinagracia, J.; Wiedow, C.; Morgan, E. Development, Management and Utilization of a Kiwifruit (*Actinidia* spp.) In Vitro Collection: A New Zealand Perspective. *Plants* **2023**, *12*, 2009. https://doi.org/10.3390/plants12102009

Academic Editors: Anna De Carlo and Waed Tarraf

Received: 6 March 2023
Revised: 9 May 2023
Accepted: 11 May 2023
Published: 17 May 2023

1. Introduction to New Zealand Kiwifruit In Vitro Collection

1.1. History of Kiwifruit In Vitro Collection in New Zealand

Cultivars of the genus *Actinidia* (kiwifruit) are an important export (NZD 4 billion) for New Zealand, resulting in significant investment in breeding programs to ensure the industry remains competitive [1]. Breeding draws on the diversity of a species or genus, requiring access to large numbers of individuals that are valued for the traits they offer. The New Zealand Institute for Plant and Food Research Limited (PFR) supports The Kiwifruit Breeding Centre Limited (KBC) in a large kiwifruit breeding program that includes more than 20 species; while thousands of genotypes of these species are maintained in the field, around 450 genotypes are held in vitro with the germplasm collection comprising a subset of these plants.

The arrival of *Pseudomonas syringae* pv. *Actinidiae*-biovar 3 (Psa-3) (previously known as Psa-V) in the North Island of New Zealand in 2010 [2] had a devastating effect on the health of field collections. More than 80% of diploid and about half of the tetraploid *A. chinensis* plants growing in the PFR research orchard in Te Puke were removed because of Psa-3 infection by 2013 [3]. This prompted the urgent establishment of an in vitro repository so that plants could be held in an environment in which they were less exposed to biotic threats. The in vitro repository is located in Palmerston North, away from field-based collections to reduce risks that might arise through natural disasters (Figure 1a). The value of the collection has been accentuated by the expensive and long quarantine procedures for importation, and because of difficulties in accessing material from some countries of origin. With field grown germplasm under constant threat from biotic and abiotic factors, and the challenges with importing planting material, we need to ensure the conservation of New Zealand's existing kiwifruit genetic resources. Ex situ conservation using in vitro technologies, including cryopreservation, provides an alternative and complementary storage method for this germplasm. However, PFR's role is also to ensure our plant material is available to users. Hence, the in vitro collection needs to be fit for purpose for PFR's and our collaborator's objectives.

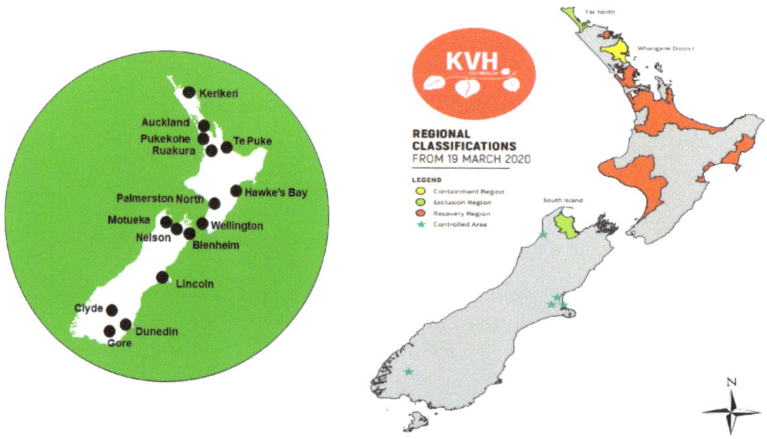

(a) (b)

Figure 1. (**a**) Plant & Food Research (PFR) locations within New Zealand. PFR kiwifruit collection is grown in orchard blocks at Kerikeri, Ruakura, Te Puke, Motueka and Clyde; (**b**) Map showing the boundaries and status of regions in respect of the presence of *Pseudomonas syringae* pv. *Actinidiae*-biovar 3 (Psa-3) (adapted with permission from KVH [4]). Containment region (yellow): Limited Psa-3 infections. Movement of plant material is restricted; Exclusion region (green): No orchards with Psa-3 identified, or Psa-3 has recently been identified for the first time. Material is unrestricted; Recovery region (red): Psa-3 is already widespread. Material is restricted; Controlled areas (stars): only within the South Island. Exclusion regions and movement restrictions apply.

Challenges associated with in vitro collections include establishment, propagation and maintenance, as well as re-establishment of plants to the in vivo environment. Consideration must be given to the health status of plants, as the Psa-3 bacterium has been shown to persist within inoculated in vitro growing kiwifruit plants through multiple culture cycles [5]. Additional key elements that will make the germplasm collection successful are accurate curation, tracking and management of the plants

1.2. Restricted Kiwifruit Plant Material Handling and Movement Limitations

In addition to plants being conserved and curated, they may require steps to remove or verify the absence of pathogens so the plants can be moved (within and beyond New Zealand). Additionally, plants may be held because they host particular pathogens that cannot otherwise be maintained. Following the arrival of Psa-3 in New Zealand, regulations were put in place to restrict the movement of kiwifruit plants due to the uneven distribution of the disease within growing areas. While some regions remain Psa-3-free, others have widespread infections, resulting in strict control of plant movements by New Zealand Kiwifruit Vine Health (KVH) and the New Zealand Ministry for Primary Industries (MPI). KVH, established in 2010 to lead the industry response to the Psa-3 incursion, has been responsible for managing biosecurity readiness, response and operations for the kiwifruit industry since 2012, in conjunction with MPI. Plant material can be moved relatively easily from exclusion to recovery regions (Figure 1b) where Psa-3 is widespread. However, moving plants from recovery to exclusion regions requires extensive testing and quarantine procedures. Although these measures have successfully protected growers in Psa-3-free regions, they resulted in limited access to new cultivars for growers in these regions. To move plants from a recovery region to an exclusion region, they must follow a pathway mandated and approved by both KVH and MPI. In New Zealand, handling of restricted kiwifruit material requires adherence to rules regulating Physical Containment (PC2) facilities. The kiwifruit PC2 laboratory in Palmerston North, operated by PFR since 2011, is used to process field-sourced and potentially Psa-3-infected shoots for initiating in vitro cultures. In 2018, MPI finalized the reopening of the *Actinidia* Import Health Standard (IHS), allowing the importation of kiwifruit plant material for the first time since the Psa-3 incursion in 2010.

This review will focus on development, management and utilization of the PFR-maintained kiwifruit in vitro collection, application of relevant in vitro technologies, and the challenges associated with these activities.

2. Use of the In Vitro Kiwifruit Collection for Germplasm Conservation

2.1. Establishing and Maintaining the In Vitro Kiwifruit Collection

2.1.1. Plant Initiation in Tissue Culture

Initiation of field-sourced plant materials into tissue culture is a vital step for establishment of an in vitro collection. Efficient establishment of in vitro cultures is affected by genotype, health of the field-grown mother plant, preparation of explant, surface sterilization procedures, medium composition and growth conditions. For our collection, these have all been improved over the years, with many factors found to contribute to initiation success. When the in vitro system was first being established at PFR (and in the context of trying to rapidly rescue many different genotypes from field collections), budwood explants for initiation were obtained directly from the field. Problems reported in subsequent cultures included: high contamination rates and growth problems, such as apical shoot collapse or slow growth with a success rate of c. 10% [6]. The initiation protocol was later improved by sprouting new shoots from canes excised from winter-dormant mother plants in a "clean" indoor environment [7] and modifying the surface sterilization protocol [8]. The original medium onto which the explants were initiated was based on Murashige and Skoog (MS) salts [9] and supplemented with zeatin. Although good for some genotypes, others, particularly males, were reported to stop growing and die after developing symptoms of shoot-tip browning, shoot dieback, leaf necrosis and significant callus growth [6]. To accommodate a wider genotype range, zeatin was replaced with meta-Topolin (mT) in the initiation and proliferation medium comprising half-strength MS macro-elements, full-strength MS microelements, B5 vitamins [10] and 3% (w/v) sucrose, supplemented with 0.66 mg/L mT, 0.05 mg/L Indole-3-butyric acid (IBA), 0.1 mg/L gibberellic acid and 0.75% (w/v) agar [11]. These and other modifications in our initiation protocol have led to a success rate of over 70% in establishing in vitro cultures of a wide range of genotypes from different species, this includes removal of plants shown to carry endogenous organisms.

After successful initiation of the explants (growing from axillary buds) in mT medium, young plants are transferred to proliferation medium for six weeks, after which these plants undergo a bacteriological screening (described in Section 2.1.2 below). Any plants exhibiting bacteriological or fungal contamination are discarded, and this is noted in the Germplasm Management System (GMS). Once the plants have cleared bacteriological screening, they are maintained on proliferation medium. A 4–6-week culture cycle is used. The plants are grown under standard culture conditions of a temperature of 24 ± 1 °C, 16 h photoperiod and a light intensity of 35–45 µmol m^{-2} s^{-1}, provided by Philips® cool white fluorescent tubes (Philips, China). To induce the growth of roots, healthy plantlets are transferred onto a medium that comprises half-strength MS macro-elements, full-strength MS microelements, Linsmaier and Skoog (LS) [12] vitamins, 3% (w/v) sucrose, 0.6 mg/L

IBA and 0.75% (*w*/*v*) agar. The duration needed for root initiation varies based on the genotype, generally ranging from two to four weeks.

2.1.2. Bacteriological Screening

The importance of plant health status with particular reference to Psa-3 has been mentioned previously. Many of the in vitro cultures were initiated from plants growing in the field. However, even apparently healthy-looking plants have been demonstrated to carry the Psa-3 bacterium [13,14]. Additionally, plants carry a microbiome that reflects their cultivation history and they may harbor endogenous pathogens that cannot be detected using standard plant tissue culture media [15]. This risk is minimized by screening for contaminating organisms following the onset of growth on the tissue culture-initiated explants. For this, the basal part of the plant is excised and transferred into a Petri plate containing a medium comprising 20 g/L potato dextrose agar (PDA) with 5 g/L peptone for one week to encourage proliferation of endogenous pathogens. Plants that show no contamination are retained, but then, depending on the purpose of the in vitro cultures, they might be subjected to further screening for Psa-3 using polymerase chain reaction (PCR) testing in conjunction with incorporation of 3 g/L peptone into growing medium to confirm the plants are Psa-3-free [13,16].

2.1.3. Optimization of Tissue Culture Medium

Perhaps the most significant challenge at present is how to cope with the differences in in vitro performance of the diverse range of kiwifruit genotypes. This is an issue for all germplasm collections that will invariably lack the resources to optimize conditions for each of the large numbers of genotypes they hold. It is generally accepted that genotypes of the same species can respond quite differently to the same in vitro conditions. The same is true for kiwifruit. Although many genotypes will grow in vitro on the same medium and in the same conditions [17,18], different genotypes, even siblings, may have different requirements for optimum in vitro growth. For this discussion, we will focus only on nutrient salt formulations for kiwifruit in vitro culture.

A range of media have been used to grow kiwifruit in in vitro culture since the publication by Harada [19]; however, there has been little optimization of the nutrient medium. Though MS salts are frequently used, other formulations have been used in different applications [17]. In trying to improve establishment of in vitro cultures, Debenham et al. [6] tested genotypes of *A. chinensis* and *A. polygama* on Long and Preece medium (LP) [20], Woody Plant Medium (WPM) [21] and B5 and MS macronutrient salt formulations. Though differences in plant performance were noted, e.g., smaller and paler leaves on LP and B5 media, these were not considered limiting to culture initiation success rates [6]. Half-strength MS medium was also recommended by some research groups [18,22], which we use frequently for our kiwifruit cultures now.

Work to identify an improved salt formulation for in vitro culture of *A. arguta* has been reported. For example, Hameg et al. [23] evaluated the response of an *A. arguta* genotype to six different nutrient formulations. They reported that the medium of Standardi [24] gave the best or equal best results and B5 the poorest results, though they noted that Standardi [24] medium resulted in some undesirable callus growth. A "Design of Experiments" [25] approach was used to formulate a new nutrient salt combination (R medium) for an *A. arguta* genotype [26]. We have begun to evaluate the "R" medium.

2.2. Current Kiwifruit In Vitro Collection Status

The PFR kiwifruit in vitro collection currently features more than 450 genotypes from various species. During the four-year period from 2019 to 2023, we have successfully initiated 370 genotypes of kiwifruit into in vitro culture. These genotypes represent *A. chinensis* var. *chinensis* (54%), *A. chinensis* var. *deliciosa* (19%), *A. melanandra* (1%), *A. arguta* (2%) species and hybrids (12%). Additionally, 31% and 49% of the plants initiated are male and female, respectively (Figure 2a,b). In addition to its diverse range of species and genotypes, this collection also preserves genotypes that no longer exist in the field, potentially adding further value to the collections.

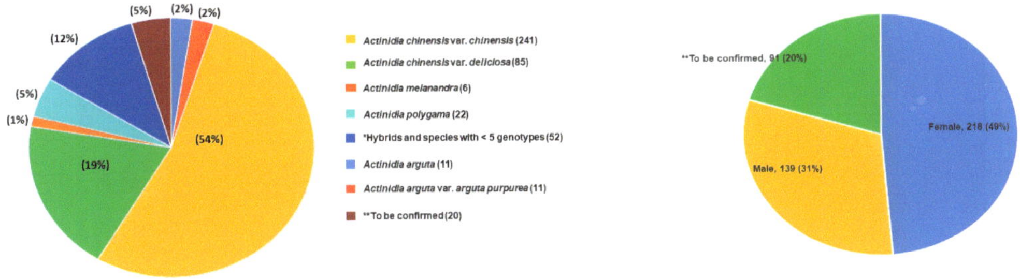

(**a**) (**b**)

Figure 2. The diversity of the genotypes in the PFR kiwifruit in vitro collection. (**a**) Number and percentage of genotypes for each species. (**b**) Genotypic distribution by sex in number and percentage. * This collection includes one genotype each from *A. callosa*, *A. chrysantha*, *A. chinensis* var. *deliciosa coloris*, *A. latifolia* and *A. valvata*; two genotypes each from *A. hemsleyana*, *A. indochinensis* and *A. setosa*; three genotypes of *A. kolomikta*; and five genotypes from *A. macrosperma* species, respectively. ** Gap analyses are underway to confirm the species and sex of the unidentified accessions.

2.3. Medium-Term Storage

Following the establishment of cultures, plants may be moved to medium-term storage (MTS) for conservation in a system that requires less maintenance of plants. Under the MTS protocol, plants are held in vials of half-strength MS medium supplemented with 3% sucrose and solidified using 0.75% agar under conditions of low temperature (5 °C) and low light (1–3 µmol m^{-2} s^{-1}, 8 h photoperiod). MTS methods reduce plant metabolic activity through physical, chemical, or nutrient limitations to plant growth [27]. This approach can reduce the frequency of subculture events from monthly to yearly, providing savings in labor and lowering risks of handling errors and genetic instability. Plants stored under MTS conditions are easily available for propagation or other purposes.

Kiwifruit accessions under MTS conditions undergo two distinct stages, i.e., slow growth storage and rejuvenation. Sixteen healthy plants of a genotype are placed in eight clear screw-top vials with growth medium described above. They are grown in standard culture conditions for four weeks, then undergo a 7-day acclimation period at 22 °C for 10 h/2 °C dark for 14 h. Following acclimation, these plants are moved to the MTS environment as described above. Visual evaluations of plant quality / health are conducted once every three months. When only 20% of the plants remain viable, they are subjected to a rejuvenation step. To accomplish this, plants are removed from all eight vials and undergo two or three rounds of subcultures to generate a new set of 16 plants, which are then returned to the MTS environment. The proliferation process begins with a medium containing mT and ends with a medium containing zeatin [6], resulting in a larger number of plants in a short period of time with a standard plant size of 3 to 4 cm for MTS. The duration of a plant can be maintained under MTS before rejuvenation is species and genotype dependent. As an example, the storage life of four genotypes of *A. macrosperma* and *A. chinensis* at 5 °C are presented in Table 1. In general, plants can be held at MTS conditions without rejuvenation for between 400 and 800 days. A small proportion (about 4%) require rejuvenation within 200 days, while a low proportion can last between 1200 and 1400 days (Figure 3). All activities are recorded in a database where the vials containing plants are labelled with details of the genotypes, vial identity, and date. All plant transfers are recorded using our in-house database systems, i.e., Germplasm Management System (GMS) and Vial Management System (VMS) (Figure 4). Trays are used to store the vials and are assigned a number in VMS to easily locate the genotypes.

Table 1. Selected examples of genotypes in medium-term storage (MTS). Genotypes 11384 and 6001 are male plants of *Actinidia macrosperma* (MA) and genotypes 11348 and 11158 are female plants of *Actinidia chinensis* var *chinensis* (CK). These genotypes have undergone repeated cycles of storage and rejuvenation, with varying storage duration at 5 °C.

Genotype ID	Species	Gender	Year of Entry	Average of Days Being in MTS
11384	MA	Male	2019	955
6001	MA	Male	2018	496
11348	CK	Female	2017	349
11158	CK	Female	2016	680

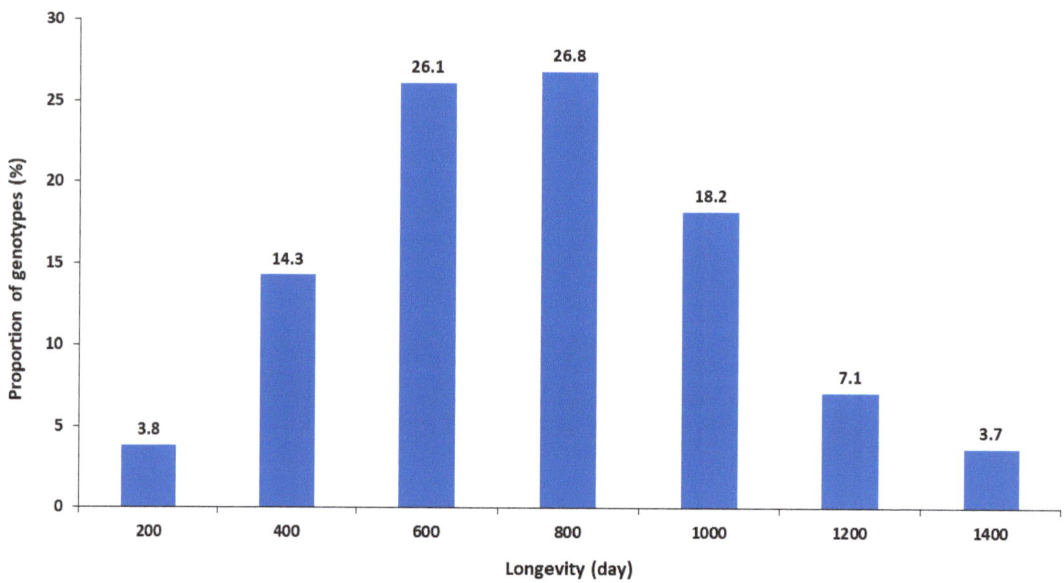

Figure 3. The proportions of kiwifruit genotypes (%) that could be maintained in medium term storage (MTS) for different periods without rejuvenation. MTS is in vitro storage of kiwifruit plants at 5 °C with longevity describing the length of time in days that the plants could be left before requiring rejuvenation.

2.4. Long-Term Conservation Using Cryopreservation

The in vitro plants provide starting material for long-term conservation using cryopreservation. Cryopreservation—the preservation of viable cells, tissues, organs and organisms at ultra-low temperatures (c. −196 °C) in the liquid or vapor (c. −150 °C) phase of liquid nitrogen. This method of preservation is growing in popularity for plant germplasm conservation due to its comparatively low maintenance cost, small space requirement and reliability [28]. Other advantages of cryopreservation over medium-term storage (MTS) or standard in vitro storage are the reduced threats of contamination and somaclonal variation [29].

Cryopreservation protocols are usually developed empirically for specific materials or explants, taking into consideration the physiological and bio-physical factors of the explants to minimize stress and maximize survival rates [30]. Cryoprotection can be applied in the form of osmo-protection alone, i.e., direct desiccation using silica gel or air drying, or in a combination of osmo-protection and chemical cryoprotection using permeating chemical cryoprotectants. The ultimate goal is to achieve a 'vitrified state' which is the critical physical state that determines the success of the post-cryopreservation survival [31].

The development of a simple and reliable cryopreservation protocol would allow wider application of this preservation technique in conservation of plant materials.

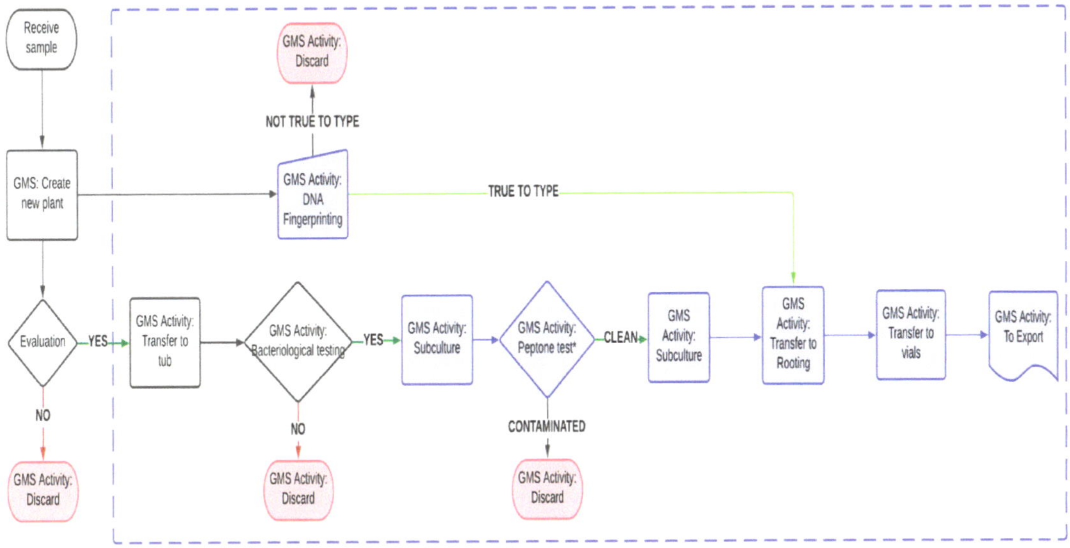

Figure 4. Flow diagram summarizing steps involved in placing a plant into medium-term storage (MTS) and tracking of plants using Germplasm Management System (GMS) and Vial Management System (VMS). When a plant is received, its details are entered into the GMS database and the plant is prepared for MTS at 5 °C. GMS provides details of every plant in the laboratory with VMS used to record plants being held in vials in MTS. The activities highlighted in blue are activities recorded in VMS. Discarded accessions are also recorded in the system (indicated in pink).

At PFR, a cryopreservation protocol has been developed for the long-term storage of kiwifruit using a droplet vitrification protocol [18]. This protocol was successfully tested on nine genotypes from five species, i.e., *A. chinensis* var. *chinensis*, *A. chinensis* var. *deliciosa*, *A. arguta*, *A. macrosperma* and *A. polygama*, utilizing shoot tips around 1 mm in size excised from two-week-old axillary shoots. These young shoots were obtained by growing nodal cuttings sourced from six-week-old in vitro grown mother plants. In our study, we observed that the age of the mother plant and size of the explant can determine the success of cryopreservation; use of younger donor plants and smaller explants significantly increasing regeneration. The response of kiwifruit plants to in vitro culture is not just species-specific, but also genotype-specific [32]. Hence, significant variation in post-cryopreservation regeneration between different genotypes was expected, and was observed to range from 59% to 88% [33,34]. Our results confirmed that different kiwifruit genotypes respond differently to the stresses imposed at various stages of the cryopreservation protocol [33,34]. We tested this protocol successfully on three additional species, i.e., *A. melanandra* (regeneration c. 70%), *A. kolomikta* (regeneration c. 68%) and *A. eriantha* × *A. chinensis* (regeneration c. 55%) (Figure 5). Since post-cryopreservation regeneration is higher than the suggested baseline of 40% for cryobanks [35], we will be implementing this cryopreservation protocol for long-term kiwifruit germplasm conservation.

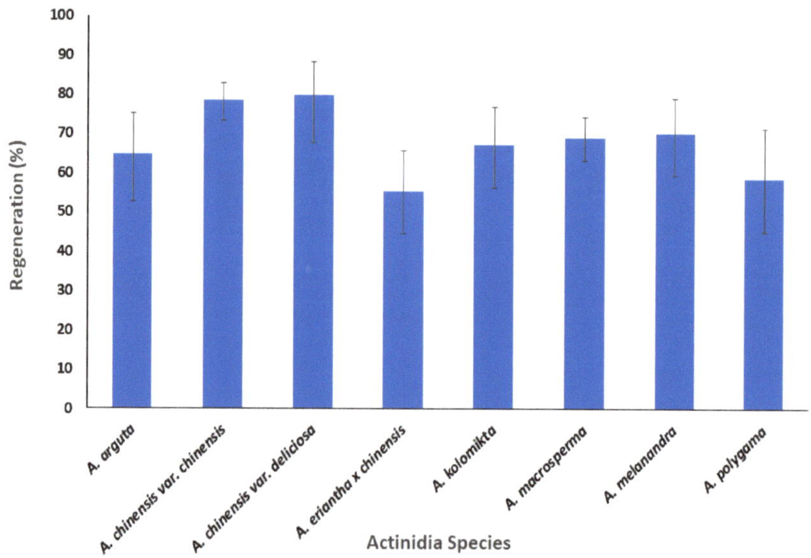

Figure 5. Post-cryopreservation regeneration percentages of eight kiwifruit species. The mean percentage of regeneration is averaged over 1–3 genotypes per species. Error bars are 95% confidence limits.

3. Use of In Vitro Collection for Plant Movements

The key role of the collection and collection facility is to ensure the availability of plants for breeding and research programs and to support industry partners. The laboratory provides a hub for distributing tissue-cultured plants to a range of users in New Zealand and offshore. A goal is to ensure supply of Psa-3 free plants. Any such movements of plant material are subject to appropriate consent/material transfer agreement from the owners of these genotypes. These plant movements comply with previously mentioned requirements from MPI and KVH, where PFR is tracking all exported plants in the GMS database (date, genotype, number of clones, etc.) audited six-monthly or annually by MPI and KVH.

3.1. Plant Movement Overseas (Export)

Before plant material leaves the facility, a true-to-type quality control assessment is carried out using DNA fingerprinting. Molecular techniques are commonly used in plant breeding, intellectual property (IP) protection and cultivar verification, and to facilitate plant genetic resources management and conservation [36]. Standard lab protocols have been adjusted and developed at PFR over the past few years for DNA profiling and are described in Rowland et al. [37] and Knäbel et al. [38], using a modified cetyltrimethylammonium bromide (CTAB) method [39,40] and in-house developed simple sequence repeat (SSR) markers that are species-specific for kiwifruit [41].

Using SSR, amplified fragments of DNA are scored as illustrated in Figure 6, where DNA fragments and alleles are displayed as peaks with their sizes in base pairs (bp). The resulting molecular characterization/scoring output shows patterns of molecular similarity indicating if the studied plant is clonal to a reference sample as shown in Figure 6.

Following true-to-type testing, plants that have passed both phases (bacteriological and DNA profiling) are prepared for shipment and export following packaging one clonal plant per vial, and labelled with a GMS barcode to provide a unique identifier. Typically, 10–12 vials of rooted plants will be provided per genotype. Vials are triple-packed and sent according to IATA shipping standard 650 [42]. Figure 7 summarizes the flow of activities (corresponding to GMS activities) for overseas exports. For example, when a plant is received, its details are placed into the GMS database and it undergoes a series of activities. These activities correlate to specific GMS activities that allow for tracking plants in specific stages. Activities for export of plants are highlighted in violet (DNA fingerprinting, peptone testing, rooting and export). Discarded accessions are also recorded in GMS (indicated in pink). The number of peptone tests carried out corresponds to the specific movement pathway. For example, for plants being sent within New Zealand, a PCR test and a minimum of three peptone tests are required (Figure 7).

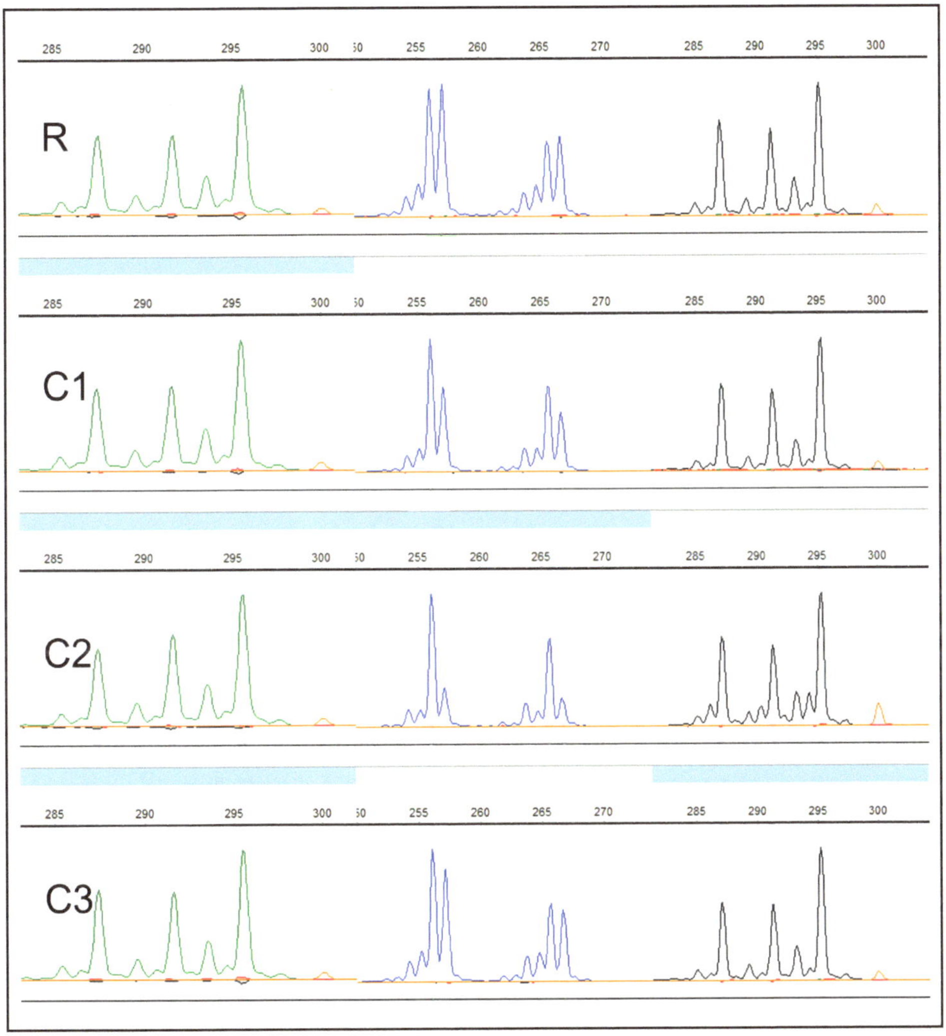

Figure 6. SSR-amplified fragments of a kiwifruit genotype. An example output of GeneMarker® v 2.2.0 software (© SoftGenetics, LLC, State College, PA, USA) showing R = reference sample displaying identical allele/peak pattern as clone 1, 2 and 3 (C1–3) and suggest true to type.

3.2. Plant Movement within New Zealand

It is sometimes desirable to move plant material bred in a recovery region (North Island) to an exclusion region (South Island) (Figure 1) to ensure commercial growers are accessing new cultivars (and to support breeding activities in other regions). In this instance, the plants go through periods of greenhouse and outdoor growing quarantine before being released into exclusion regions (Figure 1) [43]. Candidate plants must be stage 3 tissue culture plants (rooted plants) of confirmed Psa-3-free status. This is achieved by identifying Psa-3-free in vitro grown mother plants from which further plants are propagated. Psa-3-free status is first indicated by PCR testing of individual plants. These plants are subsequently propagated, with progeny undergoing repeated screenings (at least three times) on peptone-supplemented medium. If any plant derived from a mother plant is subsequently found to have Psa-3, all plants of the lineage are required to be destroyed. These steps are intended to provide industry and regulatory authorities a high degree of confidence that the in vitro plants have minimal chance of carrying Psa-3 and can be moved from containment to a Psa-3 quarantine greenhouse within the South Island. Plants are held in a Psa-3 quarantine facility for up to two years before release.

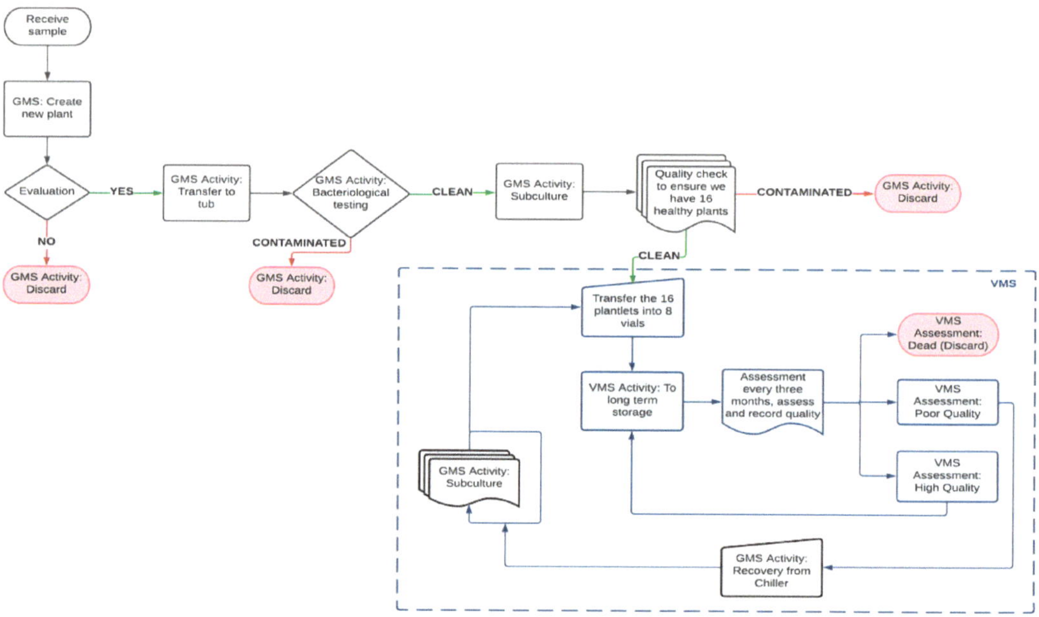

Figure 7. Summarized workflow for preparation of in vitro plants for overseas trials and plant movement within New Zealand at the PFR in vitro facility.

In addition to the South Island pathway, there is also a pathway for the movement of candidate plants within North Island recovery regions. The candidate plants undergo PCR screening to identify disease-free mother plants followed by subsequent screening on a peptone-supplemented medium to ensure Psa-3-free status. After achieving "in vitro," Psa-3-free status candidate plants may, with prior approval from KVH and MPI, be transferred from a containment facility to a non-containment facility.

3.3. Plant Movement into New Zealand (Import)

The significant loss caused by the Psa-3 incursion has resulted in New Zealand adopting a very cautious approach to new importations of *Actinidia*. All plant material must go through a tissue culture step or be imported as tissue cultures. If dormant cuttings are imported, they must be used to generate tissue cultures in a quarantine facility [44,45]. Although the current facility is not involved in imports of kiwifruit plants into New Zealand, this point is highlighted, since the requirement for the importation pathway mandates in vitro steps which places further pressure on the development of improved in vitro protocols.

4. Use of the In Vitro Collection to Maintain a Virus Reference Collection

There is value in maintaining reference collections, not just of plants but also of their pathogens. In some cases, the host range of the pathogen may be limited, or the pathogen may not be widely dispersed in the environment, and it may be convenient to maintain infected in vitro cultures. This is a method we use to maintain access to the betaflex virus, *Betaflexiviridae*, with a small number of different kiwifruit genotypes known to be infected held in culture.

5. Management and Challenges Associated with In Vitro Collection

Since 2014, all data from the PFR field kiwifruit germplasm and breeding collections have been held in Ebrida, a plant-breeding software package developed by Agri Information Partners (Wageningen, The Netherlands) and adapted for use in PFR (www.e-brida.nl, accessed on 2 February 2023). This database contains information on accessions with their taxon, passport data and ploidy, block plans, listing individual vines with their unique ID(s), planting position(s), year planted, sex, phenotypic and/or genotypic data. The Ebrida database ensures kiwifruit germplasm and breeding information is available to approved users of the collection, including breeders, germplasm curators, biometricians and molecular geneticists.

5.1. Database and Sample Tracking Systems

For management and sample tracking of the in vitro collection, GMS is used. The GMS is a user-friendly, searchable database designed to track and record all details of genotypes and plants in the in vitro collection. Plants are tracked from initiation (including passport data) to disposal with all details recorded in the database. A discarded sample is also recorded in the system, so it can be traced to determine if a genotype still exists in the collection. The system assigns unique "Activity Event ID" codes to each container and can link each container back to its previous "Activity Event ID", allowing for a complete history of events related to each plant to be recorded, right back to the plant in the field [46] (Supplementary Figure S1). One important component of the GMS is the VMS, which is specifically designed to store and organize data related to genotypes conserved in MTS. The VMS is used to maintain accurate and up-to-date records of all vials and genotypes in MTS, including information on the availability, the location of a vial and quality assessments of the conserved samples. Having accurate genotype information and ensuring the availability of all data upon sample receipt has been crucial for establishing a functional database. It is crucial to establish, follow and audit systems to ensure the best practices for sample collection, storage and analysis are practiced.

5.2. Somaclonal Variation

Somaclonal variation can be detected at morphological, cytological, cytochemical, biochemical, and molecular levels [47]. In in vitro cultures kiwifruit, the analysis of genetic stability over the long-term especially in slow growth storage is still lacking. Both genotypic and phenotypic assessments are needed to confirm if somaclonal variation is indeed a problem. Though extreme levels of somaclonal variation between 100% [48] and 90% in banana [49] have been reported, typically an average of 15–20% can be expected [50], depending on the number of culture cycles and culture conditions. The levels of somaclonal variation can vary with genotype, type of tissue, explant source, medium components and the duration of the culture cycle [51]. Some level of somaclonal variation may be inevitable in tissue culture; however, manipulating, controlling and minimizing the putative inductive factors will reduce the level of occurrence. The PFR kiwifruit in vitro collection is being studied for somaclonal variation. We are examining morphological changes in a subset of frequently subcultured genotypes from our MTS collection and using DNA sequencing to detect somaclonal variation at the molecular level.

5.3. Backing up Collections

To mitigate the risk of losing the in vitro collection due to environmental catastrophes or human errors, duplication or back up of the collection is recommended [52]. The backup collection should be in a different location to reduce risks to the collection from natural disasters such as earthquake, flood or fire. For our kiwifruit in vitro collection, we are planning to back up the collections in MTS and in cryostorage at different PFR sites. These sites are selected to provide the same storage environment as the original collections. Our database, particularly VMS, has been upgraded to capture the information on backup collections. The backup repository is only responsible for receiving and storing the plant materials and will not be responsible for managing the collection, as all preparation of material (initiation, proliferation, rejuvenation) and database entry will be carried out in our original repository.

5.4. Specialized Operations and Resources Requirements for In Vitro Collection

Setting up an in vitro repository requires tissue culture laboratory, medium preparation room, growth chambers, controlled cabinets for medium term storage, cryopreservation facility and an efficient inventory system. Management of an in vitro facility entails cost, particularly expenses associated with electricity, culture medium and labor charges, in addition to the equipment cost. The cost of electricity, which is mainly utilized for autoclaving, artificial lighting in tissue culture rooms, air filtration of laminar air hoods and air conditioning can amount up to 60% of tissue culture production costs [53]. Culture medium is indispensable to a tissue culture facility, but preparation can be expensive. The cost of labor accounts for 60–70% of the medium cost in a tissue culture laboratory [54].

In addition to the infrastructure requirement, highly skilled staff are the backbone of an in vitro collection facility. The staff are trained in aseptic technique, require knowledge of how to modify protocols and must have manual dexterity with aseptic tools, especially when handling fragile plant tissues. Micropropagation is labor-intensive work [54,55]. Since tissue culture media are high in nutrients and plants are often held in a warm environment, microbial contamination can result in significant losses. Prevention of microbial contamination is crucial in micropropagation facilities [56] and requires highly skilled technicians. Ahloowalia and Savangikar [57] and Datta [58] pointed out that using a highly skilled work force can significantly increase laboratory efficiency and reduce costs.

6. Conclusions

In this review, we reported the ongoing development of the kiwifruit in vitro collection held at PFR. The in vitro collection complements field collections as repositories for New Zealand kiwifruit genetic resources. The large in vitro collection needs careful curation and management procedures to address the challenges posed by the genetic diversity it holds. The facilities and staff also support end users of plant material by facilitating plant movements within and from New Zealand. The facility provides Psa-3-free plants, reference collections and plants used as resources for research and new cultivar development. The specialized facilities, resources and skilled staff are key to the ongoing management and growth of this collection. We also discussed the need to work with regulators who manage New Zealand's biosecurity and compliance requirements to ensure plants can be made available to users.

Supplementary Materials: The following supporting information can be downloaded at: https://www.mdpi.com/article/10.3390/plants12102009/s1, Figure S1: Example of activities recorded in GMS for one genotype. The workflow starts at initiation, with subsequent activities until being sent for testing. Each 'Activity Event ID' number (illustrated as a circle) represents one activity.

Author Contributions: Conceptualization, J.N.; writing—original draft preparation, J.N., A.E., L.M., J.D., C.W. and E.M.; writing—review and editing, J.N., A.E., L.M., J.D., C.W. and E.M. All authors have read and agreed to the published version of the manuscript.

Funding: This research was funded by The New Zealand Institute for Plant and Food Research Limited (PFR) Kiwifruit Royalty Investment Programme (KRIP) for Germplasm Program.

Data Availability Statement: The data presented in this study are available on request from the corresponding author.

Acknowledgments: The authors would like to express their gratitude to Andrew Mullan and Belinda Diepenheim for laboratory management and tissue culture medium preparation and recipe advice; Alison Duffy, Thomas Paterson and Guilhem Morra for checking and updating the kiwifruit collection list and database assistance; Andrew McLachlan for statistical advice; Tianchi Wang and Joy Tyson for providing an internal review of the manuscript.

Conflicts of Interest: The authors declare no conflict of interest.

References

1. Zespri-Annual-Report. 2021–22. Available online: Zespri-Annual-Report-2021-22.pdf (accessed on 27 January 2023).
2. Everett, K.R.; Taylor, R.K.; Romberg, M.K.; Rees-George, J.; Fullerton, R.A.; Vanneste, J.L.; Manning, M.A. First report of Pseudomonas syringae pv. actinidiae causing kiwifruit bacterial canker in New Zealand. *Australas. Plant Dis. Notes* **2011**, *6*, 67–71. [CrossRef]
3. Datson, P.; Nardozza, S.; Manako, K.; Herrick, J.; Martinez-Sanchez, M.; Curtis, C.; Montefiori, M. Monitoring the Actinidia Germplasm for Resistance to Pseudomonas syringae pv. actinidiae. In Proceedings of the 1st International Symposium on Bacterial Canker of Kiwifruit, Mt Maunganui, New Zealand, 19–22 November 2015; pp. 181–184.
4. Psa Regional Classification. Available online: http://www.kvh.org.nz/assets/documents/Protocols-and-movement-controls-tab/Maps-and-Stats/KVH-Regional-Classifications-02022021.pdf (accessed on 27 January 2023).
5. Minardi, P.; Ardizzi, S.; Lucchese, C. Micropropagation of kiwifruit: One-year endophytic colonization of pseudomonas syringae pv. actinidiae in shoots of actinidia deliciosa cv. hayward. *J. Plant Pathol.* **2012**, *94*, S4.70.
6. Debenham, M.; Seelye, J.; Mullan, A. An in vitro repository for clonal kiwifruit. In Proceedings of the XXIX International Horticultural Congress on Horticulture: Sustaining Lives, Livelihoods and Landscapes (IHC2014): 1113, Brisbane, Autralia, 17–22 August 2014; pp. 93–98.
7. Prado, M.J.; Herrera, M.T.; Vazquez, R.A.; Romo, S.; Gonzalez, M.V. Micropropagation of two selected male kiwifruit and analysis of genetic variation with AFLP markers. *HortScience* **2005**, *40*, 740–746. [CrossRef]
8. Debenham, M.; Pathirana, R. Establishment and management of an in vitro repository of kiwifruit (Actinidia spp.) germplasm. In *Meta-Topolin: A Growth Regulator for Plant Biotechnology and Agriculture*; Springer: Berlin/Heidelberg, Germany, 2021; pp. 279–291.
9. Murashige, T.; Skoog, F. A Revised medium for rapid growth and bio assays with tobacco tissue cultures. *Physiol. Plant.* **1962**, *15*, 473–497. [CrossRef]
10. Gamborg, O.L.; Miller, R.A.; Ojima, K. Nutrient requirements of suspension cultures of soybean root cells. *Exp. Cell Res.* **1968**, *50*, 151–158. [CrossRef]
11. Saeiahagh, H.; Mousavi, M.; Wiedow, C.; Bassett, H.B.; Pathirana, R. Effect of cytokinins and sucrose concentration on the efficiency of micropropagation of Zes006' *Actinidia chinensis* var. *chinensis*, a red-fleshed kiwifruit cultivar. *Plant Cell Tissue Organ Cult.* **2019**, *138*, 1–10. [CrossRef]
12. Linsmaier, E.M.; Skoog, F. Organic growth factor requirements of tobacco tissue cultures. *Physiol. Plant.* **1965**, *18*, 100–127. [CrossRef]
13. Tyson, J.L.; Manning, M.A.; Mellow, K.D.; Vergara, M.J. Pseudomonas syringae pv. actinidiae survival in point-inoculated kiwifruit vines. *N. Z. Plant Prot.* **2018**, *71*, 45–50. [CrossRef]
14. Snelgar, B.; Blattmann, P.; Tyson, J.; Manning, M.; Curtis, C. *On-Orchard Management of Pseudomonas Syringae pv. Actinidiae Infection and Symptom Expression: Part C: Girdlingpossible Positive and Negative Effects on Psa*; A report prepared for ZESPRI Group Limited; The New Zealand Institute for Plant & Food Research Limited: Tauranga, New Zealand, 2012.
15. Orlikowska, T.; Nowak, K.; Reed, B. Bacteria in the plant tissue culture environment. *Plant Cell Tissue Organ Cult.* **2017**, *128*, 487–508. [CrossRef]
16. Tyson, J.L.; Vergara, M.J.; Butler, R.C.; Seelye, J.F.; Morgan, E.R. Survival, growth and detection of Pseudomonas syringae pv. actinidiae in Actinidia in vitro cultures. *N. Z. J. Crop Hortic. Sci.* **2018**, *46*, 319–333. [CrossRef]
17. Kumar, S.; Sharma, D.R. In vitro propagation of kiwifruit. *J. Horticult. Sci. Biotechnol.* **2002**, *77*, 503–508. [CrossRef]
18. Pathirana, R.; Mathew, L.; McLachlan, A. A simplified method for high recovery of kiwifruit (*Actinidia* spp.) shoot tips after droplet vitrification cryopreservation suitable for long-term conservation. *Plant Cell Tissue Organ Cult.* **2021**, *144*, 97–102. [CrossRef]
19. Harada, H. In vitro organ-culture of actinidia-chinensis pl as a technique for vegetative multiplication. *J. Hortic. Sci.* **1975**, *50*, 81–83. [CrossRef]
20. Quoirin, M.; Lepoivre, P. Improved media for in vitro culture of Prunus sp. In Proceedings of the Symposium on Tissue Culture for Horticultural Purposes 78, Ghent, Belgium, 6–9 September 1977; pp. 437–442.
21. Lloyd, G.; McCown, B. Commercially-feasible micropropagation of mountain laurel, Kalmia latifolia, by use of shoot-tip culture. *Comb. Proc. Int. Plant Propagators' Soc.* **1981**, *30*, 421–427.

22. Kim, M.; Kim, S.-C.; Moon, D.-Y.; Song, K.J. Rapid shoot propagation from micro-cross sections of kiwifruit (*Actinidia deliciosa* cv. 'Hayward'). *J. Plant Biol.* **2007**, *50*, 681–686. [CrossRef]
23. Hameg, R.; Arteta, T.; Gallego, P.P.; Barreal, M.E. Selecting an efficient proliferation medium for Actinidia arguta 'Issai' explants. In Proceedings of the 9th International Symposium on Kiwifruit, Porto, Portugal, 6–9 September 2018; pp. 565–572.
24. Standardi, A. Micropropagazione dell'Actinidia chinensis Pl. mediante coltura "in vitro" di apici meristematici. *Frutticoltura* **1980**, *43*, 23–27.
25. Niedz, R.P.; Evens, T.J. Regulating plant tissue growth by mineral nutrition. *Vitr. Cell. Dev. Biol.-Plant* **2007**, *43*, 370–381. [CrossRef]
26. Hameg, R.; Arteta, T.A.; Landin, M.; Gallego, P.P.; Barreal, M.E. Modeling and Optimizing Culture Medium Mineral Composition for in vitro Propagation of *Actinidia arguta*. *Front. Plant Sci.* **2020**, *11*, 905. [CrossRef]
27. Rajasekharan, P.; Sahijram, L. In vitro conservation of plant germplasm. *Plant Biol. Biotechnol. Vol. II Plant Genom. Biotechnol.* **2015**, *2*, 417–443.
28. Sakai, A. Plant cryopreservation. In *Life in the Frozen State*, 1st ed.; Fuller, B.J., Lane, N., Benson, E.B., Eds.; CRC Press: Boca Raton, FL, USA, 2004; pp. 355–372. [CrossRef]
29. Engelmann, F. Use of biotechnologies for the conservation of plant biodiversity. *Vitr. Cell. Dev. Biol. Plant* **2011**, *47*, 5–16. [CrossRef]
30. Nadarajan, J.; Pritchard, H.W. Biophysical characteristics of successful oilseed embryo cryoprotection and cryopreservation using vacuum infiltration vitrification: An innovation in plant cell preservation. *PLoS ONE* **2014**, *9*, e96169. [CrossRef] [PubMed]
31. Benson, E.E. Cryopreservation of phytodiversity: A critical appraisal of theory & practice. *Crit. Rev. Plant Sci.* **2008**, *27*, 141–219.
32. Wu, J. Establishment of in vitro culture systems for breeding new types of kiwifruit and for Actinidia genomics studies. In Proceedings of the VII International Symposium on Production and Establishment of Micropropagated Plants 1224, Lavras, Brazil, 24–28 April 2017; pp. 195–202.
33. Mathew, L.; McLachlan, A.; Jibran, R.; Burritt, D.J.; Pathirana, R. Cold, antioxidant and osmotic pre-treatments maintain the structural integrity of meristematic cells and improve plant regeneration in cryopreserved kiwifruit shoot tips. *Protoplasma* **2018**, *255*, 1065–1077. [CrossRef]
34. Ren, L.; Wang, M.-R.; Wang, Q.-C. ROS-induced oxidative stress in plant cryopreservation: Occurrence and alleviation. *Planta* **2021**, *254*, 1–18. [CrossRef] [PubMed]
35. Reed, B.M. Implementing cryogenic storage of clonally propagated plants. *Cryo-Letters* **2001**, *22*, 97–104. [PubMed]
36. Sivaraj, N.; Pandravada, S.; Venkateswaran, K.; Ramamoorthy, S.; Anitha, K. Plant Genetic Resources Management in the Framework of Policy Developments. In *Plant Genetic Resources, Inventory, Collection and Conservation*; Springer: Berlin/Heidelberg, Germany, 2022; pp. 389–410.
37. Rowland, L.J.; Ogden, E.L.; Bassil, N.; Buck, E.J.; McCallum, S.; Graham, J.; Brown, A.; Wiedow, C.; Campbell, A.M.; Haynes, K.G. Construction of a genetic linkage map of an interspecific diploid blueberry population and identification of QTL for chilling requirement and cold hardiness. *Mol. Breed.* **2014**, *34*, 2033–2048. [CrossRef]
38. Knäbel, M.; Friend, A.P.; Palmer, J.W.; Diack, R.; Wiedow, C.; Alspach, P.; Deng, C.; Gardiner, S.E.; Tustin, D.S.; Schaffer, R. Genetic control of pear rootstock-induced dwarfing and precocity is linked to a chromosomal region syntenic to the apple Dw1 loci. *BMC Plant Biol.* **2015**, *15*, 230. [CrossRef]
39. Doyle, J.J.; Doyle, J.L. A rapid DNA isolation procedure for small quantities of fresh leaf tissue. *Phytochem. Bull.* **1987**, *19*, 11–15.
40. Kobayashi, N. A simple and efficient DNA extraction method for plants, especially woody plants. *Plant Tissue Cult. Biotech.* **1998**, *4*, 76–80.
41. Wiedow, C.; Debenham, M.; Bassett, H.; Corpe, S.; Seelye, J.; Pathirana, R.; Gardiner, S.E. Quality Control in a Kiwifruit Germplasm Collection. In Proceedings of the Plant & Food Research: 4th International Symposium on Molecular Markers in Horticulture, Napier, New Zealand, 7–10 March 2017.
42. IATA Dangerous Goods Regulations: Packing Instructions 650. Available online: https://www.iata.org/contentassets/b08040a138dc4442a4f066e6fb99fe2a/dgr-62-en-pi650.pdf (accessed on 7 February 2023).
43. Tissue Culture Standard for the Movement of Actinida Plant Material into Exclusion and Containment Regions. Available online: https://kvh.org.nz/assets/documents/Protocols-and-movement-controls-tab/KVH-Protocols/Tissue-Culture-standard-for-the-movement-of-Actinidia-plant-material-Version-2-June-2021.pdf (accessed on 23 February 2023).
44. Standard for Transitional Facilities for Biological Products. Available online: https://www.mpi.govt.nz/dmsdocument/1640-biological-products-transitional-facilities-standard#:~:text=Biological%20products%20shall%20not%20be,in%20writing%20to%20the%20supervisor (accessed on 28 January 2023).
45. Actinidia Plants for Planting IHS. Available online: https://www.mpi.govt.nz/dmsdocument/29894-Actinidia-Plants-for-Planting-Import-Health-Standard#:~:text=(5)%20For%20Actinidia%20plants%20for,occur%20following%20inspection%20and%20certification (accessed on 2 February 2023).
46. Seelye, J.F.; Corpe, S.; Debenham, M.C. A management system for tracking high health in vitro kiwifruit germplasm. In Proceedings of the 29th International Horticultural Congress on Horticulture Sustaining Lives, Livelihoods and Landscapes (IHC)/International Symposium on Micropropagation and In Vitro Techniques, Brisbane, Australia, 17–22 August 2016; pp. 113–118.
47. Wang, Q.-M.; Wang, L. An evolutionary view of plant tissue culture: Somaclonal variation and selection. *Plant Cell Rep.* **2012**, *31*, 1535–1547. [CrossRef]

48. Orton, T. Somaclonal variation: Theoretical and practical considerations. In Proceedings of the Gene Manipulation in Plant Improvement: 16th Stadler Genetics Symposium, Columbia, MO, USA, 19–21 March 1984; pp. 427–468.
49. Smith, M.; Drew, R. Growth and yield characteristics of dwarf off-types recovered from tissue-cultured bananas. *Aust. J. Exp. Agric.* **1990**, *30*, 575–578. [CrossRef]
50. Skirvin, R.M.; McPheeters, K.D.; Norton, M. Sources and frequency of somaclonal variation. *HortScience* **1994**, *29*, 1232–1237. [CrossRef]
51. Pierik, R. In vitro culture of higher plants as a tool in the propagation of horticultural crops. In Proceedings of the International Symposium on Propagation of Ornamental Plants 226, Geisenheim, Germany, 23–27 August 1987; pp. 25–40.
52. Engels, J. *A Guide to Effective Management of Germplasm Collections*; Bioversity International: Rome, Italy, 2003.
53. Espinosa-Leal, C.A.; Puente-Garza, C.A.; García-Lara, S. In vitro plant tissue culture: Means for production of biological active compounds. *Planta* **2018**, *248*, 1–18. [CrossRef] [PubMed]
54. Chen, C. Cost analysis of plant micropropagation of Phalaenopsis. *Plant Cell Tissue Organ Cult. (PCTOC)* **2016**, *126*, 167–175. [CrossRef]
55. Murphy, R.; Adelberg, J. Physical factors increased quantity and quality of micropropagated shoots of Cannabis sativa L. in a repeated harvest system with ex vitro rooting. *Vitr. Cell. Dev. Biol. Plant* **2021**, *57*, 923–931. [CrossRef]
56. Reed, B.M.; Tanprasert, P. Detection and control of bacterial contaminants of plant tissue cultures. A review of recent literature. *Plant Tissue Cult. Biotechnol.* **1995**, *1*, 137–142.
57. Ahloowalia, B.; Savangikar, V. Low cost options for energy and labour. In *Low Cost Options for Tissue Culture Technology in Developing Countries, Proceedings of a technical meeting IAEA-TECDOC-1384, Vienna, Austria, 26–30 August 2002*; International Atomic Energy Agency: Vienna, Austria, 2004; pp. 41–46.
58. Datta, S.; Chakraborty, D.; Janakiram, T. Low cost tissue culture: An overview. *J. Plant Sci. Res.* **2017**, *33*, 181–199.

plants

MDPI

Article

Response of Rowan Berry (*Sorbus redliana*) Shoot Culture to Slow Growth Storage Conditions

Nóra Mendler-Drienyovszki * and Katalin Magyar-Tábori

Research Institute of Nyíregyháza, Institutes for Agricultural Research and Educational Farm (IAREF), University of Debrecen, P.O. Box 12, 4400 Nyiregyhaza, Hungary
* Correspondence: mendlernedn@gmail.com

Abstract: Slow growth storage can preserve the genetic resources of endangered species such as those of genus *Sorbus*. Our aim was to study the storability of rowan berry in vitro cultures, their morpho-physiological changes, and regeneration ability after different storage conditions (4 ± 0.5 °C, dark; and 22 ± 2 °C, 16/8 h light/dark). The cold storage lasted for 52 weeks, and observations were made every four weeks. Cultures showed 100% survival under cold storage, and those taken from the storage showed 100% regeneration capacity after the passages. A dormancy period lasting about 20 weeks was observed, followed by intensive shoot growth until the 48th week, which led to the exhaustion of the cultures. The changes could be traced to the reduction of the chlorophyll content and the F_v/F_m value, as well as in the discoloration of the lower leaves and the appearance of necrotic tissues. Long, etiolated shoots (89.3 mm) were obtained at the end of cold storage. Shoot cultures stored in a growth chamber as control (22 ± 2 °C, 16/8 h light/dark) senesced and died after 16 weeks. Explants from stored shoots were subcultured for four weeks. The number and length of newly developed shoots were significantly higher on explants from cold storage compared to those from control cultures if the storage was longer than one week.

Keywords: *Sorbus redliana* 'Burokvölgy'; in vitro conservation; midterm preservation; tissue culture; cold storage; endangered species

check for updates

Citation: Mendler-Drienyovszki, N.; Magyar-Tábori, K. Response of Rowan Berry (*Sorbus redliana*) Shoot Culture to Slow Growth Storage Conditions. *Plants* **2023**, *12*, 1287. https://doi.org/10.3390/plants12061287

Academic Editors: Anna De Carlo and Waed Tarraf

Received: 1 February 2023
Revised: 10 March 2023
Accepted: 10 March 2023
Published: 12 March 2023

1. Introduction

Sorbus redliana (common name: rowan berry) belongs to genus *Sorbus* (*Rosaceae*, *Maloideae*) [1]. *Sorbus* species are found throughout the northern temperate zone, where they are native to the natural forests. The genus is extremely rich in species and forms, and includes several interspecific hybrids [1,2]. *Sorbus* species and cultivars are important ornamental trees and medicinal plants [3]. They are deciduous trees or shrubs with lobed [4] or pinnately compound leaves [5]. They have pleasing umbel inflorescence and typically bloom in May. Their small apples are colored from white to yellow through red to brown. *Sorbus* trees maintain a long-lasting decorative appearance with their beautifully colored foliage and fruits which last into autumn [5].

Although a very small proportion of berry yield is currently utilized as raw materials in the products intended for human consumption, the bioactive compounds detected in *Sorbus* plants may make it more useful in the future for medicines, cosmetics, and functional foods [3]. *Sorbus* species can also be valuable raw materials for the wood industry [6].

Sorbus redliana is a rowan berry tree with a spherical crown and low plant height, which has branching from the base and grows into a large bush or a small tree of up to 5 m. The leaves are simple, ovate, wedge-shouldered, and dentate with a bright coloration and a downy lower surface. It is a spontaneous hybrid of *S. aria* and *S. torminalis*. Its white flowers bloom in May, with the ripe fruits being cinnabar-red and slightly warty. The autumn foliage color is yellow to brown. *S. redliana* 'Burokvölgy' is a Hungarian cultivar, native to one valley of the Bakony Mountains. It has tolerance to drought and is a decorative ornamental tree of gardens and streets [7].

Many *Sorbus* species have only a small number of individuals and are considered either critically endangered or vulnerable due to the harmful effects of climate change and habitat loss [8,9]. The special reproduction method of the genus *Sorbus*, including the easy creation of hybrids and polyploids (especially triploids) [10] often leads to the formation of small, fragmented populations that are particularly sensitive to adverse environmental effects, e.g., grazing, harmful forest management, etc. [9]. The purpose of gene conservation is to preserve genetic diversity and adaptability, which is a prerequisite for the long-term survival of species and populations [11,12]. The search for and preservation of valuable sources of disease resistance, such as fire blight-resistant genotypes in members of the *Rosaceae* family [13], is a vital role of current conservation efforts. One of the most common means of preserving valuable genetic material for seed-propagated crops is storage in the form of seeds at temperatures between −15 and −20 °C or between 0 and 4 °C [14].

Those species that are propagated vegetatively because they do not produce seeds or those with offspring whose genetic identity cannot be ensured with seeds may be best suited to storage with biotechnological methods, including tissue cultures used for micropropagation, slow growth methods, or cryopreservation [14–16]. These methods are particularly suitable for the conservation of genetic resources of the forest trees and horticultural plants [17], and the pathogen-free in vitro cultures can also be used effectively for the distribution of germplasm [18].

Slow growth storage (SGS) is a cost- and space-efficient method, suitable for midterm storage for months to up to several years [19]. The growth of in vitro cultures can be slowed down by changing the culture medium (e.g., less nutrients, sugar, modified growth regulator content, addition of osmotically active agents) and/or modified environmental factors (temperature and/or illumination reduction, even complete darkness) [15], or a combination of these variables [19]. This storage method most frequently takes the form of in vitro organ cultures—or even more recently alginate-encapsulated shoot explants [20] being kept at a low temperature (about 4 °C for temperate plant species and about 10–15 °C for tropical crops) [14]. With SGS, the number of subcultures is significantly reduced, and depending on the species, subcultures can even be extended for several years [15].

Guidelines describing the conservation of *Sorbus* species such as *S. domestica* and *S. cuneifolia* suggest that at least 50 individuals should be protected in situ, as these are among the most endangered forest tree species in the world [21–23]. Ex situ methods are of great importance, and these include conservation in the form of seeds (*S. domestica*) [24], the cryopreservation of shoot tips [25], in vitro tissue culture [26,27], and culture in cold storage conditions [28]. However, the morphological and physiological response of each plant species must be investigated to develop optimal in vitro culture and storage conditions [29].

The most suitable medium for *S. domestica* is the Schenck and Hildebrandt [30] medium supplemented with 22.2 µM of 6-benzylaminopurine (BA) [26]; the most suitable for *S. aucuparia* is the MS [31] medium with 0.9–1.8 µM of BA and 0.5 µM of indole-3-butyric acid (IBA) [32]; and the most suitable for *S. rotundifolia* is the MS medium containing 1.6 µM of IBA, 1.4 µM of 6-benzyladenine riboside (BAR), and 0.6 µM of gibberellic acid (GA_3) [33]. An application of $\frac{1}{2}$ MS medium supplemented with 3.3 µM BA, 0.5 µM IBA, and 0.3 µM GA_3 has also been used effectively for *S. rotundifolia* [34]. Application of meta-topolin (2.1 µM) with IBA (0.5 µM) in MS medium was shown to be useful for micropropagation of *S. abscondita* and *S. omissa*, whereas *S. gemella* can be propagated on MS medium containing only meta-topolin (2.1 µM) as plant growth regulator [35]. However, a three- or four-week subculture period must be considered for micropropagated plants [36,37]; therefore, in order to preserve genetic resources more efficiently, it is advisable to use techniques that can significantly reduce the number of subcultures of in vitro cultures [37]. We studied the storage ability of in vitro shoot cultures of rowan berry (*Sorbus redliana* 'Burokvölgy') by monitoring their morpho-physiological changes during cold storage under 4 ± 0.5 °C and dark (4 °CD) and their viability after different lengths of cold storage compared to those left in a growth chamber under 22 ± 2 °C, 16/8 h light/dark conditions (22 °CL).

2. Results

2.1. Morphological Traits of In Vitro Shoot Cultures during Storage

Samples to evaluate the morpho-physiological stage of *S. redliana* 'Burokvölgy' in vitro shoot cultures were observed on the first day of storage (three weeks old) and every four weeks during storage. Both the shoot growth (shoot length: SL) and shoot proliferation rate (number of newly developed shoots per explant: SN) were affected by environmental conditions during storage of in vitro shoot cultures.

2.1.1. Shoot Length

No significant shoot growth occurred during the first 24 weeks of the cold storage (at 4 °C and darkness: 4 °CD), and the length of shoots was between 28 ± 5 and 31 ± 7 mm. Shoot elongation started after 24 weeks of cold storage. A period of significant shoot elongation was then observed, which lasted until week 48. The average shoot length reached 89 ± 16 mm. In contrast, intensive shoot growth was detected during the first four weeks of storage at 22 °CL, but no significant changes were observed after that. The average length of the shoots stored at 22 °CL reached just 43 ± 11 mm between weeks 4 and 16, and these shoots were significantly longer than those stored at 4 °CD during the same period. However, the shoots stored at 4 °CD reached a significantly longer shoot length (51 ± 19 mm) from the 32nd week compared to the highest results obtained in the growth chamber (43 ± 11 mm) (Figure 1).

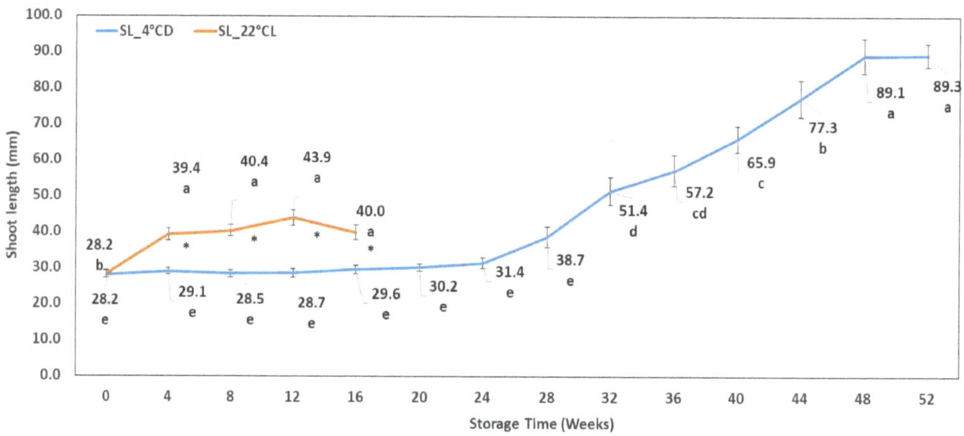

Figure 1. Shoot length (SL) during cold storage (SL_4 °CD) and during storage in the growth chamber (SL_22 °CL). Line graphs present mean values ± SE of SL. Lower case indicates the significant differences between the SL measured after different storage times. * indicates the significant differences between the SL measured after different storage conditions.

2.1.2. Shoot Number

In the case of the shoot cultures stored at 4 °CD, the degree of shoot proliferation was not proven to be significant during most of the storage period. However, at the 36th and 52nd week, we detected significantly higher shoot numbers compared to the initial state.

In shoot cultures stored at 22 °CL, the number of shoots increased significantly in the first four weeks and then did not change significantly in the period between weeks 4 and 16, which was similar to the trend in shoot length. Moreover, in the period between the 4th and 16th week, the degree of shoot proliferation in the cultures stored at 22 °CL was significantly higher compared to those stored at 4 °CD (Figure 2).

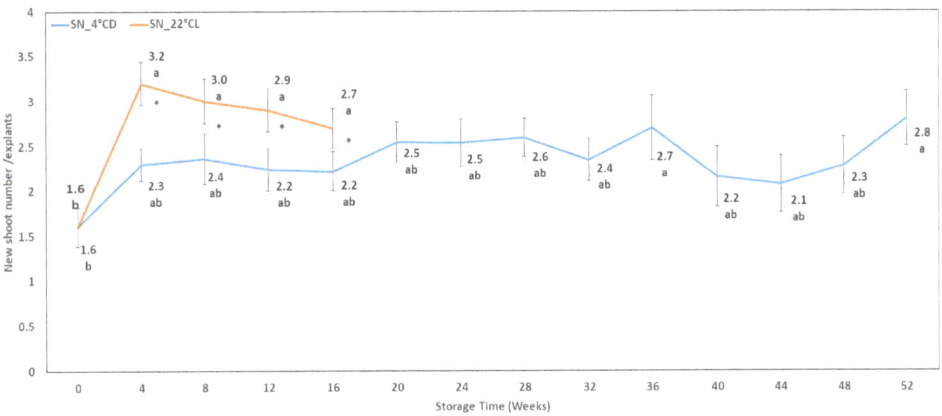

Figure 2. Number of newly developed shoots (SN) during cold storage (SN_4 °CD) and during storage in the growth chamber (SN_22 °CL). Line graphs present mean values ± SE of SN. Lower case indicates the significant differences between the SN measured after different storage times. * indicates the significant differences between the SN measured after different storage conditions.

2.2. Physiological Traits of In Vitro Shoot Cultures during Storage

The rate of hyperhydration was high after 48 weeks of cold storage (up to 60%) and after 12 weeks of storage at 22 °CL (up to 25%). In the case of the shoot cultures stored at 4 °CD, the yellowing of the leaves started after the 24th week, and after the 32nd week, browned tissues also appeared. Around the 28th week, in the phase of intensive shoot growth, etiolated shoots appeared, and after the 36th week, most shoots were already etiolated (Figure 3). After 48 weeks of storage, tissue death appeared on 30% of the shoot cultures. The shoots stored at 22 °CL showed signs of senescence around the 12th week, and extensive browning and tissue death were already common in the 16th week (Figure 4) (Table 1). In general, the etiolation was located on the upper part of shoots, while yellow and brown discoloration tended to occur on the lower part of shoots. All these changes were also reflected in the changes in chlorophyll content.

Figure 3. Shoot cultures of *Sorbus redliana* 'Burokvölgy' after storage at 4 °CD and darkness for four weeks (**A**), 16 weeks (**B**), 24 weeks (**C**), 36 weeks (**D**), 40 weeks (**E**), 44 weeks (**F**), 48 weeks (**G**), and 52 weeks (**H–J**).

Figure 4. Shoot cultures of *Sorbus redliana* 'Burokvölgy' after storage at 22 °CL and 16 h light for four weeks (**A**), 8 weeks (**B**), 12 weeks (**C**), and 16 weeks (**D**).

Table 1. The rate of physiological changes observed for the in vitro shoot culture of *S. redliana* 'Burokvölgy' stored under cold condition (4 °CD) and the growth chamber (22 °CL).

Storage Weeks	Hyperhydration (%)	Discoloured Leaves (%)	Necrotic Tissues (%)	Etiolated Shoots (%)
		4 °CD		
0	6.2	0.0	0.0	0.0
4	15.9	0.0	0.0	0.0
8	11.2	0.0	0.0	0.0
12	24.5	0.0	0.0	0.0
16	16.5	0.0	0.0	0.0
20	21.4	0.0	0.0	0.0
24	25.6	0.0	0.0	0.0
28	34.2	4.6	4.6	29.1
32	37.8	45.9	0.0	51.6
36	38.5	45.8	0.0	71.5
40	29.2	58.2	0.0	90.0
44	51.1	64.5	5.2	100.0
48	60.7	73.8	28.7	100.0
52	46.6	87.1	31.4	87.4
		22 °CL		
0	3.5	0.0	0.0	0.0
4	12.7	0.0	0.0	0.0
8	14.0	14.4	0.0	0.0
12	25.0	79.3	16.5	0.0
16	27.2	100.0	85.0	0.0

2.2.1. Chlorophyll Content of Leaves

The chlorophyll content of the leaves was measured every 4 weeks during the experiment. At the start of the experiment, the chlorophyll content of the leafy shoots (three weeks old) was 0.62 and 0.27 mg g^{-1} FW (chl *a* and chl *b*).

Both the chl *a* and chl *b* content decreased significantly during storage regardless of the storage conditions. The chl *a* and chl *b* content of shoots stored at 4 °CD decreased significantly in the first four weeks and then remained roughly at the same level, and no further significant decrease occurred until the end of the 20th week. This period was followed by a phase in which the chl *a* and chl *b* content decreased significantly again. A similar periodic change was observed in the quantitative changes of the total chlorophyll content. In contrast, in the case of shoot cultures stored at 22 °CL, a very significant decrease in both chl *a* and chl *b* content occurred during the first 8 weeks, and the same trend could be seen in the changes in the total chl content.

Comparing the storage conditions, we found that the chl *a* and chl *b* content of the shoot cultures was lower during storage at 22 °CL than during storage at 4 °CD, and in the eighth week, the difference was significant. The shoot cultures stored under the 22 °CL condition had a significantly lower chl *a*/chl *b* ratio than did those cultures stored at 4 °CD (Table 2).

Table 2. Chlorophyll content (Chl *a*, Chl *b*, and Chl*a*/*b*) in the in vitro leaves of *S. redliana* 'Burokvölgy' under storage conditions of 4 °CD and 22 °CL. Values are expressed as the mean ± SE of chlorophyll content. Lower case indicates the significant differences between the chlorophyll content measured after different storage times. * indicates the significant differences between the chlorophyll content of leaves from different storage conditions.

Weeks in Storage	Chl *a*		Chl *b*		Chl *a*/*b*	
	mg g^{-1} Fresh Weight					
	4 °CD	22 °CL	4 °CD	22 °CL	4 °CD	22 °CL
0	0.62 ± 0.06 a	0.62 ± 0.06 a	0.27 ± 0.02 a	0.27 ± 0.03 a	2.29 ± 0.03 ab	2.29 ± 0.03 a
4	0.56 ± 0.03 b	0.47 ± 0.03 b	0.24 ± 0.01 b	0.22 ± 0.01 b	2.38 ± 0.03 ab	2.16 ± 0.02 a
8	0.49 ± 0.02 bc *	0.28 ± 0.02 bc	0.20 ± 0.01 b *	0.13 ± 0.01 c	2.45 ± 0.03 ab *	2.09 ± 0.03 a
12	0.54 ± 0.02 b *	0.31 ± 0.03 bc	0.21 ± 0.01 b *	0.14 ± 0.01 c	2.54 ± 0.02 ab *	2.29 ± 0.04 a
16	0.53 ± 0.02 b *	0.22 ± 0.02 c	0.21 ± 0.01 b *	0.10 ± 0.01 c	2.50 ± 0.03 a *	2.30 ± 0.03 a
20	0.46 ± 0.02 bc	-	0.17 ± 0.01 bc	-	2.63 ± 0.02 ab	-
24	0.33 ± 0.02 cd	-	0.13 ± 0.01 cd	-	2.59 ± 0.04 ab	-
28	0.21 ± 0.02 de	-	0.08 ± 0.01 de	-	2.48 ± 0.04 ab	-
32	0.27 ± 0.04 de	-	0.11 ± 0.01 de	-	2.54 ± 0.04 ab	-
36	0.21 ± 0.02 de	-	0.08 ± 0.01 de	-	2.57 ± 0.04 ab	-
40	0.18 ± 0.02 de	-	0.07 ± 0.01 de	-	2.56 ± 0.04 ab	-
44	0.09 ± 0.01 e	-	0.04 ± 0.004 e	-	2.27 ± 0.03 b	-
48	0.15 ± 0.02 e	-	0.06 ± 0.01 de	-	2.47 ± 0.05 ab	-
52	0.14 ± 0.02 e	-	0.06 ± 0.01 de	-	2.50 ± 0.05 ab	-

2.2.2. Chlorophyll Fluorescence Results

The average F_v/F_m value measured on the leaves of the starting material (three-week-old shoot cultures) was 0.68. Later, values of F_v/F_m varied between 0.33 and 0.65 as measured on shoots stored at 4 °CD and between 0.61 and 0.71 as measured on cultures stored at 22 °CL.

In the case of shoot cultures stored at 4 °CD, the F_v/F_m value decreased significantly in the first eight weeks in respect to the initial value, an increased tendency was detected around the 20th week, and then decreased values were measured from the 28th week. In contrast, no significant changes could be detected in F_v/F_m value in shoot cultures kept under the 22 °CL condition. Moreover, in the case of cultures stored at 22 °CL, significantly higher values were measured in the storage period between the 8th and 16th week compared to those stored at 4 °CD (Figure 5).

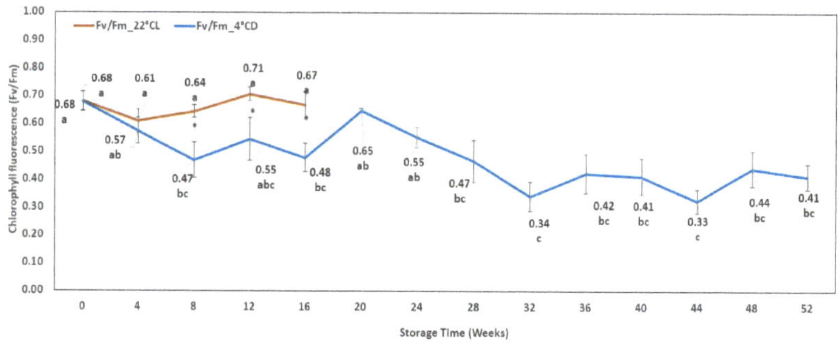

Figure 5. F_v/F_m values measured on shoots after different periods of cold storage (F_v/F_m _4 °CD) and of storage in the growth chamber (F_v/F_m _22 °CL). Line graphs present mean values ± SE of F_v/F_m. Lower case indicates the significant differences between the F_v/F_m measured after different storage times. * indicates significant differences between the F_v/F_m values measured after different storage conditions.

2.3. *Number of Usable Explants*

Shoot cultures were taken from the starting materials and during storage every four weeks for the purpose of subculturing. The number of new explants obtained per original shoot explant (number of usable explants) was recorded. In the case of shoot cultures stored at 4 °CD, it was almost the same between the 8th and the 32nd week, but it significantly decreased at the 40th week relative to the 36th week. During this period, extremely variable results were observed: there were already long shoots but not in a very good physiological condition, they were etiolated, and a certain amount of tissue death was already noticeable. In the case of shoot cultures kept at 22 °CL, a significant increase was detected in the first four weeks, and then a significant decrease was observed between the 8th and 12th week (Figure 6).

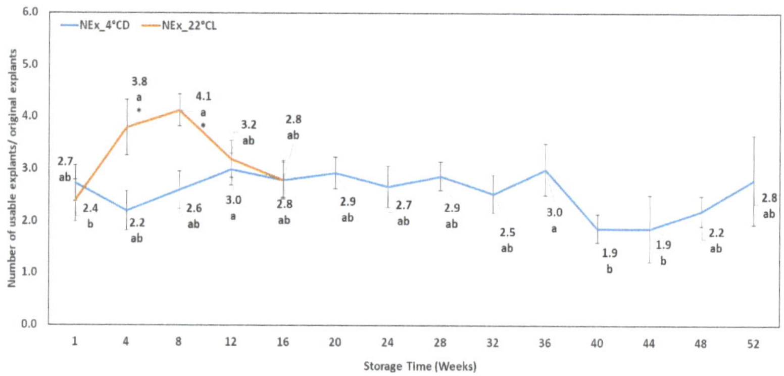

Figure 6. The number of usable explants after different periods of cold storage (NEx_4 °CD) and of storage in the growth chamber (NEx _22 °CL). Line graphs present mean values ± SE of the number of usable explants. Lower case indicates the significant differences between the number of usable explants counted after different storage times. * indicates the significant differences between the number of usable explants after different storage conditions.

Comparing the two storage conditions, we found that at around the 4th and 8th week, significantly more explants could be obtained from the cultures stored at 22 °CL. In the case of cold storage, the usable explant number per shoot clump varied between 1.9 ± 1.1 and 3.0 ± 1.2, while it varied between 2.4 ± 1.6 and 4.1 ± 1.2 in 22 °CL.

2.4. Shoot Proliferation Parameters of Explants from Different Storage Conditions

Explants collected from stored in vitro cultures were cultured under growing conditions (16 h light with 57 µmol m^{-2} s^{-1} PPFD, 22 ± 2 °C) to regenerate new shoots. The length and the number of newly developed shoots was observed after a 4-week period. Both shoot parameters were affected by previous environmental conditions during storage. Lengths of newly developed shoots varied between 23 and 33 mm and between 18 and 23 mm on explants from storage at 4 °CD and 22 °CL, respectively. The number of new shoots varied between 2.7 and 4.7 mm and between 1.9 and 2.8 mm on explants from storage at 4 °CD and 22 °CL, respectively.

2.4.1. Length of Shoots at the End of Subculture Period

In the case of explants stored at 4 °CD, we observed that longer shoots (29 mm on average) developed after a storage period of 4–36 weeks compared to a shorter or longer storage (both were 23 mm on average). However, significantly longer shoots developed only when storage duration was 16 to 20 weeks compared to those stored for one week or longer than 40 weeks. No significant differences could be detected after a different period of storage at 22 °CL, whereas the length of new shoots developed on explants from cold storage (4 °CD) was significantly greater than that developed on explants from storage at 22 °CL (Figure 7).

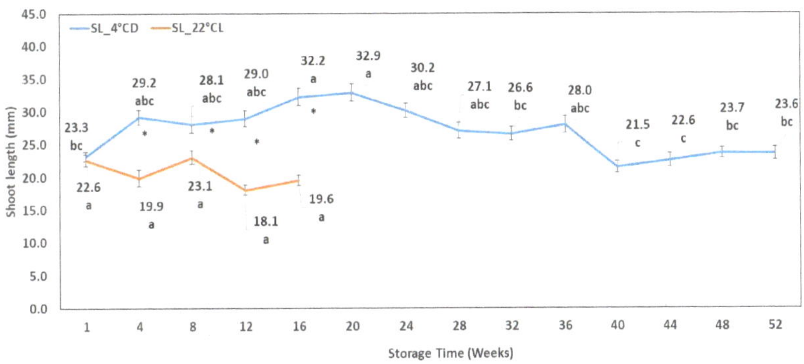

Figure 7. Shoot length (SL) of new shoots developed on subcultured explants from cold storage (SL_4 °CD) and from the growth chamber (SL_22 °CL). Line graphs present mean values ± SE of the SL of new shoots developed from subculture. Lower case indicates the significant differences between the SL measured after different storage times. * indicates the significant differences between the SL measured after different storage conditions.

2.4.2. The number of New Shoots at the End of the Subculture Period

In the case of explants from shoot cultures stored at 4 °CD, an increase could be seen after the first four-week storage period, but it was not statistically significant. However, after the 20th week, the proliferative capacity of the explants increased significantly, but later a significant decrease could be observed starting from the 36th week.

In the case of shoots stored at 22 °CL, a significant increase in the shoot proliferation capacity of explants was not observed until the 12th week. The number of newly developed shoots collected at week 16 was significantly lower than that of explants from 12-week storage. Comparing the shoot proliferation capacity of shoot explants from different storage conditions, we observed that the number of new shoots was significantly lower on explants from 22 °CL than that on those from 4 °CD storage (Figure 8).

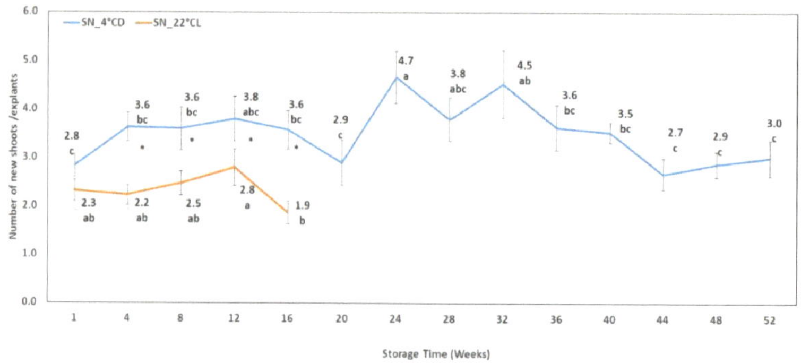

Figure 8. Shoot number (SN) of new shoots developed on subcultured explants from cold storage (SN_4 °CD) and in the growth chamber (SN_22 °CL). Line graphs present mean values ± SE of SN of new shoots developed from subculture. Lower case indicates the significant differences between the SN measured after different storage times. * indicates the significant differences between the SN measured after different storage conditions.

However, the physiological state of the subcultured shoot cultures was completely adequate. Although necrotic tissues could be detected on shoots, the slow growth storage condition (4 °C and darkness) prolonged up to 52 weeks did not affect the survival rate (100%) or the success of regrowth during the first subculture of *Sorbus redliana* 'Burokvölgy' (Figure 9).

Figure 9. Four-week-old shoot cultures of *Sorbus redliana* 'Burokvölgy' developed during the first subculture after storage at 4 °CD for four weeks (**A**), at 4 °CD for 16 weeks (**B**), at 4 °CD for 36 weeks (**C**), at 4 °CD for 52 weeks (**D**), at 22 °CL for four weeks (**E**), and at 22 °CL for 16 weeks (**F**).

3. Discussion

Although maintenance of valuable genotypes in field collections is the most common way to preserve woody plants, the application of in vitro techniques has also been proposed [38]. Responses to cold storage of in vitro cultures of plant species and varieties can be highly varied, and even if they can survive the storage, differences in their regrowth capacity are common.

In our experiments, in vitro shoot cultures of *Sorbus redliana* 'Burokvölgy' were stored at 4 °CD for 52 weeks. In general, woody plants of the temperate zone tolerate low storage temperatures well because they can go dormant to survive the winter [39]. Dark conditions can enhance the stability of in vitro culture via the inhibition of development of new tissues, as was observed during a 60-day storage of potato and ginger cultures [18]. Darkness can also enhance the vigor of shoots when stored up to 12 weeks, but in a longer storage period, illumination was found to significantly increase the regrowth ability of chokecherry shoots (*Prunus virginiana*) [40].

Accordingly, we found that all in vitro shoot cultures survived the cold storage at least 52 weeks; however, after an initial dormant period of about 5 months, the shoots began to grow, and the newly developed parts of these shoots were strongly etiolated and elongated.

In another study, newly developed lateral shoots were observed on in vitro shoot cultures of apple (*Malus pumila* Mill., 'Starkspur Red') stored at 4 °C after 8 months [41]. In contrast, we did not find an increase in the number of shoots because the growth of lateral buds might have been inhibited further.

As mentioned above, the possibility of slow growth storage in the case of temperate woody plants is based on the ability to adapt to the cold by entering dormancy [28]. This dormant period is usually observed during the first 20 weeks of cold storage, after which intensive shoot growth begins, lasting until about the 40th week, and is followed by a slightly reduced growth period [26]. Thus, growth of woody plants with in vitro cultures during cold storage is a common phenomenon. Similarly, in our experiments, the dormant period of *Sorbus* lasted to 24 weeks, starting approximately between the 4th and 8th week of cold storage, and after this, an intensive shoot growth was observed, which appeared to slow after 48 weeks.

Although we obtained explants from each cold storage period for the first subculture with very high regeneration ability, longer new shoots and higher multiplication rates were detected when shoot cultures were stored from 4 to up to 36 weeks. In a previous study, decreased proliferation was detected in dwarf apple (*Malus domestica*) P2 rootstocks in subculture after 24, 30, and 36 weeks of cold storage (4 °C, dark) [42]. In contrast, in the case of two woody plants (chokecherry and saskatoon berry), the most vigorous regrowth was observed after a 12-month storage period [40].

Moreover, the length of new shoots developed on explants from *Prunus avium* in vitro cultures stored at 2 °C increased as the storage period increased up to nine months [43]. However, the number of newly developed shoots increased only on those explants collected from cold storage lasting shorter than six months, and a longer cold storage resulted in decreased shoot number in the first subculture [43].

In contrast, we did not discover there to be significant differences during the first subculture when explants were obtained from shoot cultures stored for different period durations in the growth chamber (22 °CL). However, the multiplication rates were significantly higher on those explants from cold storage compared to those from the growth chamber when the storage period was longer than four weeks. In contrast, another study reported no significant differences were in the propagation rate of cold-stored *Taraxacum pieninicum* explants compared to that of a nonstored control [44]. Because our control cultures were stored in a growth chamber until they fully aged, the differences may be due to the aged explant sources stored in the growth chamber (22 °CL).

Although all cultures survived the one-year cold storage and proved to be an adequate source of explants for further propagation, signs of damage became visible during the second half of storage. In the shoot cultures stored at 4 °CD, yellow leaves first appeared on the 24th week, and intensive shoot growth began around the 28th week, which resulted in etiolated shoots. By the 32nd week, some brown leaves were also visible, and at the 36th week, all shoots were already etiolated. Necrosis could be observed on approximately 30% of in vitro shoot cultures, especially in the lower part of the shoots after the 48th week. In a study of chokecherry (*Prunus virginiana*), yellow discoloration of lower leaves had no effect on the regrowth capacity [40].

Changes related to aging and damage were also reflected in changes in chlorophyll content. In vivo, during the study of 937 plant species, the total chlorophyll content was found to be very variable, with values ranging from 1.45 to 19.2 mg/g [45]. In our experiments, even the highest total chlorophyll content measured in the absolute control plants (3-week-old *Sorbus* shoot cultures before storage) did not exceed the value of 0.89 mg/g. Both storage conditions resulted in a decrease in chlorophyll content. During storage at 4 °CD, all chl *a* and chl *b* content decreased significantly by the 20th week. In contrast, during storage at 22 °CL, the chl *a* content fell significantly only at the end of the 16th week, whereas the chl *b* chlorophyll content had already decreased significantly at the 8th week. However, from the 8th week, all chlorophyll parameters showed significantly lower values in the 22 °CL storage compared to the 4 °CD storage. The decrease in chlorophyll content during storage at 22 °CL was probably attributable to aging, while the appearance of etiolation during storage at 4 °CD could have caused the decrease in chl content. The chl *a*/*b* ratio did not show significant changes over time in any of the storage conditions. However, significantly lower values were observed in plants stored at 22 °CL starting from the 8th week compared those stored at 4 °CD. Since the light absorption peaks of chl *a* and chl *b* are different (chl *a*: red; chl *b*: blue), the changes in the chl *a*/*b* ratio may indicate the plant's adaptive response to altered light conditions [45]. In response to a decrease in light intensity, the number of light-harvesting complexes can increase [46]; thus, the chl *b* biosynthesis is expected to become more intense to maximize the light-harvesting capacity due to decreasing illumination [47]. However, in our experiments, low temperature probably inhibited chl biosynthesis as has been observed in many plant species [48].

The correlation between the chlorophyll content of the leaves and the chlorophyll fluorescence parameters is a known phenomenon in the plant kingdom and was also proven in the case of intact *Sorbus aucuparia* plants [49]. Measurement of the photosystem II chlorophyll fluorescence is a very common way to monitor the plant functions since PS II is very sensitive to several stress effects [50]. In general, the F_v/F_m value of the intact C3 type of plant is about 0.83 under optimal conditions [51], and it was found that the F_v/F_m ratio from in vivo plants without any kind of stress varied between 0.7 and 0.8 [52]. Indeed, this value (0.827) was also obtained on intact *Sorbus aucuparia* plants [49]. However, in vitro shoot cultures of *Sorbus redliana* 'Burokvölgy' never reached the 0.8 F_v/F_m value, and even the highest value was only 0.706 (measured in control plants stored in growth chamber: 22 °CL), suggesting that in vitro cultivation is an inherently stressful environment [53]. Moreover, we detected a significant decrease in F_v/F_m ratio during the cold storage period, except for around week 20 when we measured an increase. This happened around the time when the rest period ended and shoot growth started. In contrast, research indicates that in vitro shoot cultures stored at 22 °CL do not show any significant changes: the cold stress results in a decreased F_v/F_m ratio under in vivo conditions in woody plants, such as grape [52] and cherry [54], and in herbaceous plants, such as corn [55] and rice [48].

Further studies may help to increase the efficiency and possibly contribute to increasing the storage time. The negative morpho-physiological changes occurring from the second half of the cold storage can perhaps be attributed to the fact that the intensive shoot growth after the end of the rest period resulted in the exhaustion of the cultures. Shoot cultures of *Sorbus*, 'Krasnaja Krupnaja', grown on medium with high sucrose content (60 g L^{-1}) have been successfully stored for up to five years [28]. Further studies including altered medium may help to increase the storage time. Testing other temperatures may also lead to an increase in storage efficiency as in the case of a gooseberry study, in which

2 °C was found to be much more suitable for the storage of shoot cultures than was 4 °C, which was reflected in a lower proportion of necrotic shoots [56].

4. Materials and Methods

4.1. Establishment and Maintainance of S. redliana In Vitro Cultures and Experimental Conditions

Axillary buds and shoot tips were collected from a dormant mother tree in the arboretum of the Hungarian University of Agriculture and Life Sciences in 2015. Surface sterilization started with soaking in tap water with Tween 80 for 2 h, followed by $HgCl_2$ (0.5%) and ethanol (70%) treatment for five minutes, and finished by a rinse in sterile distilled water (three times). The medium for initiation consisted of Benczúr-Márta (BM) macro elements [57], Heller microelements [58] supplemented with Murashige-Skoog (MS) [31] vitamins, and 1.4 μM of 6-benzyladenine riboside (BAR) [59].

The maintenance and propagation of in vitro shoot cultures of *S. redliana* 'Burokvölgy' for the experiments were performed using MS medium, which was supplemented with 2.8 μM of BAR, 0.6 μM of gibberellic acid (GA_3), 1.48 μM of indole-3-butyric acid (IBA), 100 mg L^{-1} of myo-inositol, 3% sucrose, and 0.7% agar-agar. Before autoclaving, the pH was adjusted to 5.8. Five shoot explants (stem segments about 15 mm with at least one node) per 400 mL Kilner jar were cultured in a growth chamber (16/8 h light/dark with 57 μmol m^{-2} s^{-1} PPFD, 22 ± 2 °C) for 3 weeks. Then, 100 jars were transferred into another room (4 °CD) for cold storage periods of 4–52 weeks, while the remaining 56 jars were left in the growth chamber as controls (22 °CL) for a further 16 weeks (total 19 weeks) until they were completely senesced. Samples were taken every four weeks to monitor changes induced by different storage conditions.

4.2. Morphological Observations on Stored Shoot Cultures

The number of new shoots per explant were counted, and their length (mm) was measured and recorded at the beginning of storage when the cultures were 3 weeks old and every four weeks thereafter.

4.3. Measurements to Detect Physiological Changes in Stored Cultures

In addition to the above-mentioned measurements, visual observation was conducted after every period of storage (every four weeks), and the explants were transferred to fresh medium. The main qualitative parameters of stored shoots were studied, including hyperhydration, intensity of green color, the presence of discoloration or tissue necrosis, and etiolation. Recording of changes consisted of the presence or lack of these symptoms independent of their extension, and then the percent (%) of the newly developed shoots was calculated.

4.3.1. Measurement of the Chlorophyll Content in Leaves

A spectrophotometric method [60] was used to determine the quantity of chlorophyll a and b (chl *a* and chl *b*) of the leaves. The absorbance of solutions was measured at 653, 666, and 750 nm with a 6705 UV/Vis spectrophotometer (Jenway 6705, VWR International Kft, Debrecen, Hungary). Calculated chl content (mg g^{-1} fresh weight) was obtained after determination of chl *a* and chl *b* values as follows (after excluding the effect of solution turbidity measured at 750 nm) [60]:

$$\text{chl } a = 17.12 \text{ A666} - 8.68 \text{ A653}$$

$$\text{chl } b = 32.23 \text{ A653} - 14.55 \text{ A666}$$

$$\text{chl } a + \text{chl } b = 2.57 \text{ A666} + 23.6 \text{ A653}$$

4.3.2. Measurement of Chlorophyll Fluorescence

Chl fluorescence was measured with an OS5p modulated fluorometer (Opti-Sciences, Hudson, NH, USA) on fully developed leaves. One leaf per plant was selected, and this was repeated for ten independent plants per treatment (on 2 plantlets per jar, from 5 jars). After a 30-min-long dark adaptation of leaves (by using leaf clips), the F_v/F_m protocol was used to measure maximum quantum yield (estimate of the maximum ratio of absorbed quanta used in the PSII reaction center) [61]. Maximum (F_m) and minimum (F_o) were measured in dark-adapted leaves after 0.8 s for the saturation pulse (35 W halogen lamp with 690 nm short pass filter according to the OS5p User's Guide). The potential quantum efficiency of PSII (F_v/F_m ratio), which is an indicator of photosynthetic performance, was calculated by the built-in software (Opti-Sciences Inc. 603-883-4400) of the fluorometer [62].

4.3.3. Number of Usable Explants after Storage

Usable shoot explants from another 3 jars were counted to help calculate the required number of shoot cultures for storage in the future. These explants (stem segments about 15 mm with at least one node) were subcultured. Moreover, only those shoots which did not show yellowing, browning, or necrosis on their upper part were subcultured. In the case of etiolated shoots, the weak, thinned upper part of shoots was not used for subculture. Before statistical analysis, the number of usable explants was calculated as follows:

NEx = the number of usable explants per jar/the number of initial explants per jar

4.4. Morphological Observations on Subcultured Shoot Cultures

Shoot explants were subcultured onto fresh medium and transferred to the growth chamber for a 28-day period to test the ability of regrowth (the medium and growing conditions were the same as those used for the shoot multiplication). The observations were the same as those described in Section 4.2.

4.5. Data Collection and Analysis

The shoots included in morphological observations (shoot length, number of new shoots) were developed on a total of 25 explants from 5 jars. Samples were collected as five replications from five different jars (2 subsamples per jar) for measurements (chlorophyll content, chlorophyll fluorescence). Statistical analysis of the data was carried out using one-way analysis of variance (ANOVA) and followed by Tukey's test to reveal significant differences between the means at a level of $p \leq 0.05$ with SPSS software version 21.0 (SPSS®) for Windows.

5. Conclusions

This is the first report on slow growth storage of *S. redliana*. In the current study, it was found that in vitro shoot cultures of *S. redliana* can be stored safely at 4 °C in the dark for 52 weeks on MS medium supplemented with 3% sucrose and 2.8 μM of BAR, 0.6 μM of GA_3, and 1.48 μM of IBA. During this period, the survival of stored shoots was 100%, and thus stored explants can be regrown at any time with an efficient multiplication rate. However, considering the performance of subcultured explants, we advise waiting for the end of the dormant period. Moreover, after week 32, the number of usable explants becomes unreliable, and after the 48th week, some damage may already be detectable on the stored plants, with the performance of the first subcultured explants being decreased. However, in vitro shoot cultures were completely senescent at 19 weeks (they were three weeks old at the beginning of the 16-week storage period) under normal conditions (22 °CL). As the most frequently used subculture period four weeks, it is possible to avoid at least three subcultures with significant savings in raw materials, costs, and working time.

Author Contributions: Conceptualization, N.M.-D. and K.M.-T.; methodology, N.M.-D. and K.M.-T.; investigation, N.M.-D. and K.M.-T.; writing—original draft preparation, N.M.-D.; writing—review and editing, K.M.-T. All authors have read and agreed to the published version of the manuscript.

Funding: This research received no external funding.

Institutional Review Board Statement: Not applicable.

Informed Consent Statement: Not applicable.

Data Availability Statement: Data are contained within the article.

Acknowledgments: The authors would like to thank Zsuzsanna Márföldiné Gyurján and to Alexandra Hanász for their valuable technical assistance as well as Hazel Brown for her English language editing.

Conflicts of Interest: The authors declare no conflict of interest.

References

1. Phipps, J.B.; Robertson, K.R.; Smith, P.G.; Rohrer, J.R. A checklist of the subfamily *Maloideae* (*Rosaceae*). *Can. J. Bot.* **1990**, *68*, 2209–2269. [CrossRef]
2. Campbell, C.S.; Donoghue, M.J.; Baldwin, B.G.; Martin, F.; Wojciechowski, M.F. Phylogenetic relationships in *Maloideae* (*Rosaceae*): Evidence from sequences of the internal transcribed spacers of nuclear ribosomal DNA and its congruence with morphology. *Am. J. Bot.* **1995**, *82*, 903–918. [CrossRef]
3. Sarv, V.; Venskutonis, P.R.; Bhat, R. The Sorbus spp.—Underutilised Plants for Foods and Nutraceuticals: Review on Polyphenolic Phytochemicals and Antioxidant Potential. *Antioxidants* **2020**, *9*, 813. [CrossRef]
4. Hebda, A.; Kempf, M.; Wachowiak, W.; Pluciński, B.; Kauzal, P.; Tomasz Zwijacz-Kozica, T. Hybridization and introgression of native and foreign *Sorbus* tree species in unique environments of protected mountainous areas. *AoB Plants* **2021**, *13*, plaa070. [CrossRef]
5. Robertson, K.R.; Phipps, J.B.; Rohrer, J.R.; Smith, P.G. A Synopsis of Genera in *Maloideae* (*Rosaceae*). *Syst. Bot.* **1991**, *16*, 376–394. [CrossRef]
6. Wolf, H.; Arenhövel, W.; Behm, A.; Franke, A.; Kleinschmit, J.; Rogge, M.; Schneck, D.; Schneck, V.; Schulzke, R.; Tabel, U. Conservation and breeding of wild fruit tree species in forestry. *Acta Hortic.* **2000**, *538*, 57–62. [CrossRef]
7. Ördögh, M.; Jámbor-Benczúr, E.; Tilly Mándy, A.; Lelik, L. The effects of growth regulators in proliferation of *Sorbus redliana* 'Burokvölgy'. *Int. J. Hortic. Sci.* **2006**, *12*, 77–83. [CrossRef]
8. Enescu, C.M.; de Rigo, D.; Houston Durrant, T.; Caudullo, G. Sorbus domestica in Europe: Distribution, habitat, usage and threats. In *European Atlas of Forest Tree Species;* San-Miguel-Ayanz, J., de Rigo, D., Caudullo, G., Houston Durrant, T., Mauri, A., Eds.; European Commission: Luxembourg, 2016; p. e019db5.
9. Rivers, M.C.; Beech, E.; Bazos, I.; Bogunić, F.; Buira, A.; Caković, D.; Carapeto, A.; Carta, A.; Cornier, B.; Fenu, G.; et al. *European Red List of Trees;* IUCN: Cambridge, UK; Brussels, Belgium, 2019; p. 60. [CrossRef]
10. Hamston, T.J.; de Vere, N.; King, R.A.; Pellicer, J.; Fay, M.F.; Cresswell, J.E.; Stevens, J.R. Apomixis and Hybridization Drives Reticulate Evolution and Phyletic Differentiation in *Sorbus* L.: Implications for Conservation. *Front. Plant Sci.* **2018**, *9*, 1796. [CrossRef] [PubMed]
11. Aravanopoulos, F.A. Conservation and Monitoring of Tree Genetic Resources in Temperate Forests. *Curr. For. Rep.* **2016**, *2*, 119–129. [CrossRef]
12. Lambardi, M.; Ozudogru, A. Advances in the safe storage of micropropagated woody plants at low temperature. *Acta Hortic.* **2013**, *988*, 29–42. [CrossRef]
13. Postman, J.D. The USDA Quince and Pear Genebank in Oregon, a World Source of Fire Blight Resistance. *Acta Hort.* **2008**, *793*, 357–362. [CrossRef]
14. Keller, E.R.J.; Senula, A.; Leunufna, S.; Grübe, M. Slow growth storage and cryopreservation—Tools to facilitate germplasm maintenance of vegetatively propagated crops in living plant collections. *Int. J. Refrig.* **2006**, *29*, 411–417. [CrossRef]
15. Cruz-Cruz, C.A.; González-Arnao, M.T.; Engelmann, F. Biotechnology and Conservation of Plant Biodiversity. *Resources* **2013**, *2*, 73–95. [CrossRef]
16. Hussain, A.; Ahmed, I.; Nazir, H.; Ullah, I. Plant Tissue Culture: Current Status and Opportunities. In *Recent Advances in Plant In Vitro Culture;* IntechOpen: London, UK, 2012. [CrossRef]
17. Agrawal, A.; Singh, S.; Malhotra, E.V.; Meena, D.P.S.; Tyagi, R.K. In Vitro Conservation and Cryopreservation of Clonally Propagated Horticultural Species. In *Conservation and Utilization of Horticultural Genetic Resources;* Rajasekharan, P., Rao, V., Eds.; Springer: Singapore, 2019. [CrossRef]
18. Li, J.; He, M.; Xu, X.; Huang, T.; Tian, H.; Zhang, W. In Vitro Techniques for Shipping of Micropropagated Plant Materials. *Horticulturae* **2022**, *8*, 609. [CrossRef]
19. Chauhan, R.; Singh, V.; Quaraishi, A. In vitro conservation through slow-growth storage. In *Synthetic Seeds;* Faisal, M., Alatar, A.A., Eds.; Springer Nature: Cham, Switzerland, 2019; pp. 397–415. [CrossRef]
20. Hatzilazarou, S.; Kostas, S.; Joachim, M.; Economou, A. Regeneration of *Viburnum dentatum* L. from Alginate-Encapsulated Shoot Explants after Short-Term Cold Storage and Assessment of Genetic Stability Using ISSR Analysis. *Agronomy* **2020**, *10*, 1660. [CrossRef]
21. Pritchard, H.W.; Moat, J.F.; Ferraz, J.B.S.; Marks, T.R.; Camargo, J.L.C.; Nadarajan, J.; Ferraz, I.D.K. Innovative approaches to the preservation of forest trees. *Forest Ecol. Manag.* **2014**, *333*, 88–98. [CrossRef]
22. Rotach, P. *EUFORGEN Technical Guidelines for Genetic Conservation and Use for Service Tree (Sorbus domestica);* International Plant Genetic Resources Institute: Rome, Italy, 2003; 6p.
23. Rich, T.C.G.; Houston, L.; Goodwin, A.; Morgan, V.; Bird, S.; Jones, R.; May, R.; Shiel, D.; Stockdale, R. Conservation status of *Sorbus cuneifolia* (*Rosaceae*), Llangollen whitebeam. *Br. Ir. Bot.* **2019**, *1*, 231–242. [CrossRef]

24. Prknová, H. Long-term storage of service tree (*Sorbus domestica* L.) seeds and induction of their germination. *J. For. Sci.* **2015**, *61*, 417–421. [CrossRef]
25. Reed, B.M. The in vitro genebank of temperate fruit and nut crops at the National Clonal Germplasm Repository-Corvallis. In *Management of Field and In Vitro Germplasm Collections*; International Plant Genetic Resources Institute: Rome, Italy, 1999; pp. 132–135.
26. Arrillaga, I.; Marzo, T.; Segura, J. Micropropagation of juvenile and adult *Sorbus domestica* L. *Plant Cell Tiss. Org.* **1991**, *27*, 341–348. [CrossRef]
27. Máchová, P.; Malá, J.; Cvrčková, H.; Dostál, J.; Buriánek, V. In vitro reproduction of rare and endemic species of rowan tree. *J. For. Sci.* **2013**, *59*, 386–390. [CrossRef]
28. Kļaviņa, D.; Ievinsh, G. Growth of tissue culture and changes in oxidative enzyme activity of *Sorbus* and tayberry cultivars during cold storage. *Acta Univ. Latv.* **2008**, *745*, 179–186.
29. Kulak, V.; Longboat, S.; Brunet, N.D.; Shukla, M.; Saxena, P. In Vitro Technology in Plant Conservation: Relevance to Biocultural Diversity. *Plants* **2022**, *11*, 503. [CrossRef]
30. Schenk, R.V.; Hildebrandt, A.C. Medium and Techniques for Induction and Growth of Monocotyledonous and Dicotyledonous Plant Cell Cultures. *Can. J. Bot.* **1972**, *50*, 199–204. [CrossRef]
31. Murashige, T.; Skoog, F. A revised medium for rapid growth and bioassays with tobacco tissue cultures. *Phys. Plant* **1962**, *15*, 473–497. [CrossRef]
32. Chalupa, V. In vitro propagation of mature trees of *Sorbus aucuparia* L. and field performance of micropropagated trees. *J. For. Sci.* **2002**, *48*, 529–535. [CrossRef]
33. Dobránszki, J.; Magyar-Tábori, K.; Jevcsák, M.; Ördögh, M.; Benczúr, E.J. Improving the in vitro rooting of micro-shoots of *Sorbus rotundifolia* 'Bükk szépe' by the sequential application of Humus®FW and Wuxal®Super organic and chemical fertilisers. *J. Hortic. Sci. Biotech.* **2012**, *87*, 509–513. [CrossRef]
34. Jámbor-Benczúr, E.; Ónadi, K.; Kissimon, J.; Horváth, G. The effect of carbon source on in vitro multiplication, photosynthesis and anatomical structure of *Sorbus rotundifolia* L. *Acta Hortic.* **1997**, *447*, 157–159.
35. Šedivá, J.; Velebil, J.; Zahradník, D. Micropropagation as a Tool for the Conservation of Autochthonous *Sorbus* Species of Czechia. *Plants* **2023**, *12*, 488. [CrossRef] [PubMed]
36. Lambardi, M.; De Carlo, A. Application of Tissue Culture to the Germplasm Conservation of Temperate Broad-Leaf Trees. In *Micropropagation of Woody Trees and Fruits*; Forestry Sciences; Jain, S.M., Ishii, K., Eds.; Springer: Dordrecht, Germany, 2003; p. 75. [CrossRef]
37. Benelli, C.; Tarraf, W.; Izgu, T.; De Carlo, A. In Vitro Conservation through Slow Growth Storage Technique of Fruit Species: An Overview of the Last 10 Years. *Plants* **2022**, *11*, 3188. [CrossRef]
38. González-Arnao, M.T.; Dolce, N.; González-Benito, M.E.; Martínez, C.R.C.; Cruz-Cruz, C.A. Approaches for In Vitro Conservation of Woody Plants Germplasm Chapter 13. In *Biodiversity and Conservation of Woody Plants*; Sustainable Development and, Biodiversity; Ahuja, M.R., Jain, S.M., Eds.; Springer International Publishing AG: Cham, Switzerland, 2017; Volume 17, pp. 355–419. [CrossRef]
39. Hu, Z.; Lin, S.; Wang, H.; Dai, J. Seasonal variations of cold hardiness and dormancy depth in five temperate woody plants in China. *Front. For. Glob. Change* **2022**, *5*, 1061191. [CrossRef]
40. Pruski, K.; Kozai, T.; Lewis, T.; Astatkie, T.; Nowak, J. Sucrose and light effects on in vitro cultures of potato, chokocherry and saskatoon berry during low temperature storage. *Plant Cell Tiss. Org.* **2000**, *63*, 215–221. [CrossRef]
41. Negri, V.; Tosti, N.; Standardi, A. Slow-growth storage of single node shoots of apple genotypes. *Plant Cell Tiss. Org.* **2000**, *62*, 159–162. [CrossRef]
42. Orlikowska, T. Effect of in vitro storage at 4 °C on survival and proliferation of two apple rootstocks. *Plant Cell Tiss. Org. Cult.* **1992**, *31*, 1–7. [CrossRef]
43. Janeiro, L.; Vieitez, A.; Ballester, A. Cold storage of in vitro cultures of wild cherry, chestnut and oak. *Ann. Sci. For.* **1995**, *52*, 287–293. [CrossRef]
44. Kamińska, M.; Skrzypek, E.; Wilmowicz, E.; Tretyn, A.; Trejgell, A. Effect of light conditions and ABA on cold storage and post-storage propagation of *Taraxacum pieninicum*. *Plant Cell Tiss. Org.* **2016**, *127*, 25–34. [CrossRef]
45. Li, Y.; Liu, C.C.; Zhang, J.H.; Yang, H.; Xu, L.; Wang, Q.F.; Sack, L.; Wu, X.Q.; Hou, J.H.; He, N.P. Variation in leaf chlorophyll concentration from tropical to cold-temperate forests: Association with gross primary productivity. *Ecol. Indic.* **2018**, *85*, 383–389. [CrossRef]
46. Walters, R.G. Towards an understanding of photosynthetic acclimation. *J. Exp. Bot.* **2005**, *56*, 435–447. [CrossRef]
47. Dale, M.P.; Causton, D.R. Use of the chlorophyll *a*/*b* ratio as a bioassay for the light environment of a plant. *Funct. Ecol.* **1992**, *6*, 190–196. [CrossRef]
48. Zhao, Y.; Han, Q.; Ding, C.; Huang, Y.; Liao, J.; Chen, T.; Feng, S.; Zhou, L.; Zhang, Z.; Chen, Y.; et al. Effect of Low Temperature on Chlorophyll Biosynthesis and Chloroplast Biogenesis of Rice Seedlings during Greening. *Int. J. Mol. Sci.* **2020**, *21*, 1390. [CrossRef] [PubMed]
49. Ptushenko, V.V.; Ptushenko, O.S.; Tikhonov, A.N. Chlorophyll Fluorescence Induction, Chlorophyll Content, and Chromaticity Characteristics of Leaves as Indicators of Photosynthetic Apparatus Senescence in Arboreous Plants. *Biochemistry* **2014**, *79*, 260–272. [CrossRef]

50. Lichtenthaler, H.K.; Buschmann, C.; Knapp, M. How to correctly determine the different chlorophyll fluorescence parameters and the chlorophyll fluorescence decrease ratio RFd of leaves with PAM fluorometer. *Photosynthetica* **2005**, *43*, 379–393. [CrossRef]
51. Björkman, O.; Demmig, B. Photon yield of O_2 evolution and chlorophyll fluorescence characteristics at 77 K among vascular plants of diverse origins. *Planta* **1987**, *170*, 489–504. [CrossRef]
52. Aazami, M.A.; Asghari-Aruq, M.; Hassanpouraghdam, M.B.; Ercisli, S.; Baron, M.; Sochor, J. Low Temperature Stress Mediates the Antioxidants Pool and Chlorophyll Fluorescence in *Vitis vinifera* L. Cultivars. *Plants* **2021**, *10*, 1877. [CrossRef]
53. Desjardins, Y.; Dubuc, J.-F.; Badr, A. In vitro culture of plants: A stressful activity! *Acta Hortic.* **2009**, *812*, 29–50. [CrossRef]
54. Vosnjak, M.; Sircelj, H.; Hudina, M.; Usenik, V. Response of chloroplast pigments, sugars and phenolics of sweet cherry leaves to chilling. *Sci. Rep.* **2021**, *11*, 7210. [CrossRef] [PubMed]
55. Riva-Roveda, L.; Escale, B.; Giauffret, C.; Périlleux, C. Maize plants can enter a standby mode to cope with chilling stress. *BMC Plant Biol.* **2016**, *16*, 212. [CrossRef] [PubMed]
56. Kucharska, D.; Orlikowska, T.; Maciorowski, R.; Kunka, M.; Niewiadomska-Wnuk, A. Storage of proliferating gooseberry cultures under slow growth conditions. *Hort. Sci.* **2021**, *48*, 134–140. [CrossRef]
57. Jámbor-Benczúr, E.; Márta-Riffer, A. In vitro propagation of *Philodendron tuxtlanum* bunting with benzylaminopurine. *Acta Agron. Hung.* **1990**, *39*, 341–348.
58. Heller, R. Recherces sur la Nutrition Minérale des Tissus Cultivés in Vitro. *Ann. Sci. Nat. Bot. Veg.* **1953**, *2*, 1–225.
59. Jámbor-Benczúr, E.; Géczi, J.; Kissné, I.; Szafián Zs Nagy, T. Results in micropropagation of *Sorbus redliana* Karp. I. Culture initiation and multiplication. *Publ. Untv. Hortic. Ind. Aliment.* **1998**, *LVII*, 77–82.
60. Felföldy, L. A biológiai vízminősítés. Budapest: Vízgazdálkodási Intézet. *Vízügyi Hidrobiol.* **1987**, *16*, 141–148.
61. Kitajima, M.; Butler, W.L. Quenching of chlorophyll fluorescence and primary photochemistry in chloroplasts by dibromothymoquinone. *Biochim. Biophys. Acta.* **1975**, *376*, 105–115. [CrossRef] [PubMed]
62. Maxwell, K.; Johnson, G.N. Chlorophyll fluorescence—A practical guide. *J. Exp. Bot.* **2000**, *51*, 659–668. [CrossRef] [PubMed]

plants

Article

Cretan Dittany (*Origanum dictamnus* L.), a Valuable Local Endemic Plant: In Vitro Regeneration Potential of Different Type of Explants for Conservation and Sustainable Exploitation

Virginia Sarropoulou, Eleni Maloupa and Katerina Grigoriadou *

Hellenic Agricultural Organization (HAO)-DIMITRA, Institute of Plant Breeding and Genetic Resources, Balkan Botanic Garden of Kroussia, Thermi, 57001 Thessaloniki, Greece
* Correspondence: katgrigoriadou@elgo.gr; Tel.: +30-2310-471613

Abstract: *Origanum dictamnus* L. is a medicinal local endemic to the Island of Crete, Greece. Its propagation through biotechnological tissue culture techniques is essential due to its augmented multi-industrial sector demand. For direct organogenesis, among different culture media variants (MS, Gamborg B5), and cytokinins [6-benzyladenine (BA), kinetin (Kin), 2-isopentenyl adenine (2-iP)], the MS + added with BA (2.2 μM) was the most effective treatment for shoots and roots formation. For indirect organogenesis, all explant types (leaves, petioles, roots) showed a 100% callusing rate after 2 months in all media variants tested; ODK1: 20 μM thidiazuron (TDZ) + 5 μM indole-3-butyric acid (IBA) or ODK2: 0.5 μM kinetin + 5 μM 2,4-dichlorophenoxy acetic acid (2,4-D). The leaves and petiole explants assured a low rate of shoot regeneration (20%) in ODK1. Afterwards, leaf-, petiole-and root-callus derived from both media were transferred to four new media plant growth regulators—free or with BA + IBA + gibberellic acid (GA_3). After 10 months from callus transferring, the petiole callus gave rise to roots (20–75%) while the leaf callus exhibited 10–30% shoot or 30% root regeneration. In this study, indirect organogenesis of *O. dictamnus* was carried out for the first time, thus various organs can be used for plant regeneration, and the developed protocol may be applicable in the horticulture industry.

Keywords: aromatic–medicinal plant; basal culture media; Cretan dittany; explant types; ex situ conservation; Greek flora; in vitro organogenesis; plant growth regulators

Citation: Sarropoulou, V.; Maloupa, E.; Grigoriadou, K. Cretan Dittany (*Origanum dictamnus* L.), a Valuable Local Endemic Plant: In Vitro Regeneration Potential of Different Type of Explants for Conservation and Sustainable Exploitation. *Plants* **2023**, *12*, 182. https://doi.org/10.3390/plants12010182

Academic Editor: Anna De Carlo

Received: 2 December 2022
Revised: 24 December 2022
Accepted: 29 December 2022
Published: 1 January 2023

1. Introduction

Origanum dictamnus L. (Lamiaceae) or Cretan dittany is a native and range-restricted Greek chamaephyte endemic to the island of Crete (Iraklio, Mt Psiloreitis) that grows in habitats such as cliffs, rocks, walls, ravines, and boulders [1]. The conservation status of this taxon based on the IUCN category has been declared as Near Threatened, being under protection by the Greek Presidential Decree 67/81 [1]. *O. dictamnus* represents the most promising case of an endemic neglected and underutilized plant due to the very high potential (94.4%) in the medicinal–cosmetic sector coupled with very high feasibility for sustainable exploitation (91.67%), outlining the extant value chain and the sustainable commercial exploitation associated with this taxon, as already achieved mostly in Crete but also abroad [2]. Finalized monographs and the European Medicines Agency have contented and endorsed the medicinal properties and recommendations of this taxon [3]. Dittany infusion was a complex mixture consisting of several flavones, flavonols, and hydroxycinnamic acid derivatives, partially related to its antioxidant/antiradical activity, being very active as an antiglycative agent, enhancing the functional properties of the Cretan tea beverage, to be used as a food supplement useful in chronic and degenerative disorders [4].

Many species of the genus *Origanum* growing as wild populations in the Mediterranean basin have frequently become the object of predatory exploitation because of overharvest practices seriously threatening the sustainability of these resources, therefore cultivation of these species is highly recommended to lessen their overexploitation [5]. This suggests that the plant could be used as an ornamental in the landscaping industry, or used practically in the pharmaceutical or food industry after the development of protocols for its multiplication through traditional or modern biotechnological methods. Conventional propagation methods are known to face difficulties related to low seed germination rates and moderate rooting percentages in the case of cuttings, pointing out the importance of in vitro propagation to be used as an alternative for large scale multiplication of *Origanum* species, allowing their cultivation and further sustainable use of biological diversity [6]. Plant tissue culture can act as a possible alternative, which may allow rapid propagation for commercial purposes [7] and the development of in vitro cell cultures for accelerating agricultural processes in producing valuable secondary metabolites and other useful phytochemicals [8]. However, there have been limitations in tissue culture systems including the availability and quality of plant materials and exact time of year needed for culture initiation, thus, various parts of a plant such as leaves, petioles, and roots could be used as better explant sources with greater potential due to their abundance and easiness of production as compared to shoot tips [9]. Organogenesis is a process of organ formation such as leaves, shoots, or roots facilitating their regeneration potential from cells and tissues, including two types; the direct one in which cultured explants regenerate shoots without the intermediate stage of callus formation and the indirect one where shoot regeneration occurs only after callus formation [10].

Based on the literature, there is no report on the application of tissue cultures for indirect organogenesis of *O. dictamnus* L. In this context, the main aim of this research was to study the direct and indirect organogenesis potential of the Cretan dittany. In terms of direct organogenesis, the specific objectives were (1) to identify an improved medium, thus two basal culture media were included: Murashige and Skoog (MS) [11], and Gamborg B5 (B5) [12] and (2) to optimize the existing micropropagation protocol and to maximize the ex vitro survival success of shoot node explants according to a previous study conducted by Sarropoulou et al. [13], herein under the effect of different cytokinin types and concentrations. In the framework of indirect organogenesis, this study was oriented (1) to evaluate callus induction, shoot and/or root regeneration potential of different tissue parts (leaves, petioles, roots) under the effect of different plant growth regulators (PGRs) and (2) to elucidate whether leaf-, petiole- and root-callus explants as induced in the previous stage could be further differentiate into new callus, shoots, and/or roots.

2. Results

2.1. In Vitro Shoot Proliferation—Rooting and Ex Vitro Acclimatization

Among the 14 treatments tested, the MS medium containing 2.2 μM BA exhibited the highest proliferation rate (3.15) and root number (16.5 roots/rooted explant), while the B5 medium enriched with 2.2 μM BA gave the highest percentage in the formation of new multiple shoots (90%); as reported in the previous work [13]. The MS medium enriched with either 2.2 μM BA or 2.32 μM Kin gave higher shoot lengths (2.66–2.78 cm) and root lengths (1.73–2.07 cm) as compared to the other 12 treatments. Statistically non-significant differences were observed in the number of new shoots/explants among the 14 treatments (Figure 1a–h).

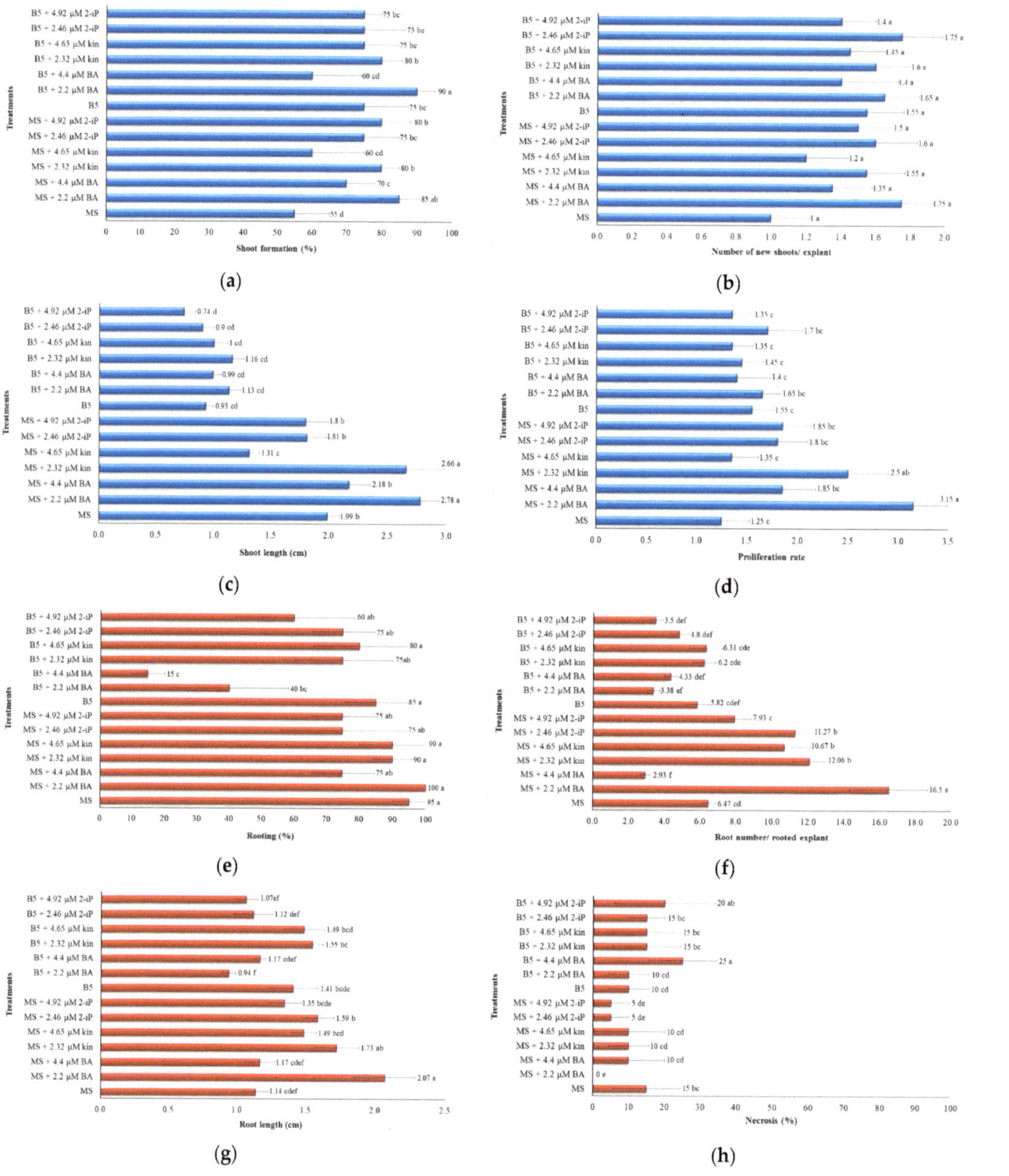

Figure 1. Effect of two basal culture media (MS, B5) supplemented with 0.25 µM IBA, 0.3 µM GA_3, 20 g L^{-1} sucrose and 6 g L^{-1} Plant Agar (pH: 5.8) in combination with three cytokinin types (BA, Kin, 2-iP) at three concentrations on shoot proliferation and rooting parameters in *O. dictamnus* explants after 30 days of in vitro culture: (**a**) Shoot formation (%) (graph horizontal scale bar: 1 unit = 10%); (**b**) Number of new shoots/explant (graph horizontal scale bar: 1 unit = 0.2); (**c**) Shoot length (cm) (graph horizontal scale bar: 1 unit = 0.5 cm); (**d**) Proliferation rate (graph horizontal scale bar: 1 unit = 0.5); (**e**) Rooting (%) (graph horizontal scale bar: 1 unit = 10%); (**f**) Root number/rooted explant (graph horizontal scale bar: 1 unit = 2); (**g**) Root length (cm) (graph horizontal scale bar: 1 unit = 0.5 cm); (**h**) Necrosis (%) (graph horizontal scale bar: 1 unit = 10%).

Taking simultaneously into consideration all proliferation and rooting parameters, the MS medium was the ideal one, BA the preferred cytokinin type, and 2.2 μM its optimum concentration, promoting better direct organogenesis of explants in a single 30-day culture stage (3.15 proliferation rate, 85% shoot formation, 2.78 cm shoot length, 100% rooting, 16.5 roots/rooted explant 2.07 cm long; as shown in the previous work [13] without occurrence of necrosis (0%) (Figure 2a,b). A survival rate of 100% was achieved in the case of plants derived in vitro from the B5 medium enriched with 2.46 μM 2-iP. Additionally, survival rates of 80–100% were recorded for rooted plants derived in vitro from B5 medium and 66.67–86.67% for those derived from the MS one, exhibiting a final 80.4% ex vitro survival rate (mean value) after 2 months in the greenhouse mist regardless of previous in vitro treatment (medium type, cytokinin type, cytokinin concentration) (Figure 2c,d).

Figure 2. Micropropagation and ex vitro acclimatization of in vitro rooted plantlets of *O. dictamnus* explants: (**a**,**b**) Shoot proliferation and rooting after 30 days of in vitro culture in MS basal nutrient medium supplemented with 2.2 μM BA, 0.25 μM IBA, 0.3 μM GA_3, 20 g L^{-1} sucrose and 6 g L^{-1} Plant Agar (pH: 5.8) inside and outside vessels, respectively (scale bar: 1 cm); (**c**) Effect of two culture media (MS, B5) supplemented with 0.25 μM IBA, 0.3 μM GA_3, 20 g L^{-1} sucrose and 6 g L^{-1} Plant Agar (pH: 5.8) in combination with three cytokinin types (BA, Kin, 2-iP) at three concentrations on survival rate (%) (graph horizontal scale bar: 1 unit = 10%); (**d**) Acclimatized plants in a peat: perlite (1:1) substrate in the greenhouse mist (scale bar: 1 cm).

2.2. In Vitro Culture of Different Plant Tissue Explants (Leaves, Petioles, Roots)

After 2 months of culture, callus formation was noticed in all treatments. Regeneration of shoots at a rate of 20% on leaf and petiole callus in ODK1 was observed, while no shoot regeneration occurred in the other treatments. Leaf, petiole and root explants cultured in ODK2 medium did not result in shoot regeneration. No treatment led to callus, shoot and root induction at the same time regardless of initial explant type (Figure 3a–n, Table 1).

(a) (b) (c) (d)
(e) (f) (g) (h)
(i) (j) (k)
(l) (m) (n)

Figure 3. In vitro culture of three explant types (leaves, petioles, roots) of *O. dictamnus* L. in different media after two months: (**a–d**) Root, callus and shoot formation from leaf explants in ODK1; (**e–h**) Callus and shoot regeneration from petioles in ODK1; (**i**,**j**) Callus formation on root explants in ODK1; (**k**) Callus formation in root explants in ODK2; (**l–n**) Callus formation on leaf, petiole and root explants, respectively in ODK2. (Scale bar: 1 cm).

Table 1. Effect of three different explant types (leaves, petioles, roots) of *O. dictamnus* L. and two different media on callus formation, shoot and root regeneration (%) after a 2-month period.

	Treatments					Response (after 2 Months)		
Explant Type	Culture Medium	TDZ (μM)	Kin (μM)	IBA (μM)	2,4-D (μM)	Callus Formation (%)	Shoot Regeneration (%)	Root Regeneration (%)
Leaves	ODK1	20	-	5	-	100	20.0 ± 6.3 a	0
	ODK2	-	0.5	-	5	100	0.0 ± 0.0 b	0
Petioles	ODK1	20	-	5	-	100	20.0 ± 11.0 a	0
	ODK2	-	0.5	-	5	100	0.0 ± 0.0 b	0
Roots	ODK1	20	-	5	-	100	0.0 ± 0.0 b	0
	ODK2	-	0.5	-	5	100	0.0 ± 0.0 b	0

Means ± standard error (S.E.) with different letters within the "Shoot Regeneration (%)" column are statistically significant different from each other according to the Duncan's multiple range test at $p \leq 0.05$.

2.3. *In Vitro Culture of Different Callus Explants (Leaf-, Petiole-, Root-Callus)*

In the following stage, after 10 months of culture, formation of new secondary callus at a 30% rate on initial primary root callus was observed after transferring the explants from ODK2 (0.5 μM Kin + 5 μM 2,4-D) to ODR0 (PGR's-free) medium while petiole callus exhibited 60% and 75% new callus formation rates when transferred from ODK2 to ODR3 (2.2 μM BA + 0.25 μM IBA + 0.3 μM GA_3) and ODR0 media, accordingly. Shoot regeneration from leaf callus was evident at a rate of 10% and 30% when leaf callus transferred from ODK1 (20 μM TDZ + 5 μM IBA) to ODR0 and ODR2 (1.1 μM BA + 0.25 μM IBA + 0.3 μM GA_3) medium, respectively. Root regeneration from leaf callus (30%) was recorded when explants derived from ODK2 and cultured afterwards in ODR0. Petiole callus originated from ODK2 were afterwards differentiated and gave regeneration to roots at a 20–75% rate in ODR0 (75%), ODR1: 0.25 μM IBA + 0.3 μM GA_3 (20%), ODR2 (20%) and ODR3 (70%) media. Among the three callus explant types, petiole callus had the best root regeneration performance followed by leaf callus. Root callus did not have any shoot and/or root organogenetic response, while leaf callus led to shoot or root regeneration when derived from ODK1 or ODK2, respectively (Figure 4a–g, Table 2).

Table 2. Effect of explant type (leaf callus, petiole callus, root callus) of *O. dictamnus* L. and different media (ODR0, ODR1, ODR2, ODR3) on callus, shoot and root regeneration (%) (10 months).

Medium Prior Callusing	Explant Type	Culture Medium Code Number	BA (μM)	IBA (μM)	GA_3 (μM)	New Callus Formation (%)	Shoot Regeneration (%)	Root Regeneration (%)
ODK1 (leaves, petioles, roots)	Leaf callus	ODR0	0	0	0	0.0 ± 0.0 d	10.0 ± 5.8 b	0.0 ± 0.0 d
		ODR1	0	0.25	0.3	0.0 ± 0.0 d	0.0 ± 0.0 c	0.0 ± 0.0 d
		ODR2	1.11	0.25	0.3	0.0 ± 0.0 d	30.0 ± 11.5 a	0.0 ± 0.0 d
		ODR3	2.22	0.25	0.3	0.0 ± 0.0 d	0.0 ± 0.0 c	0.0 ± 0.0 d
	Petiole callus	ODR0	0	0	0	0.0 ± 0.0 d	0.0 ± 0.0 c	0.0 ± 0.0 d
		ODR1	0	0.25	0.3	0.0 ± 0.0 d	0.0 ± 0.0 c	0.0 ± 0.0 d
		ODR2	1.11	0.25	0.3	0.0 ± 0.0 d	0.0 ± 0.0 c	0.0 ± 0.0 d
		ODR3	2.22	0.25	0.3	0.0 ± 0.0 d	0.0 ± 0.0 c	0.0 ± 0.0 d
	Root callus	ODR0	0	0	0	0.0 ± 0.0 d	0.0 ± 0.0 c	0.0 ± 0.0 d
		ODR1	0	0.25	0.3	0.0 ± 0.0 d	0.0 ± 0.0 c	0.0 ± 0.0 d
		ODR2	1.11	0.25	0.3	0.0 ± 0.0 d	0.0 ± 0.0 c	0.0 ± 0.0 d
		ODR3	2.22	0.25	0.3	0.0 ± 0.0 d	0.0 ± 0.0 c	0.0 ± 0.0 d

Table 2. *Cont.*

Medium Prior Callusing	Explant Type	Culture Medium Code Number	BA (μM)	IBA (μM)	GA_3 (μM)	New Callus Formation (%)	Shoot Regeneration (%)	Root Regeneration (%)
ODK2 (leaves, petioles, roots)	Leaf callus	ODR0	0	0	0	0.0 ± 0.0 d	0.0 ± 0.0 c	30.0 ± 10.0 b
		ODR1	0	0.25	0.3	0.0 ± 0.0 d	0.0 ± 0.0 c	0.0 ± 0.0 d
		ODR2	1.11	0.25	0.3	0.0 ± 0.0 d	0.0 ± 0.0 c	0.0 ± 0.0 d
		ODR3	2.22	0.25	0.3	0.0 ± 0.0 d	0.0 ± 0.0 c	0.0 ± 0.0 d
	Petiole callus	ODR0	0	0	0	75.0 ± 14.4 a	0.0 ± 0.0 c	75.0 ± 2.9 a
		ODR1	0	0.25	0.3	0.0 ± 0.0 d	0.0 ± 0.0 c	20.0 ± 5.8 c
		ODR2	1.11	0.25	0.3	0.0 ± 0.0 d	0.0 ± 0.0 c	20.0 ± 5.8 c
		ODR3	2.22	0.25	0.3	60.0 ± 11.5 b	0.0 ± 0.0 c	70.0 ± 5.8 a
	Root callus	ODR0	0	0	0	30.0 ± 5.8 c	0.0 ± 0.0 c	0.0 ± 0.0 d
		ODR1	0	0.25	0.3	0.0 ± 0.0 d	0.0 ± 0.0 c	0.0 ± 0.0 d
		ODR2	1.11	0.25	0.3	0.0 ± 0.0 d	0.0 ± 0.0 c	0.0 ± 0.0 d
		ODR3	2.22	0.25	0.3	0.0 ± 0.0 d	0.0 ± 0.0 c	0.0 ± 0.0 d

Figure 4. In vitro culture of different callus explant types of *O. dictamnus* in different PGRs-supplemented MS media: (**a**–**c**) New callus and root regeneration from leaf-callus when transferred from ODK2 to ODR0; (**d**–**f**) Differentiation of petiole callus into new callus and roots after transferred from ODK2 to ODR3; (**g**) Root callus turned brown after successive subcultures in ODR0. (Scale bar: 1 cm).

3. Discussion

3.1. Micropropagation and Ex Vitro Acclimatization

Shoot proliferation and rooting in vitro is highly contingent on the composition of the basal medium in salts including type, concentration and macronutrients/micronutrients ratio, source of iron, vitamins and surrounding conditions such as photoperiod and temperature [14], therefore different media used in this research with Cretan dittany (MS, B5) have different composition and concentration of constituents, leading to different responses towards plant growth. The augmented potential requirement of *O. dictamnus* shoot node explants for proliferation, rooting and subsequent growth could explain its better performance in the MS medium than in B5. The different responses of explants to the two basal culture media studied herein could be ascribed to the different NO_3^-/NH_4^+ ratio in each medium or to the difference in total nitrogen content, which is 39.4 mM (18.79 mM KNO_3 + 20.61 mM NH_4NO_3) in MS, and 25.74 mM in B5 (24.73 mM KNO_3 + 1.01 mM NH_42SO_4) [15]. The MS proved to be the ideal basal medium enhancing shoot proliferation in other *Origanum* species as well including *O. onites* [16], *O. syriacum* L. [17], *O. majorana* L. [18], *Origanum vulgare* x *applii* [19], and *O. vulgare* [20].

Except the basal culture medium composition, the setting up of a competent in vitro direct regeneration protocol is also strongly dependent on the quest of the best cytokinin type and its best concentration as a crucial success coefficient whichever the plant species is [21]. Shoot proliferation performance of *O. dictamnus* microshoots under study was better enhanced by 2.2 μM BA followed by 2.32 μM kinetin and afterwards by 2.46 μM 2-iP. The height of the in vitro grown shoots, the increase in the length of the internodes, the rate of the proliferated microshoots, and above all, the genotype are parameters which are highly affected by the individual and combined application of the type and concentration of the added cytokinin to the culture medium [22]. Differences that are due to stability, mobility, conjugation and hormone oxidation rates among the different cytokinins could explain the disparate competence of explants to different cytokinin-relative strengths during the reinforcement of the shoot induction stage [23]. The supremacy of BA in relation to kinetin and 2-iP in prompting multiple shoot production has been demonstrated, presumably firstly because the chemical stability of BA is the highest among the group of cytokinins in in vitro plant production systems whereas the majority of other cytokinins, especially of the purine-type are deemed of lower stability [24]. Secondly, the quantity of BA declined in the medium is of lower rate as compared to the other phytohormones or is not facile easily disintegrated and therefore carries on in the medium, allowing greater amounts of free or ionized BA be promptly disposable to plant tissues through the medium [25]. Thirdly, there is the higher discontinuating effect of BA on lifting the dominance of the shoot apex and switching on the lateral shoots configuration and shoot tip cultures leaf vegetative evolvement [23]. Finally, plant tissues have a higher ability to metabolize BA more readily than other synthetic growth regulators [26]. Numerous studies support the higher effectiveness of BA, as the most appropriate cytokinin type applied alone or in combination with an auxin (IBA, NAA) for in vitro proliferation in several other *Origanum* species, including *O. majorana* L. [18], *O. sipyleum* [27], *O. syriacum* and *O. ehrenbergii* [6], *O. vulgare* [28], *O. minutiflorum* [29], *O. vulgare* x *applii* [19] and *O. vulgare* [20].

Rooting performance of *O. dictamnus* microshoots under study was better stimulated by the cytokinin BA in the MS medium followed by kinetin and subsequently by 2-iP, while in B5 medium kinetin was the most preferred cytokinin, BA the least preferred and 2-iP of moderate ability, each cytokinin applied with 0.25 μM IBA + 0.3 μM GA_3. Different responses have been outlined in other *Origanum* species where rooting of microshoots was better promoted either in a full-strength MS medium PGRs-free such as in the case of *O. syriacum* and *O. ehrenbergii* [6] or in different MS or B5 strength media supplemented only with auxins (IBA, NAA, IAA) in the absence of cytokinins including *O. sipyleum* L. [27,30] and *O. acutidens* (Hand.-Mazz.) Ietswaart [7], *O. syriacum* [17,31], *O. heracleoticum* L. [32] and *O. vulgare* [33]. The percentage of rooted microshoots is a coefficient of greater importance in comparison to root elongation for plantlets' settlement and effectual hardening outcome to ex vitro conditions, nevertheless more lengthy roots permit better fastening of plantlets in the substrate and more efficient exploitation of water potential and minerals due to greater easiness and higher rate of absorption [34]. Rooted Cretan dittany plantlets grown on a peat: perlite (1:1 *v/v*) substrate mixture after 2 months in the greenhouse mist gave a mean 80.4% ex vitro survival rate, regardless of previous in vitro conditions. A wide range of survival rates (48–98%) of rooted in vitro plantlets to ex vitro conditions during acclimatization and gradual hardening process is reported for other *Origanum* species including *O. sipyleum* L. [30], *O. majorana* L. [18], and *O. syriacum* L. [17] as well as a wide range of substrate mixtures such as peat: perlite (1:1, 1:2, 1:3, 2:1 *v/v*) [6,30–32]; peat moss: vermiculite (1:1: *v/v*) [18]; peat moss: sand: vermiculite (1:1:1 *v/v*) [17]; and peat: sand: perlite (1:1:1 *v/v*) [30]. Therefore, between the two basal media (MS, B5), the three cytokinin types (BA, kinetin, and 2-iP) and the three cytokinin concentrations tested, the MS proved to be the ideal medium, BA the most effective cytokinin and 2.2 μM the optimum concentration combined with 0.25 μM IBA and 0.3 μM GA_3 promoting the best proliferation and rooting in vitro of shoot-node explants of *O. dictamnus* in a single 30-day culture stage ensuring 100% ex vitro survival success. The high ex vitro acclimatization and hardening survival rate of in vitro plantlets and their following vegetative and root system growth is of paramount importance in every plant tissue culture system since it can contribute to cost-savings, marketing trade cycle abridge, speeding up the installation of propagation material, and the further introduction of the plants into cultivation in field conditions [35].

In a previous study on the same plant species [13] as presented herein, among the two different basal media (MS and B5) and the two different BA concentrations (1.11 and 2.22 μM) applied simultaneously with 0.25 μM IBA and 0.3 μM GA_3, the MS medium + 2.22 μM BA found to be superior for shoot proliferation and rooting in vitro as well as for ex vitro survival success of the rooted microshoots. The further advancements of this study were the effect of two other cytokinins types except of BA, that of kinetin and 2-iP and the fact that all these three cytokinins were supplemented in MS and B5 media at two different and marginal concentrations, the lower of 0 μM, (cytokinin-free) and the higher of 4.4 μM, which had not been tested in the previous work of Sarropoulou et al. [13]. The findings of this study illustrate that the increase in all shoot proliferation and rooting in vitro parameters, except of shoot number in MS medium under the 2.2 μM BA + 0.25 μM IBA + 0.3 μM GA_3 combination, as well as in the ex vitro survival rate, can be the result of an optimal division and elongation reinforcement due to the role of BA as a cytokinin mainly in cell division/proliferation and less to cell expansion, while IBA as an auxin in both cell division and cell enlargement [21], compared to the other two BA concentration combinations, 0 and 4.4 μM. In parallel with our findings, in another *Origanum* species, in *O. acutidens* (Hand.-Mazz.) Ietswaart, each increase in BA concentration (0.6–2.4 mg/L) with 0.2 mg/L NAA led consistently to increased root numbers, shoot and root lengths up to 1.8 mg/L BA, however a sharp decline in all the above-mentioned parameters was observed when higher BA concentration (2.4 mg/L) was used [7] and this response could be attributed to the declined absorption rate of water and mineral components from the medium [36]. It becomes evident that 2.2 μM herein is the optimum BA concentration for a cytokinin and endogenous auxin balance leading to release of the shoot apical dominance [37], thus besides the cytokinin/auxin balance there is a strong inter-correlation between the endogenous concentrations of PGRs and the applied exogenous ones in the medium affecting morphogenesis [38]. As concerns rooting, the effect of BA on root growth was concentration-dependent [39] since root meristems were enlarged and gave rise to faster growing and more branched roots under treatment with cytokinins, especially BA which regulates the meristem activity due to the increased expression of cytokinin oxidase [40]. The augmented mitotic activity in plants BA-induced has been associated with higher radicle length and higher number of initiated roots [41].

3.2. *In Vitro Culture of Different Plant Tissue Explants (Leaves, Petioles, Roots)*

The process of plant organs regeneration is controlled by the appropriate concentration ratio between auxins and cytokinins, the two main PGRs groups that are essential to drive specific organogenesis responses from different tissue parts (leaf, petiole, root, etc.), thus the interplay actions between cytokinins and auxins are complex and roots, shoots and callus formation are highly affected by different combinations and concentrations of PGRs [42], explant type and plant species [43]. The results of this study showed 100% callus induction in all cases, irrespective of explant type, PGRs types, and concentrations. An explanation for the callusing response herein could be the increase of the endogenous auxin level on the cutting edge of each tissue part (leaf, petiole, root), which switches on cell proliferation, especially in media fortified with cytokinins [44]. In several *Origanum* spp. (*O. vulgare, O. vulgare* subsp. *hirtum, O. syriacum*) [45], the MS medium supplemented with PGRs proved to be effective for callus production from either leaf explants [45,46] or from hypocotyl explants in marjoram (*O. majorana* L.) [47].

The medium composition and explant source as a combined factor exerted a significant effect on callus initiation and plant regeneration [48]. In the studied *O. dictamnus*, a simultaneous occurrence of callus formation (100%) and shoot regeneration (20%) was obtained by leaf and petiole explants in ODK1 (20 μM TDZ + 5 μM IBA) medium. TDZ is a phenylurea that impels a high morphogenesis activity in plant tissues especially in leaves [49] and actively participates in auxin synthesis by raising the endogenous IAA level and its precursor tryptophan [50], displaying different regeneration responses dependent on genotype and initial explant type [49]. In tissue culture systems, TDZ alone or combined with IBA has been found to be effective for shoot regeneration responses [49]. TDZ has been shown to exert both cytokinin and auxin like activity as compared to other commonly used auxins and cytokinins, owing to its ability to shift the level of endogenous PGRs [51] and this ability might be due to the speed up that causes in endogenous cytokinin production rate and inhibition in the activity of cytokinin oxidase enzymes [52], ensuring better nutrients absorption and increased regeneration [53].

3.3. In Vitro Culture of Different Callus Explants (Leaf-, Petiole-, Root-Callus)

The regeneration ability of callus is dependent on plant species, explant type, medium components and endogenous concentrations of PGRs as an outcome of their uptake from extracellular sources, and their metabolism and endogenous interaction [54]. Even though plant cells convey the same genetic make-up information, their morphogenetic competence is not consistent based on differences on the one hand to the spatial and temporal distribution of the cells and on the other hand to their physiological and developmental stage [55]. In the studied *O. dictamnus*, after 10 months of culture, there was new secondary callus formation over the initial primary ones only by root callus when transferred from ODK2 (0.5 μM kinetin + 5 μM 2,4-D) to ODR0 medium (PGR's-free), and by petiole callus from ODK2 to ODR3 (2.2 μM BA + 0.25 μM IBA + 0.3 μM GA_3) or ODR0.

Previous studies have demonstrated that the cytokinin BA at a certain concentration can boost callus subculture, differentiation, and regeneration in a large number of plant species [56]. In this study, shoot regeneration occurred only through leaf derived callus when transferred from ODK1 to either ODR0 medium (PGRs-free) or ODR2 (1.1 μM BA + IBA + GA_3) medium. The low shoot regeneration rate from leaf callus (10–30%) might be due to the long-term maintenance period (10 months) of the explants in culture [57]. In vitro shoot formation may be subjected to change depending upon the explant types used [58]. Ehsandar et al. [59] pointed out the differentiation ability of calli into shoot primordia after repeated subcultures in MS medium containing BA. Leaf explants sourced calli demonstrated shoot regeneration ability versus petiole and root callus explants of *O. dictamnus* as presented herein and the reasons for this response might be (1) the higher juvenility and the lower number of vascular tissue in leaf-callus explants [60], (2) young leaves are healthy, non-senescing, nutrient-rich tissues that contain higher concentrations of endogenous hormones [61] and (3) leaf cells have higher organogenic potential than petiole and root cells [62]. Root regeneration from *O. dictamnus* leaf-callus derived explants was observed in ODK2-ODR0 (PGRs-free) transition combination media whereas petiole-callus explants differentiated into roots in all tested regeneration media, both in the absence of PGR's and enriched with BA + IBA + GA_3. The in vitro shoot regeneration of *O. dictamnus* through leaf-derived callus and petiole-derived callus and not through callus of root origin is a very important step because the vegetative parts of a plant are desirable explants for tissue culture systems due to their ability to preserve the genetic homozygosity of the parent genotype [63].

In this study with *O. dictamnus*, root callus explants in the ODR0 (PGRs-free) medium after successive subcultures became dark-brown, then black, and finally collapsed. In consistency with our findings, hypocotyl callus of marjoram (*O. majorana* L.) retained high proliferation rate for two subcultures but afterwards the callus grew slower, turned brown, and did not survive in subsequent subcultures [47]. Such retardation may be related to the oxidation of phenolic compounds [64] that inhibit the activity of some essential enzymes, thus suppressing in vitro proliferation [65]. This explanation could also justify the survival failure of *O. dictamnus* desiccated shoots regenerated from leaf-callus in ODK1 and then transferred to ODR0 or ODR2 media.

4. Materials and Methods

4.1. Plant Material and Culture Conditions

After taxonomic identification of *O. dictamnus* plants conserved ex situ in the premises of the Balkan Botanic Garden of Kroussia, Institute of Plant Breeding and Genetic Resources, Hellenic Agricultural Organization—DIMITRA, seed collection took place from mature inflorescences and these seeds were given thereafter the unique IPEN (International Plant Exchange Network) accession number GR-1-BBGK-03,2108. All information regarding the disinfection protocol (fungicide, ethanol, NaOCl) followed, the initial establishment conditions of 7-year-old seeds after long-term storage (4 °C, RH < 5%) and 37.93% maximum germination rate obtained under 16 h photoperiod and 21–23 °C temperature after 12 days of in vitro culture in MS medium enriched with 20 g L^{-1} sucrose and 6 g L^{-1} Plant Agar are elaborately described in a previous study [13]. The sub-culturing and transferring of the in vitro stock cultures to fresh nutrient medium were taking place every 4 weeks. In all experiments, culture media were sterilized through autoclave at a temperature of 121 °C for 20 min after adjustment of the pH value of the media to 5.8 using dilute solutions of 1 N KOH and 0.02% HCL in drops for balancing until reaching the exact value. Magenta vessels (Baby food jars, autoclavable, reusable, 62.4 mm × 95.8 mm, 200 mL, Sigma-Aldrich, Merck KGaA, Germany) sealed with Magenta™ B-caps were used for the placement of the nutrient medium and the culture of the explants. Each Magenta vessel was filled with 25 mL of medium. A growth chamber under controlled laboratory conditions of temperature 22 ± 2 °C, photoperiod 16 h light/8 h dark, illumination type of cool white fluorescent lamps (WFLs) (PHILIPS, 36 W/830 G13 1214 mm) and 40 μmol m^{-2} s^{-1} light quantity was used for the incubation of the in vitro plant tissue cultures.

4.2. In Vitro Direct Organogenesis and Ex Vitro Acclimatization

For direct organogenesis, internodal shoot segments 1–1.5 cm long that bore two buds were the experimental explants stem from already established in vitro cultures. The concurrent application of two basal nutrient media (MS, B5) with three cytokinin types, i.e., BA, Kin and 2-iP was evaluated. Each cytokinin was incorporated into the basal culture medium, either MS or B5 at three concentrations, i.e., BA: 0, 2.2 and 4.4 μM, Kin: 0, 2.32 and 4.65 μM, and 2-iP: 0, 2.46 and 4.92 μM. All media were filled with 0.25 μM indole-3-butyric acid (IBA) as an auxin type boosting root formation, 0.3 μM gibberellic acid (GA_3) for shoot elongation, and 20 g L^{-1} sucrose (Duchefa Biochemie, The Netherlands) as the main energy carbon source and afterwards were gelled with Plant Agar (Duchefa Biochemie, The Netherlands) at 6 g L^{-1}. The basic environmental requirements for incubation of the in vitro plant cultures inside the controlled laboratory growth chamber were set as follows: 22 ± 2 °C temperature, 16 h photoperiod, WFLs of 40 μmol m^{-2} s^{-1} light irradiance. The experiment was terminated after 30 days, wherein eight different macroscopic attributes for shoot proliferation and rooting potential were evaluated.

In early January, microshoots with well-developed root systems after washing with tap water to remove any adhering solidifying agent were then transferred in the internal heated mist of the greenhouse within multiple propagation trays each of 100 mL volume composed of peat (Terrahum, Klasmann) and perlite (Geoflor) substrate mixture at a 1:1 *v*/*v* ratio. The prevailing environmental conditions (temperature—T, relative humidity—

RH%, decreased light intensity) inside the greenhouse mist system were 18 °C base T for plantlets root system, 15 °C air T for the vegetative part of the plants and RH ranging between 80% and 100% under conditions of thermal curtains. After 60 days period time in the mist for gradual acclimatization, on early March, the survival rate of the successfully undergo hardening plants was recorded. The following step included the transplantation of the plants into bigger pots of 0.33 L volume capacity (dimensions: 8 cm length × 8 cm width × 7 cm height) supplemented with peat moss (TS2, Klasmann) and perlite mixture substrate at 3:1 v/v, respectively. The newly transplanted plants were then transferred outside the indoor mist system in the bench of the same greenhouse without heating where the prevailing environmental conditions were scheduled to be 17–24 °C temperature and 55–70% RH range providing a progressively decrement of RH about 5% per day and a gradual increment of light intensity for a period of 3 months for further growth. In the subsequent stage during the onset of summer, particularly in early June due to increased temperatures inside the greenhouse, the vegetative and root system developed plants were transplanted into 2.5 L higher volume pots filled with peat (TS2), perlite and soil (2:1:½ v/v) and placed in the natural environment, in the outdoor nursery under 50% direct sun exposure provided by a shading net for ex situ conservation purposes and future sustainable exploitation.

4.3. In Vitro Culture of Different Plant Tissue Explants (Leaves, Petioles, Roots)—Indirect Organogenesis

Three different explant types were used: (1) leaves, (2) petioles, and (3) roots, cultured in MS medium with two different PGRs type and concentration combinations; ODK1: 5 μM IBA + 20 μM thidiazuron (TDZ), and ODK2: 5 μM 2,4-dichlorophenoxy acetic acid (2,4-D) + 0.5 μM Kin. These two media (ODK1 and ODK2) were also enriched with 30 g L^{-1} sucrose and 6 g L^{-1} Plant Agar. After two months, callus, shoot and root regeneration rates (%) were recorded.

In the following stage, the callus derived from the three explant types (leaves, petioles, roots) in ODK1 and ODK2 media were then transferred in four new PGRs-supplemented MS media for further regeneration (ODR0, ODR1, ODR2, and ODR3), all enriched with 20 g L^{-1} sucrose and 6 g L^{-1} Plant Agar. The composition of the four new regeneration media was: ODR0 (PGR's-free), ODR1: 0.25 μM IBA + 0.3 μM GA_3, ODR2: 1.1 μM BA + 0.25 μM IBA + 0.3 μM GA_3, and ODR3: 2.2 μM BA + 0.25 μM IBA + 0.3 μM GA_3. Aggregates of brittle and very soft texture sponge-like callus were used as experimental explants, cultured under 24 h darkness at 21–23 °C. After 10 months, the following data were recorded: new callus formation percentage (%): the number of initial callus explants with formation of new callus over initial ones/the total number of initial callus explants × 100%, shoot regeneration percentage (%): the number of callus explants with multiple shoot induction/the total number of callus explants × 100%, and root regeneration percentage (%): the number of callus explants with root formation/the total number of callus explants × 100%. Detailed chemical compositions of the different culture media tested for indirect in vitro organogenesis subsequent stages are presented in Table 3.

4.4. Statistical Analysis

All experiments were completely randomized, and data were analyzed with the use of the SPSS version 17.0 statistical package (SPSS Inc., Illinois, New York, USA). In Tables and figure graphs, the analysis of variance (one-way ANOVA) and the Duncan multiple-range test at a 5% level were deployed for mean values ± standard errors (S.E.) variability and comparison among treatments for each parameter per experiment conducted. In tables, means ± S.E. with different letters within a column are statistically significant different from each other according to the Duncan's multiple range test at $p \leq 0.05$ (Tables 1 and 2). In figures representing results data in the form of graphs, different bars per graph denoted by different letters indicate statistically significant differences among treatments ($p \leq 0.05$) and error bars in each bar per graph are standard errors. For each figure graph, a horizontal scale bar on the X-axis is provided in its respective caption in parenthesis (Figures 1a–h and 2c). Furthermore, in figures representing photo images and not graphs, a scale bar of 1 cm is provided on each multi-plate image (Figures 2a,b,d, 3a–h and 4a–g).

Table 3. Chemical composition of the different PGRs-supplemented MS basal culture media tested in this study for indirect in vitro organogenesis subsequent stages of *O. dictamnus* L.

PGRs [1] (μM)	Culture Medium Code Number					
	ODK1	ODK2	ODR0	ODR1	ODR2	ODR3
TDZ [2]	20	-	-	-	-	-
IBA [3]	5	-	-	0.25	0.25	0.25
Kin [4]	-	0.5	-	-	-	-
2,4-D [5]	-	5	-	-	-	-
BA [6]	-	-	-	-	1.11	2.22
GA_3 [7]	-	-	-	0.3	0.3	0.3

[1] Plant Growth Regulators (PGRs), [2] Thidiazuron (TDZ), [3] Indole-3-Butyric Acid (IBA), [4] Kinetin (Kin), [5] 2,4-Dichlorophenoxy Acetic Acid (2,4-D), [6] 6-Benzyladenine (BA), [7] Gibberellic acid (GA_3).

The micropropagation experiment included 14 treatments with 12 repetitions/treatment (four explants/vessel × three vessels/treatment), therefore it was a 2 × 3 × 3 factorial one with two culture media (MS, B5), three cytokinin types (BA, kin, 2-iP) and three cytokinin concentrations.

The first stage of the indirect organogenesis experiment (six treatments) was a 3 × 2 factorial one including three explant types (leaves, petioles, roots) and two culture media (ODK1, ODK2) with 25 explants per treatment and per explant type (5 groups × 5 explants/vessel, i.e., 25 leaves, 25 petioles and 25 roots).

The second stage of indirect organogenesis experiment was a 3 × 4 × 2 factorial one consisting of 24 treatments corresponding to three callus explant types (leaf-callus, petiole-callus and root-callus), four new tested culture media (ODR0, ODR1, ODR2, ODR3), and two media initially used (ODK1, ODK2) prior to callus formation, i.e., 3 groups (vessels) × 10 calluses per vessel, thus 30 leaf calluses, 30 petiole calluses and 30 root calluses per treatment.

5. Conclusions

An efficient in vitro indirect regeneration protocol of the Cretan dittany was carried out and is reported for the first time in the present study. The unravelling and accomplishments of direct and indirect regeneration potential of various plant tissues and organs of *O. dictamnus* are deployed and underlined. In particular, from the results, it was observed that MS medium provides a better response for micropropagation and ex vitro survival of *O. dictamnus* when compared with Gamborg B5. It is clearly illustrated that between different cytokinin types and concentrations tested, there is a strong inter-correlation and interaction. Among the 14 combination treatments tested for direct organogenesis, the 2.2 μM BA + 0.25 μM IBA + 0.3 μM GA_3 was the most cost- and time-effective one within a single-one stage (30 days). Among the explant types and the culture media tested, shoot regeneration occurred only by leaf- and petiole explants cultured in medium containing 20 μM TDZ + 5 μM IBA. Petiole callus had the best root regeneration performance followed by leaf callus, and leaf callus resulted in shoot or root regeneration. *O. dictamnus* callus can be utilized for obtaining active constituents, nevertheless, significant scope exists in establishing cell cultures for enhanced production of the desired metabolite in a controlled environment. Further studies could be carried out on the analysis of secondary metabolites produced in vitro. Selected clones with higher bioactive compounds content could satisfy pharmaceutical industry needs, reducing overexploitation of this range-restricted/threatened Greek endemic from nature for plant material sourcing.

Author Contributions: Conceptualization, K.G.; methodology, V.S and K.G.; software, V.S.; validation, V.S. and K.G.; formal analysis, V.S.; investigation, K.G.; resources, E.M. and K.G.; data curation, V.S.; writing—original draft preparation, V.S.; writing—review and editing, V.S. and K.G.; visualization, V.S.; supervision, E.M. and K.G.; project administration, K.G.; funding acquisition, K.G. All authors have read and agreed to the published version of the manuscript.

Funding: This work was supported (co-financed) by the European Regional Development Fund of the European Union and Greek national funds through the Operational Program Competitiveness, Entrepreneurship and Innovation, under the call RESEARCH—CREATE—INNOVATE (grand number: T2EΔK-02927, MIS 5069915) entitled "Development and optimization of in vitro culture methods in bioreactors to produce repeatable and excellent quality plant material for extracts used in food supplements and cosmetics" (Acronym: BIOREACT).

Institutional Review Board Statement: Not applicable.

Informed Consent Statement: Not applicable.

Data Availability Statement: Not applicable.

Conflicts of Interest: The authors declare no conflict of interest. The funders had no role in the design of the study; in the collection, analyses, or interpretation of data; in the writing of the manuscript; or in the decision to publish the results.

References

1. Dimopoulos, P.; Raus, T.; Bergmeier, E.; Constantinidis, T.; Iatrou, G.; Kokkini, S.; Strid, A.; Tzanoudakis, D. Vascular plants of Greece—An annotated checklist supplement. *Willd* **2016**, *46*, 301–347. [CrossRef]
2. Krigas, N.; Tsoktouridis, G.; Anestis, I.; Khabbach, A.; Libiad, M.; Megdiche-Ksouri, W.; Ghrabi-Gammar, Z.; Lamchouri, F.; Tsiripidis, I.; Tsiafouli, M. Exploring the potential of neglected local endemic plants of three Mediterranean regions in the ornamental sector: Value chain feasibility and readiness timescale for their sustainable exploitation. *Sustainability* **2021**, *13*, 2539. [CrossRef]
3. Grigoriadou, K.; Krigas, N.; Lazari, D.; Maloupa, E. Chapter 4—Sustainable use of Mediterranean medicinal-aromatic plants. In *Feed Additives: Aromatic Plants and Herbs in Animal Nutrition and Health*; Florou-Paneri, P.E., Christaki, E., Giannenas, I., Eds.; Elsevier Academic Press: London, UK, 2020; pp. 57–74. [CrossRef]
4. Maietta, M.; Colombo, R.; Corana, F.; Papetti, A. Cretan tea (*Origanum dictamnus* L.) as a functional beverage: An investigation on antiglycative and carbonyl trapping activities. *Food Funct.* **2018**, *9*, 1545–1556. [CrossRef] [PubMed]
5. Arslan, N.; Gurbuz, B.; Gumuscu, A.; Ozcan, S.; Mirici, S.; Khawar, K.M. Cultivation of *Sternbergia fischeriana* (Herbert) Rupr., and a study on its morphological characteristics. *Pak. J. Bot.* **2002**, *34*, 411–418.
6. El-Beyrouthy, M.; Elian, G.; AbouJaoudeh, C.; Chalak, L. In vitro propagation of *Origanum syriacum* and *Origanum ehrenbergii*. *Acta Hortic.* **2015**, *1083*, 169–172. [CrossRef]
7. Yildirim, M.U. Micropropagation of *Origanum acutidens* (HAND.-MAZZ.) Ietswaart using stem node explants. *Sci. World J.* **2013**, *2013*, 276464. [CrossRef]
8. Srivastava, S.; Srivastava, A.K. Hairy root culture for mass-production of high-value secondary metabolites. *Crit. Rev. Biotechnol.* **2007**, *27*, 29–43. [CrossRef]
9. Seo, S.G.; Ryu, S.H.; Zhou, Y.; Kim, S.H. Development of an efficient protocol for high-frequency regeneration system in *Hibiscus syriacus* L. *J. Plant Biotechnol.* **2017**, *44*, 164–170. [CrossRef]
10. Babaoğlu, M.; Gürel, E.; Özcan, S. *Bitki Biyoteknolojisi I. Doku Kültürü ve Uygulamaları*; Selçuk Üniversitesi Press: Basimevi, Turkey, 2001; 374p.
11. Murashige, T.; Skoog, F. A revised method for rapid growth and bioassays with tobacco tissue cultures. *Physiol. Plant* **1962**, *15*, 472–497. [CrossRef]
12. Gamborg, O.L.; Miller, R.A.; Ojima, K. Nutrient requirements of suspension cultures of soybean root cells. *Exp. Cell Res.* **1968**, *50*, 151–158. [CrossRef]
13. Sarropoulou, V.; Maloupa, E.; Grigoriadou, K. In vitro direct organogenesis of the Cretan dittany (*Origanum dictamnus* L.), an important threatened Greek endemic species. *Not. Bot. Horti Agrobot. Cluj-Napoca* **2022**, *50*, 12715. [CrossRef]
14. Sokolov, R.S.; Atanassova, B.Y.; Iakimova, E.T. Physiological response of in vitro cultured *Magnolia* sp. to nutrient medium composition. *J. Hortic. Res.* **2014**, *22*, 49–61. [CrossRef]
15. George, E.F. *Plant Propagation by Tissue Culture. Part 1. The Technology*; Exegetics Ltd.: Edington, UK, 1993.
16. Atar, H.; Çölgeçen, H. Regeneration in *Origanum onites* L. by plant tissue culture. *Karaelmas Fen ve Müh Derg* **2019**, *9*, 177–180. [CrossRef]
17. Abdallah, S.A.S.; Yakoup, M.Y.A.; Abdalla, M.Y.H. Micropropagation of oregano (*Origanum syriacum* L.) through tissue culture technique. *J. Plant Prod. Mansoura Univ.* **2017**, *8*, 635–639. [CrossRef]

18. Korkor, A.M.; Mohamed, S.A.; Abd El-kafie, O.M.; Gohar, A.A. Adaptation of the in vitro culture of *Origanum majorana* L. for production of phenolic acids. *IOSR J. Pharm. Biol. Sci.* **2017**, *12*, 30–38. [CrossRef]
19. Goleniowski, M.E.; Flamarique, C.; Bima, P. Micropropagation of oregano (*Origanum vulgare* × *aplii*) from meristem tips. *In Vitro Cell. Dev. Biol. Plant* **2003**, *39*, 125–128. [CrossRef]
20. Sajina, A.; Geetha, S.P.; Minoo, D.; Rema, J.; Nirmalbabu, K.; Sadanandan, A.K.; Ravindran, P.N. Micropropagation of some important herbal species. In *Biotechnology of Spices, Medicinal and Aromatic Plants*; Edison, S., Ramana, A.V., Sasikumar, B., Eds.; Indian Society for Spices: Calicut, India, 1997; pp. 79–86.
21. George, E.F.; Hall, M.A.; de Klerk, G.J. Plant tissue culture procedure—Background. In *Plant Propagation by Tissue Culture*; George, E.F., Hall, M.A., Klerk, G.J.D., Eds.; Springer: Dordrecht, The Netherlands, 2008; pp. 104–204.
22. Gübbük, H.; Pekmezci, M. In vitro propagation of some new banana types (*Musa* spp.). *Turk. J. Agric. For.* **2004**, *28*, 355–361.
23. Buah, J.N.; Danso, E.; Taah, K.J.; Abole, E.A.; Bediako, E.A.; Asiedu, J.; Baidoo, R. The effects of different concentration cytokinins on the in vitro multiplication of plantain (*Musa* sp.). *Biotechnology* **2010**, *9*, 343–347. [CrossRef]
24. Klemš, M.; Balla, J.; Machačkova, I.; Eder, J.; Prochazka, S. The uptake and metabolism of 3H-benzylaminopurine in tobacco (*Nicotiana tabacum* L.) and cucumber (*Cucumis sativus* L.) explants. *Plant Growth Regul.* **2000**, *31*, 135–142. [CrossRef]
25. Rahman, M.Z.; Sharoar, M.G.; Matin, M.N.; Rahman, M.H.; Rahman, M.M.; Islam, R. High frequency plant regeneration of a dessert banana cv. Mehersagar for commercial exploitation. *Biotechnology* **2006**, *5*, 296–300. [CrossRef]
26. Kalia, R.K.; Arya, S.; Kalia, S.; Arya, I.D. Plantlet regeneration from fascicular buds of seedling shoot apices of *Pinus roxburghii* Sarg. *Biol. Plant.* **2007**, *51*, 653–659. [CrossRef]
27. Sevindik, B.; Izgu, T.; Simsek, O.; Tutunku, M.; Curuk, P.; Yilmaz, O.; Kaynak, G.; Aka Cakar, Y.; Teixeira da Silva, J.A.; Mendi, Y.Y. In vitro culture of Turkish *Origanum sipyleum* L. *Am. J. Plant Biol.* **2017**, *2*, 32–36. [CrossRef]
28. Leelavathi, D.; Kuppan, N. Callus induction and regeneration of multiple shoots from in vitro apical bud explant of *Origanum vulgare*, an important medicinal plant. *Int. J. Res. Pharm. Chem.* **2013**, *3*, 898–903.
29. Özkum, D. In vitro shoot regeneration of oregano (*Origanum minutiflorum* O. Schwarz & Davis). *Hacet. J. Biol. Chem.* **2007**, *35*, 97–100.
30. Oluk, E.A.; Çakır, A. Micropropagation of *Origanum sipyleum* L., an endemic medicinal herb of Turkey. *Afr. J. Biotechnol.* **2009**, *8*, 5769–5772. [CrossRef]
31. Arafeh, R.M.; Mahmoud, M.S.; Shibli, R.A. In vitro seed propagation of wild Syrian marjoram (*Origanum syriacum* L.). *Adv. Hort. Sci.* **2003**, *17*, 241–244.
32. Zayova, E.G.; Geneva, M.P.; Miladinova-Georgieva, K.D.; Hristozkova, M.G.; Stancheva, I.V. Impact of plant growth regulators on Greek oregano micropropagation and antioxidant activity. *Biosci. Biotechnol. Res. Asia* **2019**, *16*, 297–305. [CrossRef]
33. Kumari, N.; Saradhi, P.P. Regeneration of plants from callus cultures of *Origanum vulgare* L. *Plant Cell Rep.* **1992**, *11*, 476–479. [CrossRef]
34. de Bona, C.M.; Biasi, L.A.; Deschamps, C.; Reinhart, V. In vitro rooting of *Lavandula angustifolia*. *Rev. Bras. Agrocienc. Pelotas* **2011**, *17*, 401–404.
35. Idris, T.I.M.; Hussein, F.A.; Osman, M.A. Rooting and acclimatization of in vitro produced ginger plantlets (*Zingiber officinale* Rose). *Sudan J. Agric. Sci.* **2015**, *2*, 28–34.
36. Resende, C.F.D.; Bianchetti, R.E.; Oliveira, A.M.S.D.; Braga, V.F.; Peixoto, P.H.P. 2015. In vitro propagation and acclimatization of *Lippia rotundifolia*, an endemic species of brazilian campos Rupestres. *Rev. Ciênc. Agron.* **2015**, *46*, 582–589. [CrossRef]
37. Silva, F.A.B.; Pereira, L.A.R.; Silveira, C.E.S. Micropropagation of *Alibertia edulis* Rich. *Braz. Arch. Biol. Technol.* **2008**, *51*, 1103–1114. [CrossRef]
38. Mercier, H.; Souza, B.M.; Kraus, J.E.; Hamasaki, R.M.; Sotta, B. Endogenous auxin and cytokinin contents associated with shoot formation in leaves of pineapple cultured in vitro. *Braz. J. Plant Physiol.* **2003**, *15*, 107–112. [CrossRef]
39. Kumar, R.; Pal, M. Growth and proliferation of bamboo seedlings influenced by various growth regulators. *J. Bamboo Rattan* **2004**, *3*, 91–97. [CrossRef]
40. Werner, T.; Matykav, N.; Strnad, M.; Schmülling, T. Regulation of plant growth by cytokinin. *Proc. Natl. Acid. Sci. USA* **2001**, *98*, 10487–10492. [CrossRef] [PubMed]
41. El-Ghamery, A.A.; Mousa, M.A. Investigation on the effect of benzyladenine on the germination, radicle growth and meristematic cells of *Nigella sativa* L. and *Allium cepa* L. *Ann. Agric. Sci.* **2017**, *62*, 11–21. [CrossRef]
42. Su, Y.H.; Liu, Y.B.; Zhang, X.S. Auxin–cytokinin interaction regulates meristem development. *Mol. Plant* **2011**, *4*, 616–625. [CrossRef] [PubMed]
43. Ghasemi Bezdi, K.; Ahmadi, A. *Cell and Tissue Biotechnology in Micropropagation and Plant Breeding*; Makhtoomgholi Faraghi Publications: Gorgan, Iran, 2010.
44. Affonso, V.R.; Bizzo, H.R.; Lima, S.S.; Esquibel, M.A.; Sato, A. Solid Phase Microextraction (SPME) analysis of volatile compounds produced by in vitro shoots of *Lantana camara* L. under the influence of auxins and cytokinins. *J. Braz. Chem. Soc.* **2007**, *18*, 1504–1508. [CrossRef]
45. El-Gengaihi, S.; Taha, H.S.; Kamel, A.M. In vivo and in vitro comparative studies of *Origanum* species. *J. Food Agric. Environ.* **2006**, *4*, 127–134. [CrossRef]
46. Arafeh, R.M.; Shibli, R.A.; Al-Mahmoud, M.; Shatnawi, M.A. Callusing, cell suspension culture and secondary metabolites production in Persian oregano (*Origanum vulgare* L.) and Arabian oregano (*O. syriacum* L.). *Jordan J. Agric. Sci.* **2006**, *2*, 274–281.

47. Yasser, H.; Gehan, A.; El-Sayed, H.; Khaled, Y. In vitro cultivation of marjoram (*Origanum majorana* L.) under influence of 2,4-D (2,4-dichlorophenoxy acetic acid) as herbicide. *Life Sci. J.* **2014**, *11*, 249–257.
48. Ślusarkiewicz-Jarzina, A.; Ponitka, A.; Kaczmarek, Z. Influence of cultivar, explant source and plant growth regulator on callus induction and plant regeneration of *Cannabis sativa* L. *Acta Biol. Crac. Ser. Bot.* **2005**, *47*, 145–151.
49. Passey, A.; Barrett, K.; James, D. Adventitious shoot regeneration from seven commercial strawberry cultivars (*Fragaria × ananassa* Duch.) using a range of explant types. *Plant Cell Rep.* **2003**, *21*, 397–401. [CrossRef] [PubMed]
50. Murthy, B.N.S.; Murch, S.J.; Saxena, P.K. Thidiazuron: A potent regulator of in vitro plant morphogenesis. *In Vitro Cell Dev. Biol. Plant* **1998**, *4*, 267–275. [CrossRef]
51. Guo, B.; Abbasi, B.H.; Zeb, A.; Xu, L.L.; Wei, Y.H. Thidiazuron: A multi-dimensional plant growth regulator. *Afr. J. Biotechnol.* **2011**, *10*, 8984–9000. [CrossRef]
52. Podwyszynska, M.; Novák, O.; Doležal, K.; Strnad, M. Endogenous cytokinin dynamics in micropropagated tulips during bulb formation process influenced by TDZ and 2iP pre-treatment. *Plant Cell Tiss Organ Cult* **2014**, *9*, 331–346. [CrossRef]
53. Yu, K.W.; Murthy, H.N.; Jeong, C.S.; Hahn, E.J.; Paek, K.Y. Organic germanium stimulates the growth of ginseng adventitious roots and ginsenoside production. *Process Biochem.* **2005**, *40*, 2959–2961. [CrossRef]
54. El-Shafey, N.M.; Sayed, M.; Ahmed, E.S.; Hammouda, O.; Khodary, S.E.A. Effect of growth regulators on micropropagation, callus induction and callus flavonoid content of *Rumex pictus* Forssk. *Egypt. J. Bot.* **2019**, *59*, 269–278. [CrossRef]
55. Vardja, R.; Vardja, T. The effect of cytokinin type and concentration and the number of subcultures on the multiplication rate of some decorative plants. *Proc. Est. Acad. Sci. Biol. Ecol.* **2001**, *50*, 22–32. [CrossRef]
56. Horn, W.A.H. Micropropagation of rose (*Rosa* L.). In *HighTech and Micropropagation IV*; Bajaj, Y.P.S., Ed.; Springer: Berlin, Germany, 1992; pp. 320–342.
57. Krens, F.A.; Menzel, T.R.; Liu, C.; Dees, D.C.T.; Van Kronenburg, B.C.E. Oriental lily hybrids engineered to resist aphid attack. *Acta Hortic.* **2009**, *836*, 253–258. [CrossRef]
58. Zobayed, S.M.A.; Saxena, P.K. In vitro-grown roots: A superior explant for prolific shoot regeneration of St. John's wort (*Hypericum* perforatum L. cv. "New Stem") in a temporary immersion bioreactor. *Plant Sci.* **2003**, *165*, 463–470. [CrossRef]
59. Ehsandar, S.; Majd, A.; Choukan, R. Callus formation and regeneration of the first modified Iranian potato cultivar (Savalan). *Adv. Crop Sci.* **2013**, *3*, 201–208.
60. Dhital, S.P.; Lim, H.T.; Manandhar, H.K. Direct and efficient plant regeneration from different explant sources of potato cultivars as influenced by plant growth regulators. *Nepal J. Sci. Technol.* **2010**, *12*, 1–6. [CrossRef]
61. Dodds, J.; Roberts, L. *Experiments in Plant Tissue Culture*, 3rd ed.; Cambridge University Press: New York, NY, USA, 1995; p. 276.
62. Lim, K.B.; Kwon, S.J.; Lee, S.I.; Hwang, Y.J.; Naing, A.H. Influence of genotype, explants source, and gelling agent on in vitro shoot regeneration of chrysanthemum. *Hortic. Environ. Biotechnol.* **2012**, *53*, 329–335. [CrossRef]
63. Samanthi, P.W.; Puad, A.M.; Suhaimi, N.; Kumar, S.M.; Noraini, N.S. In vitro shoot regeneration from leaf explants of kenaf. *Ann. Plant Sci.* **2013**, *3*, 582–587.
64. Ji, H.; Zhang, H.X.; Ge, H.B. The relationship between browning ratio in PPO and phenols of pear explants. *J. Sichuan Agric. Univ.* **1998**, *16*, 310–313.
65. Arnaldos, T.L.; Munoz, R.; Ferrer, M.A.; Calderon, A.A. Changes in phenol content during strawberry (*Fragaria ananasa* cv. Chandler) callus culture. *Physiol. Plant* **2001**, *113*, 315–322. [CrossRef]

plants

MDPI

Review

In Vitro Conservation through Slow Growth Storage Technique of Fruit Species: An Overview of the Last 10 Years

Carla Benelli, Waed Tarraf *, Tolga Izgu and Anna De Carlo

Institute of BioEconomy, National Research Council (CNR/IBE), Sesto Fiorentino, 50019 Florence, Italy
* Correspondence: waed.tarraf@ibe.cnr.it

Abstract: Plant genetic resources conservation may be a potential option for the improvement of agricultural crops through modern biotechnologies, and in vitro conservation is a tool available to safeguard plant biodiversity. Ex situ conservation of plant genetic resources using the in vitro procedures is in progress in many countries. The slow growth storage (SGS) technique is a valid in vitro approach to preserve several vegetatively propagated species by controlling the growth and development of plantlets, economizing storage space and labor and reducing costs. Moreover, SGS prolongs the timing between subcultures, lowers the risk of losing germplasm through handling errors, such as contamination problems, and decreases the risk of genetic instability due to the reduction in the number of subcultures. SGS is applied by considering different factors: temperature, light or darkness conditions, medium composition, including mineral or sucrose concentrations, and the presence/absence of plant growth regulators, osmotic agents and growth inhibitors. SGS protocols for some fruit species have been well defined, others require additional research. The present review focuses on the effect of several factors that influence the SGS of in vitro shoots derived from temperate and tropical fruit species during the last ten years.

Keywords: ex situ conservation; minimal growth storage; in vitro banking; storage culture conditions; shoot culture; temperate and tropical species

Citation: Benelli, C.; Tarraf, W.; Izgu, T.; De Carlo, A. In Vitro Conservation through Slow Growth Storage Technique of Fruit Species: An Overview of the Last 10 Years. *Plants* **2022**, *11*, 3188. https://doi.org/10.3390/plants11233188

Academic Editor: Andreas W. Ebert

Received: 22 September 2022
Accepted: 19 November 2022
Published: 22 November 2022

1. Introduction

Nowadays, plant biotechnology offers important options for the collection, molecular characterization, pathogen indexing and elimination, propagation, preservation and exchange of disease-free plant genetic resources. In particular, in vitro techniques can provide a potential contribution to overcome some of the issues related to plant genetic resources preservation [1].

Seed banking is the most efficient method of ex situ preservation [2], but it is restricted for some species that are characterized by the null/limited production of seeds, recalcitrant seed or low germination. The traditional method of preserving vegetatively propagated species is the maintenance of clonal field collections which may include a large number of accessions representing a wide range of genetic diversity [3,4]. However, the cuttings do not always suffice or do not respond adequately as propagules for propagation. Moreover, this method is costly and at constant risk of serious losses because of biotic and abiotic stresses. In the past, the Sharka virus on plums [5,6] and, recently, the development of *Xylella* emergency on olive trees [7] are examples of biotic stresses that highlight the need for complementary ex situ conservation strategies, such as slow growth storage. In vitro shoot cultures are used for the medium-term conservation of plant germplasm; indeed, well-established shoot cultures can have their growth and development in vitro slowed by modified culture conditions that affect the normal metabolism.

The technique is generally named "slow growth storage" (SGS) or "minimal growth storage" due to the use of different physical, chemical or nutritional parameters that limit the growth of the plantlets. It may also be called "cold storage" when low temperatures are applied instead of standard growth conditions.

Basically, SGS prolongs the timing between subcultures with respect to the regular intervals at 3–5 weeks depending on the species, enhancing the conservation safety as a result of fewer interferences with the culture system and minimizing the risk of contamination during the subculture process. In vitro plants of many different species may tolerate the same standard SGS conditions. However, there are most likely other species that may require species-specific conditions. Therefore, for under-researched species, each physical or chemical factor may need to be assessed.

Furthermore, SGS is applied in commercial micropropagation laboratories as a suitable strategy for short- to medium-term storage of plant materials in limited space, offering the market high-quality produce of commercial cultivars at a reduced cost.

SGS can take advantage of investigations on the effects of plant growth regulators and growth retardants, quality and quantity of light, temperature and light interactions, propagule type and growth stage. In the past, a few reviews [8,9] and several book chapters [4,10–14] described SGS conditions for different species, while this paper presents an overview of SGS applications for shoot conservation of temperate and tropical fruit species during the last ten years.

2. Factors Affecting SGS of Shoot Cultures of Temperate Fruit Crops

Several factors can interfere with the normal growth and quality of in vitro shoots, such as the temperature applied, the presence or absence of light and its intensity, the medium composition (e.g., macroelements, carbohydrates, plant growth regulators, osmotically active substances, growth retardants or antioxidant compounds) and the characteristics of the storage containers. All these factors can influence in vitro growth of plantlets at different degrees; they can also have synergetic effects. The interplay of all elements will determine the maximum conservation time in vitro, which may differ from species to species and frequently among cultivars of the same species.

In temperate fruit plants, as a general rule, a good protocol of conservation can lead to a long conservation time ranging from a few months to more than 4 years (Table 1) maintaining the viability and potentiality of shoots to regrow under standard culture conditions.

Table 1. Shoot conservation of temperate fruit crops in SGS from 2012 to present. Culture conditions and best results are reported for each species (terminology and values are the same as mentioned by the authors).

Species	Medium	Temperature (°C)	Light Condition	Storage Time (Months)	Survival (%)	References
Arbutus unedo	MS, 1 mg L^{-1} zeatin	18	16 h, 30 $\mu E\ s^{-1}\ m^{-2}$	6	80	[15]
Castanea sativa	WPM, 0.44 µM BA, 30 g L^{-1} sucrose	8	16 h, 30 $\mu M\ m^{-2}\ s^{-1}$	48	82	[16]
Ceratonia siliqua	MS, 0.1 mg L^{-1} BA	18	16 h, 30 $\mu E\ s^{-1}\ m^{-2}$	6	100	[15]
Citrus jambhiri	WPM, 25 g L^{-1} sucrose	22	12 h, 20 $\mu mol\ m^{-2}\ s^{-1}$	12	NR	[17]
Crataegus monogyna	$\frac{1}{2}$MS, 5 mM BA, 0.5 mM IBA, 20 g L^{-1} sucrose	4	Darkness	7–12	99	[18]
Cydonia oblonga	$\frac{1}{2}$MS, 5 mM BA, 0.5 mM IBA, 20 g L^{-1} sucrose	4	Darkness	7–12	99	[18]
Eriobotrya japonica	$\frac{1}{2}$MS, 5 mM BA, 0.5 mM IBA, 20 g L^{-1} sucrose	4	Darkness	7–12	99	[18]
Ficus carica	$\frac{1}{2}$MS, 5 mM BA, 0.5 mM IBA, 20 g L^{-1} sucrose	4	Darkness	7–12	99	[18]
Fragaria x ananassa	MS, 1 mg L^{-1} BA, 30 g L^{-1} sucrose	4	16 h, 40 $\mu mol\ m^{-2}\ s^{-1}$	7	32	[19]

Table 1. *Cont.*

Species	Medium	Temperature (°C)	Light Condition	Storage Time (Months)	Survival (%)	References
Fragaria spp.	MS	4	10 h, 10 µmol m^{-2} s^{-1}	15–18	NR	[20]
	Knop medium	4	Darkness	15	100	[21]
Malus domestica	$\frac{1}{2}$MS, 5 mM BA, 0.5 mM IBA, 20 g L^{-1} sucrose	4	Darkness	7–12	99	[18]
	MS, 1 mg L^{-1} BA, 30 g L^{-1} sucrose	4	16 h, 40 µmol m^{-2} s^{-1}	7	90	[19]
	MS, 0.44 µM BA, 130.5 mM sucrose	4	Darkness	6	NR	[22]
Malus spp.	MS (25–50% NO_3), 2% sucrose + 2% mannitol	4	10 h, 10 µmol m^{-2} s^{-1}	≥36	NR	[20]
	MS, 3% sucrose	4	10 h, 15 µmol m^{-2} s^{-1}	18–20	NR	[23]
Pistacia lentiscus	MS, 1 mg L^{-1} BA, 3% sucrose	4	Darkness	12	NR	[24]
Prunus avium	MS, 1 mg L^{-1} BA, 30 g L^{-1} sucrose	4	16 h, 40 µmol m^{-2} s^{-1}	7	79.4	[19]
Prunus avium × *P. cerasus*	MS, 1 mg L^{-1} BA, 30 g L^{-1} sucrose	4	16 h, 40 µmol m^{-2} s^{-1}	7	92	[19]
	DKW, 0.5 mg L^{-1} BA, 45 or 60 g L^{-1} sucrose	4	Darkness	16	NR	[25]
Prunus mahaleb	$\frac{1}{2}$MS media without sucrose	25	16 h, 43.4 µmol m^{-2} s^{-1}	4	74.1	[26]
Prunus spp.	$\frac{1}{2}$MS, 5 mM BA, 0.5 mM IBA, 20 g L^{-1} sucrose	4	Darkness	7–12	98.6	[18]
	MS, 0.5 mg L^{-1} BA, 0.1 mg L^{-1} IBA, 2% sucrose + 2% mannitol	4	10 h, 10 µmol m^{-2} s^{-1}	30	NR	[20]
	MS, 2.2 µM BA, 0.49 µM IBA, 20 g L^{-1} sucrose	4	Darkness	12	100	[27]
Prunus webbii	MS, 0.7 mg L^{-1} BA, 0.01 mg L^{-1} NAA, 0.1 mg L^{-1} GA_3, 3% sucrose	4	Darkness	6	42.6	[28]
Punica granatum	$\frac{1}{2}$MS, 5 mM BA, 0.5 mM IBA, 20 g L^{-1} sucrose	4	Darkness	7–12	99	[18]
Pyrus communis	$\frac{1}{2}$MS, 5 mM BA, 0.5 mM IBA, 20 g L^{-1} sucrose	4	Darkness	7–12	99	[18]
	MS, 1 mg L^{-1} BA, 30 g L^{-1} sucrose	4	16 h, 40 µmol m^{-2} s^{-1}	7	91	[19]
Pyrus spp.	MS, 30 g L^{-1} sucrose	4	Darkness	6	100	[21]
	MS, 3% sucrose, 0.5 mg L^{-1} BA, 0.1 mg L^{-1} IBA/without PGRs	4	10 h, 7 µmol m^{-2} s^{-1}	18/15	NR	[29]
	MS (25% NO_3), 2% sucrose, 2% mannitol	4	10 h, 10 µmol m^{-2} s^{-1}	36	NR	[20]
	$\frac{1}{2}$MS nitrogen	1-4	12 h, (10–20 µE m^{-2} s^{-1})/dark	12–48	NR	[30]
	MS	4	12 h, 10 µE m^{-2} s^{-1}	48	NR	[31]
Ribes nigrum	MS, 0.5 mg L^{-1} BA, 0.1 mg L^{-1} IBA, 2% sucrose + 2% mannitol	4	10 h, 10 µmol m^{-2} s^{-1}	18	NR	[20]

Table 1. *Cont.*

Species	Medium	Temperature (°C)	Light Condition	Storage Time (Months)	Survival (%)	References
Rubus spp.	MS, 0.5 mg L^{-1} BA, 0.1 mg L^{-1} IBA, 3% sucrose	4	10 h, 10 μmol m^{-2} s^{-1}	15	NR	[20]
Vaccinium myrtillus	MS, 30 g L^{-1} sucrose	4	Darkness	6	90	[32]
Vitis heyneana	MS, 0.05 mg L^{-1} IBA, 0.1 mg L^{-1} IAA, 0.5 mg L^{-1} ABA, 10 g L^{-1} mannitol	10	16 h, 40 μmol m^{-2} s^{-1}	12	47.8	[33]
Vitis vinifera	$\frac{3}{4}$ MS, 5.5% sorbitol	5	Darkness	12	88.9	[34]
	MS, 300 μM ribose	15	16 h, 3000 Lux	12	73	[35]
Ziziphus jujuba	MS, 1 mg L^{-1} BA, 0.05 mg L^{-1} IBA, 3% sucrose	4	Darkness	10	78.6	[36]

NR: not reported. MS, Murashige and Skoog medium; WPM, woody plant medium; DKW, Driver and Kuniyuki walnut medium; ABA, abscisic acid; BA, 6-Benzyladenine; GA_3, gibberellic acid; IAA, indole-3-acetic acid; IBA, indole-3-butyric acid; NAA, 1-Naphthaleneacetic acid; PGRs, plant growth regulators.

The shoot size or the number of nodes used in SGS depends on the species/genotypes and quite often also varies with the practice adopted in a specific tissue culture laboratory. Commonly, in fruit species, shoots of 2–4 cm length [14] derived from healthy in vitro cultures are important to start SGS [37]. Generally, shoots subcultured for no more than 6 cycles starting from the in vitro establishment are preferred, while shoots coming from over 12 cycles (~one year) are less suitable for SGS. Moreover, the selected shoots must be in healthy conditions [31], showing no signs of physiological disorder such as hyperhydricity, apex necrosis, chlorophyll degradation, browning or decay. For example, shoot cultures with bacterial infection or even with primary symptoms of hyperhydricity must be avoided, especially when low-temperature conditions are applied. Indeed, the thermal shock might stimulate the growth of latent bacteria, subsequently leading to substantial contamination, either during the slow growth conservation period or after regrowth and development under standard growth conditions.

The main factors determining the storage period and quality of preserved shoots will be analyzed below.

2.1. Temperature and Light Conditions

The most commonly used method for reducing growth in SGS is the 'cold storage' of shoot cultures. The incubation at a temperature lower than that required for optimum growth will reduce the metabolic activities, such as respiration, water loss, wilting and ethylene production. Reduced metabolic activities, in turn, will ensure the secure preservation of shoot cultures, resulting in the restricted growth of the plantlets [13].

Although the chosen temperature usually depends on the sensitivity of the species, it is reported that the suitable storage temperature ranges from 2 to 5 °C for temperate fruit species, while for tropical and subtropical species, from 10 to 15 °C [38].

Cold storage is often combined with the reduction of light intensity or total darkness. Most of the stored cultures are maintained under dark conditions, even though several studies have demonstrated that different combinations of photoperiod and light intensity lead to better SGS results (Table 1). The storage of shoots in total darkness is mainly used by commercial tissue culture laboratories, given the low costs required to equip a storage chamber.

Most of the species reported are stored at 4 °C (Table 1). Eleven fruit species were maintained at 4 °C in the dark for at least 12 months with the highest survival rate (100%) in *Prunus* spp. [27] and *Fragaria* spp. [21].

Vaccinium myrtillus shoots showed the highest survival rate (90%) and recovery rate (80%) after cold storage at 4 °C in darkness for up to 6 months [32]. Arbeloa et al. [18]

conducted the same cold storage protocol (4 °C and darkness) for 18 fruit species (*Crataegus, Cydonia, Eriobotrya, Ficus, Malus, Prunus, Punica* and *Pyrus*). All species showed a survival rate of ~99% when the storage period ranged from 7 to 12 months. After 7 months of SGS, the multiplication rate (number of shoots/shoot) in *Crataegus* was 13, while in *Prunus* spp. and *Malus* it was 6 and 4, respectively, compared to other species with lower values.

Other storage temperatures are also reported, such as *Vitis* spp. stored at 10 °C [33] and 15 °C [35] for 1 year, *Arbutus unedo* and *Ceratonia sativa* at 18 °C for 6 months [15], *Citrus jambhiri* at 22 °C for 1 year [17] and *Prunus mahaleb* at 25 °C for 4 months [26].

After the storage in darkness, to retrieve the shoots and to overcome the visible elongation and etiolation of the shoots during the storage period, it is necessary to continue the subcultures under standard growth conditions [39]. In this context, different combinations of photoperiod and light intensity were more effective than total darkness when aiming at producing high-quality shoots with a fast recovery rate during the post-conservation period.

Various temperate fruit species have been investigated in storage under a combination of low temperature and low light intensity. The shoots have been stored in photoperiods ranging from 10h to 16h and light intensity from 25 to 40 μmol m^{-2} s^{-1}, and the response of the species to these conditions was quite different: from 50% to 100% in terms of survival after SGS (Table 1).

In *Castanea sativa*, the effects of light and temperature were evaluated under the SGS [16]. Particularly, a dark condition was compared with a reduced light intensity (30 μmol m^{-2} s^{-1}) using two storage temperatures of 8 and 4 °C. The application of a low light level and temperature of 8 °C produced positive results over long preservation periods: 82% of chestnut shoots survived after 48 months of storage. At 4 °C, the survival of shoots declined dramatically, reaching approximately 56% after 12 months, and no plants could be recovered after 24 months of storage. Higher survival rates (over 90%) were obtained with shoots of *Prunus avium* × *P. cerasus* (Gisela®5) after storage at 4 °C in standard growth light conditions (intensity 40 μmol m^{-2} s^{-1} and photoperiod 16 h) [19]. In another study on *Vitis vinifera*, the application of 3000 lux (55 μmol m^{-2} s^{-1}) as light intensity combined with higher storage temperature (15 °C) allowed the maintenance of shoots for 12 months [35].

2.2. Storage Medium Composition

The components of nutrient media have a great influence on increasing the interval period between subcultures in vitro SGS. Generally, the concentration of carbohydrates, minerals, growth regulators or osmotic agents are modified in the culture medium to reduce cell division, and therefore limit the development of shoots and the formation of a callus [9].

Usually, the shoot cultures of many plants are stored in the same medium composition (macroelements, microelements and organics) used for proliferation in standard culture conditions. Among the media, the Murashige and Skoog (MS [40]) is still today the most commonly applied, although other basal micropropagation media, such as the woody plant medium (WPM [41]) and Driver and Kuniyuki walnut medium (DKW [42]), are also utilized (Table 1). These media can be applied with full or reduced strength (concentration) of their salts. In addition, specific nutrient media with special formulations are also used such as the olive medium (OM [43]) for the storage of olive shoot cultures in dark at a low temperature (4 °C) [44,45], or Knop medium [46] for the *Fragaria* in vitro storage [21].

Shoots from *Prunus, Punica, Ficus, Cydonia, Pyrus, Malus, Eriobotrya* and *Crataegus* species were successfully stored for 7 months, on MS or half-strength MS with a reduced concentration of sucrose (2%, instead of 3%) [18]. The authors noted higher multiplication rates mainly in $\frac{1}{2}$MS than in full-strength MS, most likely related to the nutritive or osmotic effects of reducing sugar and mineral nutrients. Recently, the in vitro conservation of *Citrus jambhiri* cv. 'Florida Rough' lemon for up to 12 months was achieved using full WPM compared to other concentrations ($\frac{1}{2}$ and $\frac{1}{4}$ WPM) with 25 g L^{-1} of sucrose [17].

The effectiveness of plant growth regulators (PGR) supplied in culture media during in vitro preservation is widely discussed in the literature [20,29,31]. SGS makes use of media containing PGR and PGR-free media. As shown in Table 1, the best results are obtained with the presence of hormones in the media. The hormone most widely used is 6-Benzylaminopurine (BA), either alone or in combination with indole-3-butyric acid (IBA), 1-naphthaleneacetic acid (NAA) and/or gibberellic acid (GA_3), at different concentrations ranging from 0.05 mg L^{-1} to 1.0 mg L^{-1} (Table 1). For instance, Capuana and Di Lonardo [16] investigated the presence or the absence of BA in WPM under light conditions: the survival of *Castanea sativa* shoots was 100% and 56%, with the hormone, compared to 76 and 32%, without PGR, after 4 and 12 months, respectively. In the same study, the total darkness with 0.44 μM of BA led to the highest survival rate (82%) after 48 months. In *Ziziphus jujuba* preservation, the absence of phytohormones in the growth media was optimal only for 3 months with 82.3% of survival, while for increasing the storage period to 10 months, with a survival rate of 78.6%, adding 1 mg L^{-1} BA and 0.05 mg L^{-1} IBA to the medium was required [36].

Optimizing the medium composition was a decisive factor in extending the preservation period of strawberry genotypes. The application of MS medium supplemented with 1 mg L^{-1} BA for SGS of *Fragaria* × *ananassa* [18] allowed the survival of only 20–32% of shoots for a duration of 7 months, while a period of 15 months was achieved on hormone-free Knop's medium [21] and 15–18 months on hormone-free MS medium [20].

Both carbon sources and growth regulators affected slow growth storage of *Malus* spp. The most prolonged conservation duration of ≥36 months was obtained on PGR-free media containing sucrose and mannitol (2% each). SGS with media containing PGRs reduced the storage period to 12–18 months only [19]. In the same study, the reduction of nitrogen concentration (25–50%) increased the storage period. In contrast, the study by Kabylbekova et al. [23] on apple shoots has shown that the presence or absence of PGR in the medium did not affect the storage duration. MS medium supplemented with 3% sucrose was the most favorable medium to store apple shoots in vitro for 18–20 months. Furthermore, it has been shown that apple cultures can be stored in vitro for seven months with a 90% survival rate by slowing down their growth on MS medium with 1 mg L^{-1} BA and 30 g L^{-1} sucrose [19]. The in vitro conservation of *Prunus avium* shoots was investigated by Sedlak et al. [19] using the same multiplication medium, reporting 79.4% of survival after 7 months of storage without subculture. Later, Turdiyev et al. [20] in their study on SGS of cherries obtained the longest duration (30 months) when the cultures preserved on MS contained either sucrose alone (3%) or 2% combined with 2% mannitol, in the presence of PGRs. Furthermore, Turdiyev et al. [20] observed that MS medium supplemented with mannitol only reduced the SGS period of cherries to 12 months. A previous study on the response of in vitro cultures of *Z. jujuba* on half-strength MS medium supplemented with sucrose demonstrated the possibility to conserve shoots of this crop for ten months with a survival rate of 78.6%, while the absence of sucrose in the medium reduced the survival rate to 63% for the same conservation period [36].

The same authors investigated the effect of reduced sucrose and MS salts concentrations without PGRs for the in vitro preservation protocol of *Prunus mahaleb*. In $\frac{1}{2}$MS media without sucrose, the shoots survival was 93.4% after only 3 months of conservation and 74.1% after 4 months [26]. The media composition in *Prunus domestica* and *Prunus cerasifera* did not affect the survival percentage during the cold preservation, even if several concentrations of sucrose, BAP and IBA were tested in MS-based media [27]. The in vitro conservation via SGS of *Prunus webbii* succeeded on the same medium used during proliferation phase, reporting a maximal survival rate after 3 months, while increasing the storage period to 6 and 10 months resulted in a significant decline with 42.6 and 15.6%, respectively [28].

Considering *Pyrus* spp. preservation, MS medium is mostly used for SGS. Lukoševičiūtė et al. [21] reported successful storage at 4 °C for 6 months with 75–100% of shoots surviving after SGS on standard MS media plus 3% sucrose, regardless of BAP presence, versus 23–46% on MS containing 2% mannitol. Further, Sedlak et al. [19] indicated that pear shoots could be stored for 7 months on MS medium supplemented with 3% sucrose and 1 mg L^{-1} BAP, with a survival rate of 91%; Kovalchuk et al. [29] reported pear germplasm storage on MS medium without PGRs for 6–15 months, while MS medium supplemented with PGRs extended the storage period to 9–18 months. An increase in the storage duration of *Pyrus* spp. was obtained by Turdiyev et al. [20], using MS medium with a lower nitrogen content (50%), 2% sucrose and 2% mannitol in the presence of PGRs. Reed et al. [30] proved that pear shoots can be cold-stored at 1–4 °C for several years (1–4) on a specific cold storage medium of MS with half-strength nitrogen, but without PGRs. The cold storage period was less than 2 years for *P. pyrifolia* and 4 years for *P. gharbiana*, while wild *P. communis* could be stored for approximately 2 years [31]. For the cold storage of *Rubus idaeus*, MS medium supplemented with 3% sucrose, 0.5 mg L^{-1} BAP and 0.1 mg L^{-1} IBA ensured the longest conservation time; the first dead plants were noted after 15 months of storage [20]. In the same study, the storage on media with 3% sucrose without PGRs reduced the conservation period to only 12 months while adding mannitol at different concentrations, as well as its combination with sucrose, shortened the duration to 3 months with no good quality of preserved plants. Hormone-free MS medium was efficient to maintain *Vaccinium myrtillus* under slow growth storage for 6 months, recording high survival (90%) and the shoots were of good quality showing no browning symptoms [32]. In addition, SGS of *Ribes nigrum* was prolonged up to 18 months when the shoot cultures were conserved on MS medium, supplemented with PGRs (BA and IBA), 2% sucrose and 2% mannitol compared to only 12 months on the same medium but without the addition of PGRs [20].

Besides temperature and osmotic agents for short- to medium–term storage, growth retardants are also used for in vitro germplasm conservation such as abscisic acid (ABA) and Alar [47]. ABA has different physiological effects on plants, such as the inhibition of plant elongation and cell division [48] and the accumulation of proteins, leading to an increase in the resistance to stress associated with the storage conditions [49]; therefore, it is considered a plant growth inhibitor in in vitro conservation.

Indeed, 1 mg L^{-1} ABA added to the storage medium of pear genotypes increased the duration of storage for up to 36 months in 'Mramornaya' compared to 18 months for the control or 0.5 mg L^{-1} ABA [29]. The effect of ABA (0.5 mg L^{-1}) was also evaluated by Pan et al. [33] in *Vitis heyneana* with low survival rates (26% and 23%) when the shoots were preserved at 10 °C and at 25 °C, respectively, for up to 10 months. Pre-treatments of chestnut shoots with different concentrations of ABA did not significantly influence their survival rate; an inhibitory effect on shoot proliferation was only observed at the highest ABA concentration [16].

Alar was suitable to maintain *V. vinifera* shoot tips in SGS for 12 months, with a complete regrowth ability after this storage period. The highest survival percentage (100%) was obtained with 0.4 mg L^{-1} Alar at 6 months and then declined (80 and 60%) as the duration of storage was increased to 9 and 12 months [35]. However, to improve the survival rate up to 73% after 12 months of storage, it was necessary to use MS medium containing 300 µM of ribose, without Alar.

Although the growth retardants can prolong the subculture interval, some of them may result in mutation due to their mutagenic properties and cause physiological problems if used for a longer time [50]; therefore, careful monitoring is necessary in that case.

The carbohydrates are considered a significant component for the storage medium since the use of osmotically active substances can lower the osmotic potential of the substrate, modify the growth of the shoots and affect the storage time [51].

Mainly, sucrose, mannitol, sorbitol and ribose were found to be effective in extending the storage life of in vitro grown tissues [52]. Ozudogru et al. [25] demonstrated that increasing the sucrose concentration to 45 or 60 g L^{-1} maintained the in vitro shoots of cherry rootstock 'Gisela®5' for up to 16 months, while 30 g L^{-1} sucrose-containing medium allowed the conservation of the shoot cultures for a duration of only 9 months due to the loss of material caused by hyperhydricity or decay. Monticelli et al. [22] also proved the importance of sucrose concentrations in prolonging the conservation period of apple shoots up to six months. Indeed, shoots stored on medium supplied with 130.5 mM sucrose (~44.7 g L^{-1}) showed less necrosis and the highest multiplication rates after a subculture in standard growth conditions compared to those preserved on 87 mM sucrose (~30 g L^{-1}). During the minimal growth storage of grapevine shoots, both osmotic agents and ribose and sucrose diversely affected the survival rate. The osmotic component showed the following effects: ribose at 300 μM (45 g L^{-1}) gave the highest survival percentage (73%) after 12 months under 15 °C, followed by 53% and 47% obtained from cultures stored on media containing 200 or 100 μM (30 and 15 g L^{-1}), respectively. Raising sucrose concentrations from 100 to 300 μM (from ~35 to ~103 g L^{-1}) decreased the survival percentage while no survival was obtained on medium supplied with 300 μM of sucrose [35]. Further, different osmotic substances at various concentrations in $\frac{3}{4}$MS medium without growth regulators indicated a gradual decrease in the survival rate of *Vitis vinifera* shoots as the preservation period increased. After 12 months, good percentages (77.7%) of green and healthy explants were noted on media both with 3.5 or 4.5% glucose and 2.5% mannitol, but the highest percentage (88.9%) was obtained with 5.5% sorbitol [34]. The use of mannitol as an osmotic agent extended the storage time of *Vitis heyneana* under slow growth conditions, when compared with sucrose. The best survival (47.8%) was obtained at the end of 12 months of storage on MS medium supplemented with 10 g L^{-1} mannitol compared to the 46.7% reached after only 10 months of conservation on MS medium containing 40 g L^{-1} sucrose [33].

2.3. Containers for SGS

The type of containers used to maximize the storage duration of shoot cultures is another factor that should be considered before carrying out SGS experiments [53–55]. Several culture containers are available in different shapes, materials, volume and gas-permeability. In most papers reported in Table 1, glass jars or tubes were used, while in other ones it was not specified. A few studies tested different containers comparing the traditional glass jar with other types. Koç et al. [24] used the GA-7 Magenta™ box (a polypropylene container) to store *Pistacia lentiscus* shoots, while Giannì and Sottile [27] utilized the Microbox ECo 2™ (a polycarbonate container) for the conservation of two Italian plum species.

Ozudogru et al. [25] compared the effects of gas-tight (glass jars) and gas-permeable (Star Pac™ bags) culture containers on the development of an effective in vitro SGS protocol (4 °C, darkness) for 'Gisela®5' shoots. The Star Pac™ are heat-sealable, gas-permeable plastic containers made of five cells (15 × 4 cm in size) (Figure 1). The authors evidenced an excess of CO_2 and ethylene accumulation inside the gas-tight containers during the first weeks, referring to the physical stress of shoots as a consequence of the transfer from standard culture conditions to the conservation at low temperature and darkness. However, gas-chromatographic analysis during storage showed that the stress was quickly overcome, as both CO_2 and ethylene concentrations were drastically reduced after 1 month of conservation. For this reason, it is more favorable to use a new generation of polystyrene bags that allow for a limited exchange of the main gases produced during tissue culture with the outside, avoiding the consistent accumulation that can be observed in traditional glass jars.

The Star Pac™ bags were also used by Turdiyev et al. [20] for the SGS of genotypes from different fruit species (apple, pear, plum, cherry, raspberry, black currant and strawberry). The authors reported that the duration of storage depended on various factors (genotype, temperature and light intensity) including the type of container, showing that plastic air-permeable containers (Star Pac™ bags) were more effective for in vitro conservation. The Star Pac™ bags were already used in previous years for the SGS of apple and pear [31,39] with satisfactory cold conservation of shoots. It is noted that Star Pac™ bags are used to maximize the space during the in vitro conservation of various woody and herbaceous plants and *Musa* collections in USDA centers in Corvallis and Mayaguez [56]. Overall, the container used for SGS should have a good air exchange, a desiccation barrier and avoid microbe diffusion; moreover, it should be of adequate size for the plants [31].

(**a**) (**b**)

Figure 1. Slow growth storage of apple (**a**) and pear (**b**) shoots in Starpac Pac ™ bags (Photos by Benelli Carla).

2.4. Genotype Effect

The performance of in vitro shoots is strongly affected by the genotype. In the literature, there are different examples where the results obtained with micropropagation varied from one plant variety to another [32,57]. For this reason, efforts should not only focus on developing appropriate genotype-specific protocols for tissue culture and micropropagation of plants, but also for SGS [54,58] to be applied on a wide range of accessions. Reed and De Noma [31] mentioned that storage duration can differ greatly among genotypes and from genus to genus. The synergy among all factors involved affects the outcome of SGS; even with an effective species-specific protocol, the response to SGS among genotypes within the same species can be quite different [33,59].

Gianní and Sottile [27] reported that four plum genotypes responded differently during SGS at 4 °C in darkness. The survival rate varied significantly among 'Ariddu di Core' 100%, 'Sanacore' 75% (*P. domestica*), 'Marabolo' and 'Rapparino' 25% (*P. cerasifera*). Moreover, they also observed that genotypes of *P. cerasifera* and *P. domestica* adapted well to in vitro culture, until the rooting stage.

In another study, Sedlak et al. [19] reported the conservation of six apple genotypes, three pear cultivars, two sweet cherry cultivars, three strawberry cultivars and two dwarf sweet cherry rootstocks. All genotypes were subjected to medium-term SGS for up to 7 months at 4 °C under light conditions. Their results showed that different fruit species and their cultivars had specific survival rates after SGS. In particular, the survival of the six apple genotypes varied between 18% ('Tophola') and 90% ('Fragrance'), in the three pear cultivars from 10% ('Elektra') to 91% ('Milada') and in two sweet cherry cultivars the survival was 54.5% ('Amid') and 79.4% ('Kasandra'). Even more evident is that the genotype affected the survival of the two dwarf sweet cherry rootstocks: 92% with 'P-HL-A' and 2% with 'P-HL-C'.

Lukoševičiūtė et al. [21] underlined the effect of genotype under the SGS of different accessions from *F.* × *ananassa* ('Venta', 'Melody', 'Elsanta', 'Holiday', 'Dangė', 'Nora', 'Nida', 'Jaunė', 'Saulenė', 'Catskill', 'Juni Morgon', 'Suvetar', 'Valotar', 'Vaiva', 'Jasna' and KLP8), *F. moschata, F. vesca, F. virginiana, F. virginiana glauca* and *P. communis* ('Oranzhevaya', 'Hasselpeare', 'Princesse Dagmara', 'Karalienė Jadvyga', 'Senryo', 'Muskatelka Seda', 'Koncentrat' and No. 0408) and *P. pyraster*. The authors showed that the survival rate varied in a wide range depending on the genotype: in strawberry accessions from 11% ('Suvetar') to 100% ('Nida') after 15 months and in pear accessions from 75% ('Oranzhevaya') to 100% ('Princesse Dagmara') after 6 months of storage.

The impact of the cultivar on the stored plant quality and the duration of SGS was observed by Kovalchuk et al. [29] who successfully preserved the shoots of two *P. communis* cultivars, 'Mramornaya' and 'Talgarskaya Krasavitsa', at 4 °C for 18 and 12 months, respectively. Arbeloa et al. [18] indicated various multiplication rates of different genotypes/clones of *Prunus*, *Malus*, *Pyrus*, *Ficus* and *Cydonia* after 7 months of SGS at 4 °C in darkness.

3. Conservation in SGS of Tropical Fruit Species

Tropical germplasm resources have been reduced in their natural habitats by indiscriminate human activity, climate change and a corresponding increase in the incidence of pests, diseases and viruses. The conservation, distribution and use of the natural genetic diversity of tropical species should be considered essential, and therefore the creation of germplasm banks at national and international levels should be a priority [60]. The conservation in SGS represents a valid strategy not only for temperate species but also for tropical and subtropical species; Table 2 shows the tropical species preserved by in vitro shoot storage from 2012 to the present.

Tropical plant species are generally cold-sensitive so in many cases the conservation under SGS is not conducted at a low temperature [8]. Otherwise, physiological damages induced by cold stress referred to as chilling injury [61,62] may occur, with various changes in the metabolism, protein content, composition and functioning of the membranes.

The in vitro storage temperature depends on the cold sensitivity of the species; most of them are kept at 18–25 °C, while two species are stored at 5 °C, *Phoenix dactylifera* (date palm) and *Simmondsia chinensis* (jojoba), and one at 8 °C, *Carica pubescens* (papaja) (Table 2). In date palm, 70% of the shoots remained healthy at 5 °C in the dark after storage for 12 months on MS proliferation media without hormones [63]. Jojoba shoots were stored for 9 months at 5 °C in the dark on MS media containing a hormone (1 mg L^{-1} BA); the low temperature allowed the reduction of both the number and height of shoots and no necrosis was observed [64]. In the same study, the authors have verified the possibility to store the jojoba in standard culture conditions (25 °C and a 16 h photoperiod) using osmotic regulators in the media, such as mannitol or polyethylene glycol (PEG). The PEG addition showed a stronger growth reduction, but more shoot necrosis was observed, while the mannitol presence slightly increased the induction of new shoots. They concluded that a low temperature (5 °C) is better for the storage of healthy in vitro shoots of jojoba [64]. *Carica pubescens* was stored at 8 °C and a 16 h photoperiod for 6 months on $\frac{1}{2}$MS medium with BA (2 ppm), showing a 90% survival rate and 100% of regrowth on recovery medium [65].

For the other tropical species listed in Table 2, the temperature was not modified with respect to the standard growth conditions during micropropagation, but changes concerned the medium composition to effectively reduce the shoot growth rate and, thus, the period between subcultures, without compromising their quality and health.

Recently, de Oliviera and Aloufa [66] tested the effect of osmotic compounds added to the culture medium to slow the growth of *Hancornia speciosa* shoots maintained at 25 °C (30 µmol m^{-2} s^{-1}, 16 h photoperiod). Different concentrations of sucrose and sorbitol were tested for various storage times. Sorbitol showed a more pronounced growth-reducing effect than sucrose. The combination of 15 g L^{-1} sucrose with 5 g L^{-1} sorbitol gave the best result (95% survival), allowing the conservation of the shoots for 4 months. In contrast, higher concentrations of sucrose (30 g L^{-1}) and sorbitol (10 or 20 g L^{-1}) showed toxic and stressful effects on shoot survival with thin stems, reduced or absent leaves, high oxidation incidence and greater callus formation at the base of the explants [66].

Mannitol was effective in the conservation of taro (*Colocasia esculenta* var. *globulifera*); after 24 months, 80% of the shoots survived with a concentration of 4% mannitol in the culture medium [67]. Lower concentrations of mannitol (2%) allowed for the conservation of up to 6 months, while by increasing the concentration to 6%, the storage period fell to 2 months.

The application of ABA (3 mg L^{-1}) in the storage medium for the preservation of *Vanilla planifolia* resulted in reduced shoot growth and allowed a storage period of 6 [68] or 4 months [69], with a survival rate around 90%.

An in vitro collection of 66 pineapple accessions was successfully stored for 10 years at 21 °C, with a lower light intensity (20 µmol m^{-2} s^{-1}) and a shorter photoperiod (12 h), with a transfer of stored shoot cultures on fresh medium every 24 months [70]. Although all shoots were viable after this long period of storage, in post-conservation the capacity of the recovery and propagation potential was genotype-dependent. A few accessions required only one subculture (45 days) for a full restoration as new cultures, but other accessions needed 3–5 subcultures.

Bananas and plantains are among the most important tropical fruit crops worldwide. Local and global efforts for the ex situ preservation of banana germplasm are massive. For example, at the International Transit Centre (ITC) in Leuven (Belgium), the in vitro SGS is achieved under a combination of low temperature (16 °C) and limited light (25 µmol m^{-2} s^{-1}) [71,72]. Therefore, this method is appropriate to decrease the subculture frequency, ranging between 3 and 22 months, depending on the genomic group and in vitro browning reaction [71]. In particular, in the ITC, the AAA Mutika Lujugira bananas and AAB non-plantain bananas can be stored for up to 390 days, while the wild bananas (*M. acuminata* and *M. balbisiana*) require a subculture after 275 days. Other accessions (Lady Finger-Pome, AAB) achieved 615 days of storage compared to 60 days for SF215, a AA *M. acuminata* sp. *banksii* derivative [73,74].

Kanchanapoom and Promsorn [75] reported that sucrose (1%) was a suitable carbon source for storing shoots of *Musa balbisiana* 'Kluai Hin' (BBB group) at 25 °C and a 16 h photoperiod for 6 months. Other sources of carbohydrates (glucose and sorbitol) or higher concentrations of sucrose (3 and 5%) did not positively affect the storage duration of the cultures, and also conservation under dark conditions did not improve the regrowth capacity of the shoots.

Eight media with MS or $\frac{1}{2}$MS containing different concentrations of hormones (BA and IAA) and sucrose (20, 30, and 60 g L^{-1}) were tested for the preservation of three cultivars of banana at 26 or 18 °C [76]. After 5 months, the highest survival percentage (100%) was obtained with 2.25 mg L^{-1} BA, 0.175 mg L^{-1} IAA and 30 g L^{-1} sucrose at 18 °C. The full or half concentration of the mineral salts did not affect the conservation of the three cultivars.

To reduce the development of banana shoots, it was possible to apply growth retardants such as ABA, maleic hydrazide and paclobutrazol (PBZ) [77,78]. In particular, PBZ plays a key role in the inhibition of cell elongation and internodes by affecting the biosynthesis of gibberellins and increasing ABA and chlorophyll activity [79,80]. In the banana variant 'Kepo', the addition of PBZ at 2.5 or 5.0 ppm was effective to slow the growth of plantlet height, number of leaves and ratio of leaf length to leaf width, allowing in vitro shoot storage for 6 months at 18–22 °C. No physiological damage or loss of morphogenetic ability of the shoots were observed during the medium-term storage [81].

An innovative study evaluated the effect of different spectra of light on the conservation of the banana 'Prata Catarina' group AAB under SGS [72]. Plantlets maintained with a photoperiod of 14 h (25 °C) were illuminated with red and blue light (combined or alone) or with white light. Photosynthetic photon flux density (PPFD) for all treatments was set at 50 µmol m^{-2} s^{-1}. Plantlets grown under the blue spectrum showed a positive response to SGS for about 5 months; the blue light reduced photosynthetic activity and consequently, induced a lower metabolic activity during the storage. To date, few studies have been carried out on the influence of light on in vitro conservation, thus it is desirable to increase the attention to this important parameter. For example, in banana accessions where wide differences in storage time conservation are recorded (ranging from 3 to 22 months) [71], the authors suggest the investigation of different light spectra to promote the storage period.

Further, in tropical species, the genotype effect was documented. Several responses in shoot, leaf and root formation after 6 months of in vitro preservation were noted in four banana accessions ('Valery', 'Dwarf Cavendish', 'Pelipita-Colombia' and 'Pelipita-Costa Rica') [55]. Similar behavior was indicated by Zainy et al. [76] for three banana cultivars (C1-Pisange, C2-Brazillian and C3-William) preserved for 5 months; the application of various media and temperatures gave variability in the shoot growth of all cultivars as a result of genotype effect. In pineapple, da Silva [70] found differences in the micropropagation potential among 66 accessions after a long conservation period (10 years) under SGS, related to the response of each variety/accession.

Table 2. Tropical fruit shoots conservation in SGS from 2012 to present. Culture conditions and best results are reported for each species (terminology and values are the same as mentioned by the authors).

Species	Medium	Temperature (°C)	Light Condition	Storage Time (months)	Survival (%)	References
Ananas comosus	$\frac{1}{2}$MS, 30 g L^{-1} sucrose	21	12 h, 20 μmol m^{-2} s^{-1}	24 *	100	[70]
Carica pubescens	$\frac{1}{2}$MS, 2 ppm BA	8	16 h, 18 watts	6	90	[65]
Colocasia esculenta	MS, 4% mannitol	NR	16 h	24	80	[67]
Hancornia speciosa	MS, 15 g L^{-1} sucrose, 5 g L^{-1} sorbitol	25	16 h, 30 μmol m^{-2} s^{-1}	4	95	[66]
Musa spp.	MS, 30 g L^{-1} sucrose	25	14 h, blue light (450–465 nm) 50 μmol m^{-2} s^{-1}	3.5	100	[72]
	$\frac{1}{2}$MS or MS, 2.25 mg L^{-1} BAP, 0.175 mg L^{-1} IAA, 30 g L^{-1} sucrose	18	NR	5	100	[76]
	MS, 4% mannitol	23	12 h	16	NR	[55]
Musa acuminata × *balbisiana*	MS, 2.25 mg L^{-1} BAP, 0.175 mg L^{-1} IAA, 3% sucrose, 2.5 and 5 ppm PBZ	18-22	Natural light	6	NR	[81]
Musa balbisiana	MS, 1% sucrose	25	16 h, 20 μmol m^{-2} s^{-1}	6	25	[75]
Phoenix dactylifera	MS, 10 mg L^{-1} 2,4-D, 3 mg L^{-1} 2iP, 6 mg L^{-1} ABA, 102 g L^{-1} sucrose, 3 g L^{-1} AC	15	Darkness	12	91.8	[82]
	MS	5	Darkness	12	70	[63]
Simmondsia chinensis	MS, 1 mg L^{-1} BA	5	Darkness	9	NR	[64]
Vanilla planifolia	MS, 2 mg L^{-1} BA, 3 mg L^{-1} ABA	24	16 h, 40–50 μmol m^{-2} s^{-1}	6	90	[68]
	MS, 3 mg L^{-1} ABA	22	16 h, 50 μmol m^{-2} s^{-1}	4	93.3	[69]

* Every 24 months the shoot cultures have been renewed for a period of 10 years. NR: not reported. MS, Murashige and Skoog medium; AC, activated charcoal; ABA, abscisic acid; BA, 6-Benzyladenine; BAP, 6-Benzylaminopurine; IAA, indole-3-acetic acid; 2,4-D, 2,4-Dichlorophenoxyacetic acid; 2iP, 6-(γ,γ-Dimethylallylamino)purine; PBZ, paclobutrazol.

4. Genetic Stability

Despite that in vitro conservation techniques were optimized for several plant species, genetic stability is considered an important aspect in SGS protocols of germplasm because in vitro conditions may cause somaclonal variation and epigenetic changes [69,83].

In vitro culture conditions, the long exposure of plant material to cytokinin, can increase the probability of somaclonal variations. Kamińska et al. [84] reported that slow growth storage and further regrowth on MS medium supplemented with BA may cause abnormalities with changes in the amount of DNA in the cells. Unbalanced concentrations of auxins and cytokinins and supra-optimal levels of growth regulators have been linked with somaclonal variation. However, the effect of the type and concentration of plant growth regulators on the incidence of somaclonal variation in different plant species remains a topic of debate [85].

Moreover, excessive generation of reactive oxygen species (ROS) in explants subjected to cold stress can also lead to somaclonal variation due to DNA damage. Indeed, in cold storage, the growth of explants is slowed down but prolonged exposure to low temperatures may cause stress, especially to thermophilic species [86].

In any case, SGS is considered as an effective approach to maintaining genetic resources in in vitro gene banks [12], and the analysis of the genetic stability of the in vitro preserved material can be a step to detecting the changes that may arise altering the genetic homogeneity of the germplasm [87].

The genetic investigation was applied by molecular markers such as random amplification of polymorphic DNA (RAPD), sequence-related amplified polymorphism (SRAP) [88], simple-sequence repeat (SSR), inter-simple sequence repeat (ISSR) [89,90], amplified fragment length polymorphism (AFLP) and start codon targeted (SCOT) [24,69,70,91].

The genetic stability, using ISSR markers, of 66 pineapple accessions after micropropagation and SGS for 10 years was reported by da Silva et al. [70]. No somaclonal variation was recorded in most of the accessions; only two accessions of *A. comosus* var. *bracteaus* showed genetic instability. Therefore, a subculture interval of 24 months is recommended by the authors for pineapple shoots kept under in vitro conservation.

Bautista-Aguilar et al. [69] demonstrated high genetic stability of *Vanilla* preserved for 4 months using ISSR and SSR markers with a low percentage of polymorphism detected in different accessions of *V. planifolia* (SSR 0%, ISSR 2%) and in *V. insignis* (SSR 0%, ISSR 0%).

Koç et al. [24] reported that *Pistacia lentiscus* adapted to in vitro cold storage and showed some genetic variations by AFLP after 6 months of storage at 4 °C. According to the results, genetic similarity values of the non-preserved and cold-stored plantlets ranged from 0.66 to 0.84, with a mean of 0.74. Thus, in vitro propagation and especially cold storage of *P. lentiscus* may be affected by transposons activation, which could cause genetic instability.

5. In Vitro Banking Strategy

Several countries have built up national germplasm banks for the conservation, utilization and distribution of plant material. Germplasm conservation needs to be established for preserving high-quality plant genetic sources and to set a database of available genetic material that would facilitate the knowledge and the research. Various institutions and centers in the world have produced important applicative impact with the establishment of in vitro banks, mainly of shoot cultures maintained in SGS, spread across 15 countries of 6 continents, with a conservation policy especially directed to the most strategically important plants for human nutrition. Today, over 37,000 accessions are preserved by means of SGS of shoot cultures [92]. Only three species represent over 90% of this germplasm: potato (*Solanum* spp.), with almost 19,000 accessions conserved in SGS at 6–10 °C as shoot or microtuber cultures; cassava (*Manihot esculenta*), with over 9400 accessions; and sweet potato (*Ipomea batatas*) with over 5300 accessions mainly preserved as shoot cultures at room temperature with the use of osmotically active compounds (mannitol, sorbitol and sucrose) in the storage medium [37].

Among the fruit species, the *Musa* species has the largest gene bank established in Belgium at the International Musa Germplasm Transit Centre (ITC) with >1600 accessions sourced from different national and regional field collections in 38 countries; these accessions are ex situ conserved under minimal growth conditions to reduce the growth rate of the plant tissues and the frequency of subculturing cycles [71]. Additionally, the Tropical Agriculture Research Station (USDA-ARS TARS) in Mayagüez, Puerto Rico, has been maintaining 164 accessions of *Musa* spp. in field collection and in in vitro storage [55].

A complete listing of worldwide fruit germplasm related to in vitro collections is difficult to find because the germplasm is often stored in in vitro conditions as collections for study and research purposes. Generally, there is a need to introduce new strategies for managing in vitro collections as a routine, as well as an appropriate collecting process. Such procedures should be designed specifically for each species and variety [93] involving the introduction, evaluation, characterization and distribution of high-quality germplasm and its preservation for long term availability.

Particularly, the measures implemented should be complying with safety and ethical authority regulations and they should ensure: (i) healthy material: contamination-free material, (ii) authenticity: correct identification and (iii) stability: trueness-to-type. Good laboratory practices with careful application of aseptic techniques, clear and accurate documentation and the correct procedures that decrease the risks of genetic variation are all necessary activities to ensure the in vitro storage of plant material. In order to avoid somaclonal variation in SGS, the use of shoots and apices as an explant source and the reduction of the number of subcultures can be recommended.

SGS can be a useful tool to continue the conservation of plant diversity as mentioned by Target 8 of the Global Strategy for Plant Conservation (GSPC): 'At least 75 percent of threatened plant species in ex situ collections, preferably in the country of origin, and at least 20 percent available for recovery and restoration programmes' [94]. Indeed, SGS allows the appropriate storage of stock material to be used for restoration and provides a method for propagating plants to reduce the pressure on wild populations.

For long term conservation, many accessions can be preserved using advanced tools such as cryopreservation. Cryopreservation is the storage of biological material at an ultra-low temperature of −196 °C in liquid nitrogen, and it can be considered a safety backup method for field collections to reduce the loss of plant germplasm [95]. Currently, it involves over 10,000 accessions preserved in cryo-banks established in different countries for cassava, potato, banana, apple, pear, coffee, mulberry and garlic [71,96].

6. Conclusions and Perspectives

Ex situ plant conservation depends on the species, the methods employed and the desired storage time, and these occurrences are associated with costs, risks and scientific challenges. In vitro conservation allows for the preservation of plant genetic resources, disease-free planting material, material to be provided to growers all year round and control of the genetic fidelity verified with molecular analysis, at the same time. Particularly, SGS, with less frequency of subcultures, lowers the risk of losing germplasm through handling errors, such as contamination problems, and decreases the risks of genetic instability, manual labor, and the space needed for germplasm conservation. Thus, considering all this, SGS can also be an efficient way to reduce production costs (Figure 2).

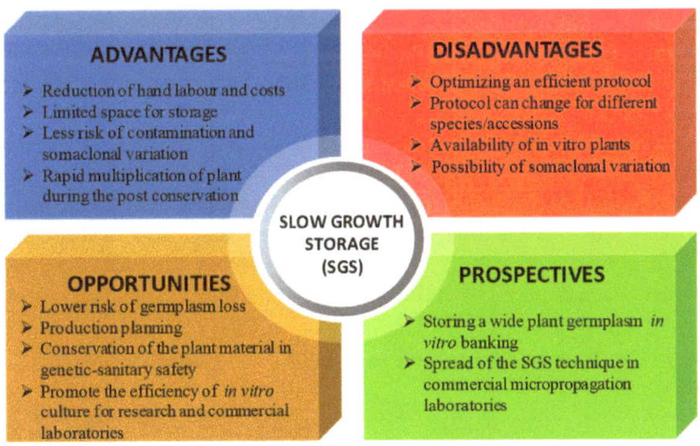

Figure 2. In vitro slow growth storage (SGS) technique as a competitive tool to make the plant conservation dynamic and applicable in present and future.

It is evident that the simplest and most successful slow growth strategies involve temperature and light limitation as reported in this review.

SGS protocols for some fruit species require additional research, but for others are well-defined. A bottleneck in the application of slow growth is the adaptation of a generic protocol to every accession of a large and various plant genebank; the possibility of one common protocol applicable as best practice across all plants is limited mainly due to variable species and genotype responses. A standard protocol can be applied to diverse genotypes, although minor improvements may be required for outlying, low response performers. Therefore, it might be more practical to develop a number of protocols as standard operating procedures which can be validated for different accessions across different genebanks [14]. For this reason, it is important to support cooperation through knowledge sharing on the best practices developed in the plant research community (public institutions, universities and private laboratories) and active genebanks.

Overall, for in vitro storage significant precautions should be taken to use healthy plants, e.g., applying tests for virus-free material, especially for vulnerable species before initiating SGS. Further research needs to improve the in vitro conservation of plant germplasm collection and to enlarge the applications and the prospects of the SGS technique (Figure 2).

Author Contributions: Conceptualization, A.D.C. and C.B.; data collection and analysis, preparation of the tables, T.I. and W.T.; writing—original draft preparation, A.D.C., W.T., T.I. and C.B.; writing—review and editing, W.T., T.I. and C.B. All authors have read and agreed to the published version of the manuscript.

Funding: This research received no external funding.

Data Availability Statement: The data presented are available in all the publications cited in this review.

Conflicts of Interest: The authors declare no conflict of interest.

References

1. Kulak, V.; Longboat, S.; Brunet, N.D.; Shukla, M.; Saxena, P. In Vitro technology in plant conservation: Relevance to biocultural diversity. *Plants* **2022**, *11*, 503. [CrossRef] [PubMed]
2. Pritchard, H. Classification of seed storage types for ex situ conservation in relation to temperature and moisture. In *Ex Situ Plant Conservation: Supporting Species Survival in the Wild*; Guerrant, E., Havens, K., Maunder, M., Eds.; Island Press: Washington, DC, USA, 2004; pp. 139–161.
3. Shii, C.T.; Ma, S.S.; Huang, P.L.; Chin, S.W. Current and future development in clonal germplasm conservation of vegetatively propagated crops. In Proceedings of the Plant Germplasm Conservation: Perspectives for the 2000s, Taiwan, China, 10–11 August 1994.
4. Reed, B.M. The in vitro genebank of temperate fruit and nut crops at the National Clonal Germplasm Repository-Corvallis. In *Management of Field and In Vitro Germplasm Collections*; Intl. Plant Genetic Resources Institute: Rome, Italy, 1999; pp. 132–135.
5. Levy, L.; Damsteegt, V.; Welliver, R. First Report of *Plum pox virus* (Sharka Disease) in *Prunus persica* in the United States. *Plant Dis.* **2000**, *84*, 202. [CrossRef] [PubMed]
6. García-Ramírez, Y.; Barrera, G.P.; Freire-Seijo, M.; Barbón, R.; Concepción-Hernández, M.; Mendoza-Rodríguez, M.F.; Torres-García, S. Effect of sucrose on physiological and biochemical changes of proliferated shoots of *Bambusa vulgaris* Schrad. Ex Wendl in temporary immersion. *Plant Cell. Tissue Organ Cult.* **2019**, *137*, 239–247. [CrossRef]
7. Martelli, G.P.; Boscia, D.; Porcelli, F.; Saponari, M. The olive quick decline syndrome in south-east Italy: A threatening phytosanitary emergency. *Eur. J. Plant Pathol.* **2016**, *144*, 235–243. [CrossRef]
8. Engelmann, F. In vitro conservation of tropical plant germplasm—A review. *Euphytica* **1991**, *57*, 227–243. [CrossRef]
9. Lambardi, M.; Ozudogru, E.A. Advances in the safe storage of micropropagated woody plants at low temperature. *Acta Hortic.* **2013**, *988*, 29–42. [CrossRef]
10. Grout, B.W.W. In vitro conservation of germplasm. In *Developments in Crop Science*; Elsevier: Amsterdam, The Netherlands, 1990; Volume 19, pp. 394–411.
11. Engelmann, F. Management of field and in vitro germplasm collections. In Proceedings of the Consultation Meeting, Rome, Italy, 5–20 January 1996; IPGRI: Rome, Italy, 1999; p. 165.
12. Cha-um, S.; Kirdmanee, C. Minimal growth in vitro culture for preservation of plant species. *Fruit Veg. Cereal Sci. Biotechnol.* **2007**, *1*, 13–25.
13. Rajasekharan, P.E.; Sahijram, L. In Vitro conservation of plant germplasm. In *Plant Biology and Biotechnology*; Bahadur, B., Venkat Rajam, M., Sahijram, L., Krishnamurthy, K.V., Eds.; Springer: New Delhi, India, 2015; Volume II, pp. 417–443. ISBN 978-81-322-2282-8.
14. Reed, B.M.; Gupta, S.; Uchendu, E.E. In vitro genebanks for preserving tropical biodiversity. In *Conservation of Tropical Plant Species*; Springer: New York, NY, USA, 2013; pp. 77–106.
15. Mascarello, C.; Sacco, E.; Di Silvestro, D.; Pamato, M.; Ruffoni, B. Reduction of the subculture frequency in three Mediterranean species during micropropagation. *Acta Hortic.* **2017**, *1155*, 461–466. [CrossRef]
16. Capuana, M.; Di Lonardo, S. In vitro conservation of chestnut (*Castanea sativa*) by slow growth. *Vitr. Cell. Dev. Biol.-Plant* **2013**, *49*, 605–610. [CrossRef]
17. Carvalho, M.D.J.D.S.D.; Souza, A.D.S.; Santos, E.B.; Filho, W.D.S.S.; Ledo, C.A.D.S.; Costa, E.M.R.; Souza, F.V.D. In vitro conservation of 'Florida Rough' lemon plants. *Ciência Rural* **2022**, *52*, 1–9. [CrossRef]

18. Arbeloa, A.; Marín, J.A.; Andreu, P.; García, E.; Lorente, P. In vitro conservation of fruit trees by slow growth storage. *Acta Hortic.* **2017**, *1155*, 101–106. [CrossRef]
19. Sedlak, J.; Zidova, P.; Paprstein, F. Slow growth in vitro conservation of fruit crops. *Acta Hortic.* **2019**, *1234*, 119–124. [CrossRef]
20. Turdiyev, T.T.; Kovalchuk, I.Y.; Kabylbekova, B.Z.; Chukanova, N.I.; Frolov, S.N. In vitro germplasm cold storage of fruit and berry plants of Kazakhstan. *Eurasian J. Biosci.* **2020**, *1219*, 1213–1219.
21. Lukoševičiute, V.; Rugienius, R.; Staniene, G.; Blažyte, A.; Gelvonauskiene, D.; Bendokas, V.; Gelvonauskis, B.; Sasnauskas, A.; Stanys, V.; Bobinas, Č. Low temperature storage of *Fragaria* sp. and *Pyrus* sp. genetic resources in vitro. *Zemdirbyste* **2012**, *99*, 125–130.
22. Monticelli, S.; Gentile, A.; Forni, C.; Caboni, E. Slow growth in "Pisana" apricot and in "Cerina" apple, two Italian cultivars. *Acta Hortic.* **2020**, *1285*, 117–124. [CrossRef]
23. Kabylbekova, B.; Kovalchuk, I.; Chukanova, N.; Turdiyev, T. Cold storage of in vitro apple germplasm in Kazakhstan. *Acta Hortic.* **2021**, *1324*, 77–82. [CrossRef]
24. Koç, I.; Akdemir, H.; Onay, A.; Çiftçi, Y.Ö. Cold-induced genetic instability in micropropagated *Pistacia lentiscus* L. plantlets. *Acta Physiol. Plant.* **2014**, *36*, 2373–2384. [CrossRef]
25. Ozudogru, E.A.; Benelli, C.; Dradi, G.; Lambardi, M. Effect of culture container and carbohydrate content on in vitro slow growth storage of the cherry rootstock 'Gisela®5. ' *Acta Physiol. Plant.* **2017**, *39*, 2–9. [CrossRef]
26. Sota, V.; Kongjika, E. The effect of nutrient media in micropropagation and in vitro conservation of wild population of mahaleb cherry (*Prunus mahaleb* L.). *J. Microbiol. Biotech. Food Sci.* **2014**, *3*, 453–456.
27. Gianní, S.; Sottile, F. In vitro storage of plum germplasm by slow growth. *Hortic. Sci.* **2015**, *42*, 61–69. [CrossRef]
28. Sota, V.; Kongjika, E. Micropropagation and in vitro Conservation via Slow Growth Techniques of Prunus webbii (Spach) Vierh: An Endangered Plant Species in Albania. *Int. J. Biol. Life Agric. Sci.* **2019**, *13*, 280–283.
29. Kovalchuk, I.; Zhumagulova, Z.; Turdiev, T.; Reed, B.M. Growth medium alterations improve in vitro cold storage of pear germplasm. *Cryo-Letters* **2014**, *35*, 197–203. [PubMed]
30. Reed, B.M.; DeNoma, J.; Wada, S.; Postman, J. Micropropagation of pear (*Pyrus* sp.). In *Protocols for Micropropagation of Selected Economically-Important Horticultural Plant*; Humana Press: Totowa, NJ, USA, 2013; pp. 3–18. ISBN 978-1-4614-3775-8.
31. Reed, B.M.; DeNoma, J. Medium-term in vitro storage of pear as a complementary germplasm preservation technique. *Acta Hortic.* **2016**, *1113*, 251–256. [CrossRef]
32. Nin, S.; Benelli, C.; Petrucci, W.A.; Turchi, A.; Pecchioli, S.; Gori, M.; Giordani, E. In vitro propagation and conservation of wild bilberry (*Vaccinium myrtillus* L.) genotypes collected in the Tuscan Apennines (Italy). *J. Berry Res.* **2019**, *9*, 411–430. [CrossRef]
33. Pan, X.J.; Zhang, W.E.; Li, X. In vitro conservation of native Chinese wild grape (Vitis heyneana Roem. & Schult) by slow growth culture. *VITIS-J. Grapevine Res.* **2014**, *53*, 207–214.
34. AlMousa, R.; Hassan, N. Effect of osmotic regulators on In vitro conservation of grape. *J. Genet. Environ. Resour. Conserv.* **2022**, *10*, 60–66.
35. Hassanen, S.A.; Abido, A.I.A.; Aly, M.A.M.; Rayan, G.A. In vitro preservation of grapevine (*Vitis vinifera* L.) Muscat of Alexandria and Black Monukka cultivars as genetic resource. *Afr. J. Basic Appl. Sci.* **2013**, *5*, 55–63. [CrossRef]
36. Sota, V.; Kongjika, E. Slow growth in vitro conservation of *Zizyphus jujuba* Mill. *Poljopr. I Sumar.* **2014**, *60*, 27–37.
37. Benson, E.E.; Harding, K.; Debouck, D.; Dumet, D.; Escobar, R.; Mafla, G.; Panis, B.; Panta, A.; Tay, D.; Van Den, I.; et al. *Refinement and Standardization of Storage Procedures for Clonal Crops*; Global Public Goods Phase 2. Part 3: Multi-crop guidelines for developing in vitro conservation best practices for clonal crops; System-Wide Genetic Resources Programme (SGRP): Rome, Italy, 2011; p. 144, ISBN 9789290439059.
38. Lambardi, M.; De Carlo, A. Application of tissue culture to the germplasm conservation of temperate broad-leaf trees. In *Micropropagation of Woody Trees and Fruits*; Springer: Dordrecht, The Netherlands, 2003; pp. 815–840.
39. Lambardi, M.; Roncasaglia, R.; Previati, A.; De Carlo, A.; Dradi, G.; Da Re, F.; Calamai, L. In vitro slow growth storage of fruit rootstocks inside gas-tight or gas-permeable containers. *Acta Hortic.* **2006**, *725*, 483–488. [CrossRef]
40. Murashige, T.; Skoog, F. A revised medium for rapid growth and bioassays with tobacco tissue cultures. *Physiol. Plant* **1962**, *15*, 473–497. [CrossRef]
41. Lloyd, G.; McCown, B. Commercially-feasible micropropagation of mountain laurel, *Kalmia latifolia*, by use of shoot tip culture. *Int. Plant Propagators' Soc.* **1980**, *30*, 421–427.
42. Driver, J.A.; Kunijuki, A. In vitro propagation of paradox walnut rootstock. *Hortscience* **1984**, *19*, 507–550. [CrossRef]
43. Rugini, E. In vitro propagation of some olive (*Olea europaea sativa* L.) cultivars with different root-ability, and medium development using analytical data from developing shoots and embryos. *Sci. Hortic.* **1984**, *24*, 123–134. [CrossRef]
44. Lambardi, M.; Benelli, C.; De Carlo, A.; Fabbri, A.; Grassi, S.; Lynch, P.T. Medium- and long- term in vitro conservation of olive germplasm (*Olea europaea* L.). *Acta Hortic.* **2002**, *586*, 109–112. [CrossRef]
45. Gardi, T.; Micheli, M.; Piccioni, E.; Sisani, G.; Standardi, A. Cold storage of micropropagated shoots of some fruit species. *Italus Hortus* **2001**, *8*, 32–40.
46. Reski, R.; Abel, W.O. Induction of budding on chloronemata and caulonemata of the moss, *Physcomitrella patens*, using isopentenyladenine. *Planta* **1985**, *165*, 354–358. [CrossRef] [PubMed]
47. Ahmed, M.; Anjum, M.A.; Ahmed, M.J.; Sajid, G.M.; Yoqub, A.; Shafqat, M. Role of plant growth regulators in preservation of *Pyrus* germplasm in vitro. *Afr. J. Biotechnol.* **2011**, *10*, 14029–14037. [CrossRef]

48. Selvakumar, G.; Bindu, G.H.; Bhatt, R.M.; Upreti, K.K.; Paul, A.M.; Asha, A.; Shweta, K.; Sharma, M. Osmotolerant cytokinin producing microbes enhance tomato growth in deficit irrigation conditions. *Proc. Natl. Acad. Sci. India Sect. B Biol. Sci.* **2018**, *88*, 459–465. [CrossRef]
49. Rivin, C.J.; Grudt, T. Abscisic acid and the developmental regulation of embryo storage proteins in maize. *Plant Physiol.* **1991**, *95*, 358–365. [CrossRef]
50. Gupta, S.; Parakh, D.B.; Verma, S.K. Ex situ conservation of strawberry (*Fragaria* spp.) germplasm in ICAR-NBPGR. *Acta Hortic.* **2020**, *1297*, 89–98. [CrossRef]
51. Martínez-Santos, E.; Cruz-Cruz, C.A.; Spinoso-Castillo, J.L.; Bello-Bello, J.J. In vitro response of vanilla (*Vanilla planifolia* Jacks. ex Andrews) to PEG-induced osmotic stress. *Sci. Rep.* **2021**, *11*, 22611. [CrossRef]
52. Silva, T.d.S.; Nepomuceno, C.F.; Soares, T.L.; de Santana, J.R.F. In vitro conservation of *Poincianella pyramidalis* (Tul.) L.P. Queiroz under minimal growth conditions. *Ciência E Agrotecnologia* **2019**, *43*, 14519. [CrossRef]
53. Reed, B.M. Application of gas-permeable bags for in vitro cold storage of strawberry germplasm. *Plant Cell Rep.* **1991**, *10*, 431–434. [CrossRef] [PubMed]
54. Kovalchuk, I.; Lyudvikova, Y.; Volgina, M.; Reed, B.M. Medium, container and genotype all influence in vitro cold storage of apple germplasm. *Plant Cell. Tissue Organ Cult.* **2009**, *96*, 127–136. [CrossRef]
55. Irish, B.M.; Goenaga, R.; Reed, B.M. Amending storage vessel and media increase transfer interval of *Musa* spp. tissue culture plantlets. *J. Agric. Univ. Puerto Rico* **2013**, *97*, 1–13.
56. Jourdan, P. Short- and medium-term in vitro conservation and management of germplasm within the USDA's National Plant Germplasm System. *Acta Hortic.* **2018**, *1224*, 31–38. [CrossRef]
57. Khan, M.R.; Muhammad, A.; Hussain, I.; Shah, S.H.; Kumar, T.; Inam, S.M.; Zubair, H.U.; Rehman, A.; Sher, N.; Rehman, S.; et al. Rapid in vitro multiplication of sugarcane elite genotypes and detection of sugarcane mosaic virus through two steps RT-PCR. *Int. J. Agric. Biol.* **2012**, *14*, 870–878.
58. Tyagi, R.K.; Goswami, R.; Sanayaima, R.; Singh, R.; Tandon, R.; Agrawal, A. Micropropagation and slow growth conservation of cardamom (*Elettaria cardamomum* Maton). *Vitr. Cell. Dev. Biol.-Plant* **2009**, *45*, 721–729. [CrossRef]
59. Du, Y.P.; Li, W.Y.; Zhang, M.F.; Bin He, H.; Jia, G.X. The establishment of a slow-growth conservation system in vitro for two wild lily species. *Afr. J. Biotechnol.* **2012**, *11*, 1981–1990. [CrossRef]
60. Jain, S.M. Prospects of in vitro conservation of date palm genetic diversity for sustainable production. *Emir. J. Food Agric.* **2011**, *23*, 110–119. [CrossRef]
61. Lyons, J.M. Chilling injury in plants. *Annu. Rev. Plant Physiol.* **1973**, *24*, 445–466. [CrossRef]
62. Graham, D.; Patterson, B.D. Responses of plants to low, nonfreezing temperatures: Proteins, metabolism, and acclimation. *Ann. Rev. Plant Physiol.* **1982**, *33*, 347–374. [CrossRef]
63. Bekheet, S.A. In vitro conservation of date palm tissue cultures. In *Date Palm Biotechnology Protocols Volume II: Germplasm Conservation and Molecular Breeding, Methods in Molecular Biology*; Al-Khayri, J., Jain, S., Johnson, D., Eds.; Springer Science+Business Media LLC.: Berlin/Heidelberg, Germany, 2017; Volume 1638, pp. 15–24. [CrossRef]
64. Bekheet, S.A.; Matter, M.A.H.S.; Taha, E.-A.A.A. In vitro conservation of jojoba (*Simmondsia chinensis*) shootlet cultures using osmotic stress and low temperature. *Middle East J.* **2016**, *5*, 396–402.
65. Rahayu, E.S.; Habibah, N.A. Optimization of in vitro conservation protocol of *Carica pubescens* lenne & k. koch through medium concentration, temperature and irradiation duration decrease. *Biosaintifika J. Biol. Biol. Educ.* **2016**, *8*, 87–95. [CrossRef]
66. de Oliveira, K.S.; Aloufa, M.A.I. Slow growth in vitro culture for conservation of *Hancornia speciosa* gomes. *Floresta* **2022**, *52*, 7–16. [CrossRef]
67. Bhuiyan, M.; Hossain, M.; Haque, M. In vitro conservation of taro (*Colocasia esculenta* var. globulifera) as influenced by mannitol. *Bangladesh J. Agric. Res.* **2016**, *41*, 67–74. [CrossRef]
68. Bello-Bello, J.J.; García-García, G.G.; Iglesias-Andreu, L. Conservación de vainilla (*Vanilla planifolia* jacks.) bajo condiciones de lento crecimiento *in vitro*. *Rev. Fitotec. Mex.* **2015**, *38*, 165. [CrossRef]
69. Bautista-Aguilar, J.R.; Iglesias-Andreu, L.G.; Martínez-Castillo, J.; Ramírez-Mosqueda, M.A.; Ortiz-García, M.M. In vitro conservation and genetic stability in *Vanilla planifolia* Jacks. *HortScience* **2021**, *56*, 1494–1498. [CrossRef]
70. da Silva, R.L.; Ferreira, C.F.; da Silva Ledo, C.A.; de Souza, E.H.; da Silva, P.H.; de Carvalho Costa, M.A.P.; Souza, F.V.D. Viability and genetic stability of pineapple germplasm after 10 years of in vitro conservation. *Plant Cell. Tissue Organ Cult.* **2016**, *127*, 123–133. [CrossRef]
71. Panis, B.; Nagel, M.; Van den Houwe, I. Challenges and prospects for the conservation of crop genetic resources in field genebanks, in in vitro collections and/or in liquid nitrogen. *Plants* **2020**, *9*, 1634. [CrossRef]
72. Rodrigues, P.H.V.; Oliveira, E.; Demetrio, C.; Ambrosano, G.; Piedade, S.M.S. The Effect of slow-grown in vitro storage under different light spectra on banana plantlets Cv. Prata Catarina (AAB). *Res. Sq.* **2021**, *150*, 479–485.
73. Van den houwe, I.; De Smet, K.; du Montcel, H.T.; Swennen, R. Variability in storage potential of banana shoot cultures under medium term storage conditions. *Plant Cell. Tissue Organ Cult.* **1995**, *42*, 269–274. [CrossRef]
74. Van den houwe, I.; Chase, R.; Sardos, J.; Ruas, M.; Kempenaers, E.; Guignon, V.; Massart, S.; Carpentier, S.; Panis, B.; Rouard, M.; et al. Safeguarding and using global banana diversity: A holistic approach. *CABI Agric. Biosci.* **2020**, *1*, 15. [CrossRef]
75. Kanchanapoom, K.; Promssorn, N. Micropropagation and in vitro germplasm conservation of endangered *Musa balbisiana* 'Kluai Hin' (BBB group). *Afr. J. Biotechnol.* **2012**, *11*, 6464–6469. [CrossRef]

76. Zainy, A.H.S.; Erum, S.; Shinwari, Z.K.; Alam, J.; Khan, U.; Jan, S.A. Ex-situ conservation and morpho-biochemical analysis of exotic cultivars of banana. *Pak. J. Bot.* **2019**, *51*, 1703–1709. [CrossRef]
77. Sharma, P.; Jha, A.B.; Dubey, R.S.; Pessarakli, M. Reactive oxygen species, oxidative damage, and antioxidative defense mechanism in plants under stressful conditions. *J. Bot.* **2012**, *2012*, 1–26. [CrossRef]
78. Trejgell, A.; Kamińska, M.; Tretyn, A. In vitro slow growth storage of *Senecio macrophyllus* shoots. *Acta Physiol. Plant.* **2015**, *37*, 234. [CrossRef]
79. Aloni, R. The induction of vascular tissues by auxin. In *Plant Hormones*; Springer: Dordrecht, The Netherlands, 2010; pp. 485–518. ISBN 9781402026867.
80. Lolaei, A.; Mobasheri, S.; Bemana, R.; Teymori, N. Role of paclobutrazol on vegetative and sexual growth of plants. *Int. J. Agric. Crop Sci.* **2013**, *5*, 958–961.
81. Indrayanti, R.; Putri, R.E.; Sedayu, A.; Adisyahputra. Effect of paclobutrazol for in vitro medium-term storage of banana variant cv. Kepok (*Musa acuminata x balbisiana Colla*). *AIP Conf. Proc.* **2018**, *2019*, 020009. [CrossRef]
82. El-Dawayati, M.M. In vitro conservation of date palm shoot-tip explants and callus cultures under minimal growth conditions. In *Methods in Molecular Biology*; Springer: Berlin/Heidelberg, Germany, 2017; Volume 1638, pp. 49–59.
83. Bednarek, P.T.; Orłowska, R. Plant tissue culture environment as a switch-key of (epi)genetic changes. *Plant Cell. Tissue Organ Cult.* **2020**, *140*, 245–257. [CrossRef]
84. Kamińska, M.; Gołębiewski, M.; Tretyn, A.; Trejgell, A. Efficient long-term conservation of *Taraxacum pieninicum* synthetic seeds in slow growth conditions. *Plant Cell, Tissue Organ Cult.* **2018**, *132*, 469–478. [CrossRef]
85. Leva, A.R.; Petruccelli, R.; Rinaldi, L.M.R. Somaclonal variation in tissue culture: A case study with olive. In *Recent Advances in Plant In Vitro Culture*; InTech: Croatia, Balkans, 2012.
86. Cassells, A.C.; Curry, R.F. Oxidative stress and physiological, epigenetic and genetic variability in plant tissue culture: Implications for micropropagators and genetic engineers. *Plant Cell. Tissue Organ Cult.* **2001**, *64*, 145–157. [CrossRef]
87. Tyagi, P.; Khanduja, S.; Kothari, S.L. In vitro culture of *Capparis decidua* and assessment of clonal fidelity of the regenerated plants. *Biol. Plant.* **2010**, *54*, 126–130. [CrossRef]
88. Kaçar, Y.A.; Dönmez, D.; Biçen, B.; Erol, M.H.; Şimşek, Ö.; Mendi, Y.Y. Micropropagation of *Spathiphyllum* with temporary immersion bioreactor system. *Turkish J. Agric.-Food Sci. Technol.* **2020**, *8*, 1195–1200.
89. Simsek, O.; Donmez, D.; Saridas, M.A.; Paydas-Kargi, S.; Aka-Kacar, Y. Genetic relationship and polymorphism of turkish myrtles (*Myrtus communis* L.) as revealed by inter simple sequence repeat (issr). *Appl. Ecol. Environ. Res.* **2020**, *18*, 1141–1149. [CrossRef]
90. Dönmez, D. Regeneration of plants from alginate-encapsulated shoot tips of myrtle (*Myrtus communis* L.). *Erwerbs-Obstbau* **2022**, *64*, 307–314. [CrossRef]
91. Kamińska, M.; Tretyn, A.; Trejgell, A. Genetic stability assessment of *Taraxacum pieninicum* plantlets after long-term slow growth storage using ISSR and SCoT markers. *Biologia* **2020**, *75*, 599–604. [CrossRef]
92. Ruta, C.; Lambardi, M.; Ozudogru, E.A. *Biobanking of Vegetable Genetic Resources by In Vitro Conservation and Cryopreservation*; Springer: Dordrecht, The Netherlands, 2020; Volume 29, ISBN 1053102002051.
93. Guarino, L.; Ramanatha Rao, V.; Reid, R. *Collecting Plant Genetic Diversity: Technical Guidelines*; IUCN—International Union for Conservation of Nature: Gland, Switzerland, 1995.
94. CBD Progress in Implementation of the targets of the global strategy for plant conservation. Eighteenth meeting of the Subsidiary Body on Scientific, Technical and Technological Advice, Montreal, QC, Canada, 23–28 June 2014.
95. Benelli, C. Plant cryopreservation: A look at the present and the future. *Plants* **2021**, *10*, 2744. [CrossRef]
96. Panis, B. Sixty years of plant cryopreservation: From freezing hardy mulberry twigs to establishing reference crop collections for future generations. *Acta Hortic.* **2019**, *1234*, 1–8. [CrossRef]

plants

MDPI

Article

Photosynthetic Parameters and Oxidative Stress during Acclimation of Crepe-Myrtle (*Lagerstroemia speciosa* (L.) Pers.) in a *meta*-Topolin-Based Micropropagation System and Genetic Fidelity of Regenerated Plants

Naseem Ahmad [1], Nigar Fatima [1], Mohammad Faisal [2,*], Abdulrahman A. Alatar [2] and Ranjith Pathirana [3]

1 Plant Biotechnology Laboratory, Department of Botany, Aligarh Muslim University, Aligarh 202002, India; naseembot@gmail.com (N.A.); nigar.fatima77@gmail.com (N.F.)

2 Department of Botany & Microbiology, College of Science, King Saud University, P.O. Box 2455, Riyadh 11451, Saudi Arabia; aalatar@ksu.edu.sa

3 Plant & Food Research Australia Pty Ltd., Waite Research Precinct, Plant Breeding #46, Waite Road, Urrbrae, SA 5064, Australia; ranjith.pathirana@plantandfood.co.nz

* Correspondence: faisalm15@yahoo.com

Abstract: An improved and stable micropropagation system using the cytokinin, *meta*-Topolin (N6 (3-hydroxybenzylamino purine—*m*T), with nodal explants in *Lagerstroemia speciosa* L. was established. Among the different doses of *m*T, the maximum number of shoots with the highest shoot length was obtained using Murashige and Skoog's (MS) medium supplemented with 5.0 μM *m*T. The results were consistent throughout the proliferation period, when recorded at week 4, 8, and 12 of being cultured, with an average of 16.4 shoots per nodal explant, and having a mean length of 4.10 cm at week 8. Shoot proliferation rates could be further improved by a combination of 5.0 μM *m*T with 0.5 μM α-naphthalene acetic acid in MS medium; nodal explants produced an average of 24.3 shoots with a mean length of 5.74 cm after 8 weeks of being cultured. Among the five different concentrations of three auxins tested for the rooting of microshoots in MS medium, a 1.0 μM indole-3-butyric acid treatment was the best, with an average of 10.3 roots per microshoot at an average length of 3.56 cm in 93% of microshoots within 4 weeks of being transferred to this medium. A significant reduction of both chlorophyll a and b in leaves during the first week of acclimation corresponded with a high accumulation of malondialdehyde (MDH), indicating that lipid peroxidation affected chlorophyll pigments. From the second week of acclimation, photosynthetic pigment content significantly increased and MDH content decreased. The net photosynthetic rate and leaf carotenoid content showed almost linear increases throughout the acclimation period. Activity of antioxidant enzymes, namely, superoxide dismutase, catalase, and peroxidases, consistently increased throughout the acclimation period, corresponding with the accumulation of photosynthetic pigments, thus demonstrating the role of the improved antioxidant enzymatic defense system during acclimation. A comparison of parent plant DNA with that of the greenhouse acclimated plants using random amplified polymorphic DNA and inter-simple sequence repeat markers showed a monomorphic pattern indicating genetic stability and the suitability of the method for micropropagation of *L. speciosa*.

Keywords: antioxidant enzymes; carotenoids; conservation; genetic integrity; plant growth regulators; photosynthetic pigments; tissue culture

Citation: Ahmad, N.; Fatima, N.; Faisal, M.; Alatar, A.A.; Pathirana, R. Photosynthetic Parameters and Oxidative Stress during Acclimation of Crepe-Myrtle (*Lagerstroemia speciosa* (L.) Pers.) in a *meta*-Topolin-Based Micropropagation System and Genetic Fidelity of Regenerated Plants. *Plants* **2022**, *11*, 1163. https://doi.org/10.3390/plants11091163

Academic Editor: Stephen O. Amoo, Anna De Carlo and Waed Tarraf

Received: 1 March 2022
Accepted: 23 April 2022
Published: 26 April 2022

1. Introduction

Queen Crape Myrtle *Lagerstroemia speciosa* Pers. (Lythraceae) is a tropical, medium-large sized tree that can grow up to 25 m and is of commercial importance as an ornamental plant and avenue tree. Tracing its origin to "Tropical Southern Asia" [1], its aesthetic beauty, accompanied by spectacular blooming with large bright pink to lavender flowers with various shades, has made it the "Queen" among flowers where it is grown, and it is also known as "Pride of India". Having moderately hard timber, the tree is used to make decorative furniture, agricultural implements, pulp, paper, ploughs, boats, fence posts, and so on. The species is also used in biofertilizer manufacturing [2] as a fodder component for afforestation and ecological restoration [3]. The demand for plants is high in many countries because their uses are multipurpose, including plantation establishment, urban horticulture, reforestation, desert rehabilitation, use in furniture, paper, pulp, and timber industries, as well as for medicinal purposes [3,4].

Bioactive compounds in the leaves and fruits of *L. speciosa*, such as ellagitannins (lagerstroemin, reginin A, flosin B, reganin C and D) [5], alanine, methionine, lageracetal, amyl alcohol, and β-sitosterol [6], serve as raw materials in pharmaceutical industries. Additionally, the leaf extracts have anti-oxidant, anti-inflammatory, anti-hypertension, analgesic, diuretic, thrombolytic, anti-cancer [7,8], antifungal, and antiviral properties [9]. The leaves act as a febrifuge and cleansing agent to regulate metabolic processes [10]. Leaf beverages (tea) possess anti-oxidative [11], anti-inflammatory [3,6], anti-hypertensive [8], diuretic, purgative, anti-ulcer, and anti-gout properties [12,13]. Numerous commercial formulations and industrial products (for example, Swanson Banaba extract) are available, including medicinal tablets, capsules, cosmetics, lotions, hair tonics, and hypoglycemic foods. The majority of formulations contain corosolic acid derived from the plant. Corosolic acid is known as "phyto-insulin" or "botanical insulin" due to its insulin-like activities, and its high level in *L. speciosa* implies that this plant may have anti-diabetic therapeutic potential. [14]. Hypoglycemic effects of *L. speciosa* have been proven not only in animal and in vitro studies, but also in clinical studies, for example, using Glucosol^{TM} (standardized to 1% corosolic acid) in Type II diabetes patients [15].

Due to its many valuable characteristics, *L. speciosa* is has been over-exploited in its natural habitat, leading to a scarcity of mature trees [16]. A low density of plants and a scarcity of pollinators means that much of the seed is fertilized through self-pollination, resulting in aborted seed [17]. Viable seeds have a short lifespan (1–2 months), making the establishment of seedlings difficult [18]. Additionally, vegetative propagation of this plant is not widely practiced. To meet these challenges, improved micropropagation methods are required. In vitro propagation of this species is complicated by poor rooting after use of the widely used synthetic cytokinin (CK) and N6-benzylaminopurine (BAP) at the proliferation stage, requiring forced ventilation of vessels [19]. When kinetin (Kn) was used, callus formation was an impediment both at the proliferation and rooting stages, with low numbers of shoots produced [20].

meta-Topolin (*m*T) (6-(3-hydroxybenzylamino) purine), is an innate aromatic cytokinin (ArCK) originally isolated from *Populus* X *robusta* leaves [21,22]. Being the most biologically active ArCK belonging to the group "Topolins", *m*T has been found to have distinct advantages in morphogenesis, differentiation, and subsequent rooting and acclimation in the micropropagation industry [21]. It is a hydroxylated form of BAP that produces an *O*-glucoside with relatively higher activity than other derivatives, resulting in less hyperhydricity in tissue culture, thus facilitating easier rooting and acclimation of rooted plantlets [22–24].

Recently, the use of *m*T to replace more traditional CKs, such as BAP and zeatin, at the proliferation stage of micropropagation has been reported to result in better rooting and acclimation of plantlets in diverse species, including *Spathiphyllum floribundum* (Araceae) [22], *Ananas comosus* (Bromeliaceae) [25], *Eucalyptus* spp. (Myrtaceae) [26], *Cannabis sativa* (Cannabaceae) [23], *Corylus colurna* (Betulaceae) [27], *Manihot esculenta* (Euphorbiaceae) [28], and *Pterocarpus marsupium* (Fabaceae) [29]. Replacement of BAP with *m*T reduced hyperhydricity, increased the multiplication rate, and resulted in the spontaneous rooting of *Cannabis sativa* [23] and *Eucalyptus* spp. [26]. In comparison with BAP, *m*T supplementation resulted in more shoots with leaves, which had better morphological and anatomical features, and a higher amount of chlorophyll in *Sesamum indicum* (Pedaliaceae) [30], *Oxystelma esculentum* (Asclepiadaceae) [31], and *Vanilla planifolia* (Orchidaceae) [32]. In a study involving 14 contrasting genotypes of gooseberry (*Ribes grossularia*—Grossulariaceae), Kucharska, et al. [33] found that *m*T-supplementation improved the shoot proliferation to 2.8, in comparison with 1.8, which was obtained using the optimal concentration of BAP, which resulted in longer and healthier shoots in subsequent subcultures. Experiments involving multiple CKs, such as Kn and BAP [29], Kn, BAP and thidiazuron [34], and BAP, Kn, and 2-isopentyl adenine [24], have also shown the superiority of *m*T in its capacity to produce healthy shoots that are more amenable to rooting and acclimation. Furthermore, the problem of recalcitrance of kiwifruit genotypes (*Actinidia chinensis*—Actinidiaceae) to the rooting of shoots proliferated in BAP or zeatin media, could be solved by switching to *m*T in culture media for micropropagation [35], and in the management of a large in vitro collection [36].

These advantages of *m*T over some other CKs used in plant tissue culture can be attributed to the lesser oxidative stress experienced by explants cultured on *m*T-supplemented media. For example, plants grown in *m*T-supplemented media had lower H_2O_2 content than in BAP-supplemented media in *Daphne mezereum* (Thymelaeaceae) [37]. *Corylus colurna* (Betulaceae) plantlets cultured in a medium containing BAP showed a significantly higher H_2O_2 content, enhanced antioxidant enzyme (superoxide dismutase (SOD), ascorbate peroxidase, and catalase (CAT)) activities, and lower chlorophyll than shoots derived from the medium containing *m*T [27].

However, there are no reports on the use of *m*T in *L. speciosa* micropropagation or in any other species in the plant family Lythraceae; therefore, the objective of our work was to optimize the concentration of *m*T in the proliferation and multiplication stages of *L. speciosa* micropropagation, and to investigate if further improvements can be achieved when combined with different auxins. Additionally, we used molecular markers to test if the *L. speciosa* plants regenerated using *m*T in proliferation media were true-to-type, and if the chlorophyll content and oxidative stress status of plants during the acclimation period could improve our understanding of the ability of tissue cultured plants to recover and re-establish themselves.

2. Results

2.1. In Vitro Establishment and Optimizing Shoot Proliferation and Rooting of Shoots

Nodal explants collected from mature *L. speciosa* plants were successfully established within 2–3 weeks of initiation. All of the nine *meta*-Topolin (*mT*) concentrations within the range 0.1–15.0 μM in MS medium resulted in the initiation of multiple shoots from the cultured nodal segments (Table 1). There was no shoot production from nodal explants when the medium was not supplemented with *mT*. Supplementation of the medium with 5.0 μM *mT* resulted in a significantly higher number of shoots that were also the longest among the nine concentrations tested (Figure 1A). This result was consistent throughout the period of growth when measurements were taken, from week 4 to 12 (Table 1). With the increase in concentration of *mT* in the medium beyond 5.0 μM, not only did the number of shoots decrease, but also there was an increase in vitrification, resulting in difficulties with rooting and plant acclimation.

When the optimal *m*T concentration for shoot proliferation of *L. speciosa* (5.0 μM *m*T) from the first experiment was combined with different concentrations of three auxins, α-naphthaleneacetic acid (NAA) produced a better proliferation of shoot buds compared with indole-3-butyric acid (IBA) and indole-3-acetic acid (IAA) (Table 2). After 8 weeks in a culture, 0.5 μM NAA in combination with 5.0 μM *m*T resulted in a significantly higher number of shoots than any other combination (Table 2; Figure 1B,C), or with *m*T alone (Table 1), and the highest proportion of explants (95%) producing shoots. Supplementation with IBA or IAA did not have a positive effect on the proliferation of shoot buds, whereas 0.1 and 1.0 μM NAA produced slightly higher numbers of shoots (17.5 and 18.2, respectively) than 5.0 μM *m*T alone (16.4) in 85% of cultures after 8 weeks (Tables 1 and 2). The shoots produced using MS medium supplemented with 5.0 μM *m*T and 0.5 μM NAA were also significantly longer than all other combinations, except the combination with 5.0 μM *m*T and 1.0 μM NAA (Table 2).

Table 1. Effect of different concentrations of *meta*-Topolin in MS basal medium on in vitro shoot regeneration from nodal explants of *Lagerstroemia speciosa* L.

meta-Topolin Concentration (μM)	4 Weeks		8 Weeks		12 Weeks	
	Number of Shoots	Shoot Length (cm)	Number of Shoots	Shoot Length (cm)	Number of Shoots	Shoot Length (cm)
0.0	0.00 ± 0.00 ^{j}	0.00 ± 0.00 ^{g}	0.00 ± 0.00 ^{i}	0.00 ± 0.00 ^{f}	0.00 ± 0.00 ^{j}	0.00 ± 0.00 ^{f}
0.1	2.04 ± 0.04 ^{i}	0.88 ± 0.02 ^{f}	5.20 ± 0.00 ^{h}	0.92 ± 0.16 ^{e}	6.85 ± 0.03 ^{i}	1.60 ± 0.10 ^{e}
0.5	3.20 ± 0.20 ^{h}	1.32± 0.08 ^{e}	7.62 ± 0.17 ^{e}	1.62 ± 0.18 ^{d}	10.2 ± 0.18 ^{h}	3.36 ± 0.16 ^{d}
1.0	4.36 ± 0.36 ^{g}	2.50 ± 0.00 ^{c}	12.8 ±0.04 ^{c}	2.70 ± 0.33 ^{bc}	14.1 ± 0.10 ^{f}	3.86 ± 0.02 ^{e}
2.5	8.10 ± 0.10 ^{d}	3.20 ± 0.20 ^{b}	14.4 ± 0.18 ^{b}	3.08 ± 0.50 ^{b}	18.5 ± 0.15 ^{c}	4.60 ± 0.00 ^{b}
5.0	12.0 ± 0.02 ^{a}	3.76 ± 0.06 ^{a}	16.4± 0.12 ^{a}	4.10 ± 0.10 ^{a}	22.4 ± 0.02 ^{a}	5.22 ± 0.00 ^{a}
7.5	10.0 ± 0.00 ^{b}	2.52 ± 0.11 ^{c}	11.5 ± 0.15 ^{d}	3.12 ± 0.09 ^{b}	20.2 ± 0.20 ^{b}	4.84 ± 0.08 ^{b}
10.0	9.16 ± 0.04 ^{c}	2.00 ± 0.00 ^{d}	6.50 ± 0.00 ^{f}	2.30 ± 0.08 ^{cd}	15.6 ± 0.10 ^{d}	4.40 ± 0.04 ^{b}
12.5	7.10 ± 0.10 ^{e}	1.20 ± 0.00 ^{e}	5.60 ± 0.17 ^{g}	1.50 ± 0.00 ^{de}	14.7 ± 0.06 ^{e}	3.80 ± 0.40 ^{c}
15.0	5.00 ± 0.00 ^{f}	0.80 ± 0.00 ^{f}	5.00 ± 0.00 ^{h}	1.20 ± 0.20 ^{de}	11.4 ± 0.04 ^{g}	2.96 ± 0.06 ^{d}

Values represent mean ± SE of three repeated experiments, each with ten replications. Means in a column with different letters (superscript) are significantly different according to Duncan's multiple range test ($p \leq 0.05$).

Table 2. Effect of different concentrations of auxins in combination with 5 μM *meta*-Topolin in MS basal medium on shoot initiation and proliferation from nodal segments of *Lagerstroemia speciosa* L. after 8 weeks of being cultured.

Plant Growth Regulators (μM)			% Regeneration	Number of Shoots/Explant	Mean Shoot Length (cm)
IAA	IBA	NAA			
0.1			70	12.5 ± 0.19 ^{gh}	4.40 ± 0.40 ^{ef}
0.5			75	14.4 ± 0.40 ^{e}	5.00 ± 0.00 ^{cd}
1.0			60	8.70 ± 0.12 ^{j}	4.92 ± 0.02 ^{cd}
1.5			70	8.00 ± 0.00 ^{k}	4.20 ± 0.00 ^{efg}
2.0			65	7.48 ± 0.12 ^{l}	3.92 ± 0.02 ^{g}
	0.1		80	13.6 ± 0.08 ^{f}	4.68 ± 0.08 ^{de}
	0.5		85	16.0 ± 0.00 ^{d}	5.22 ± 0.02 ^{bc}
	1.0		75	14.6 ± 0.16 ^{e}	4.34 ± 0.04 ^{ef}
	1.5		60	12.9 ± 0.31 ^{g}	3.84 ± 0.05 ^{g}
	2.0		60	10.0 ± 0.00 ^{i}	2.60 ± 0.10 ^{h}
		0.1	85	17.5 ± 0.36 ^{c}	5.16 ± 0.05 ^{bc}
		0.5	95	24.3 ± 0.19 ^{a}	5.74 ± 0.06 ^{a}
		1.0	85	18.2 ± 0.20 ^{b}	5.50 ± 0.00 ^{ab}
		1.5	80	15.8 ± 0.30 ^{d}	5.20 ± 0.20 ^{bc}
		2.0	75	12.0 ± 0.00 ^{h}	5.04 ± 0.24 ^{cd}

Values represent mean ± SE of three repeated experiments each with ten replications. Means in a column with different letters (superscript) are significantly different according to Duncan's multiple range test ($p \leq 0.05$). Plant growth regulators: IAA—Indole-3-acetic acid, IBA—Indole-3-butyric acid and NAA—α-naphthaleneacetic acid.

For the rooting experiment, microshoots 3–4 cm in length were harvested from the optimal shoot proliferation medium after 8 weeks of growth and transferred to BM supplemented with one of five concentrations of IAA, NAA, or IBA. The frequency of root initiation varied with auxin concentration (0.5–2.5 μM) (Table 3). BM supplemented with 1 μM IBA produced a significantly higher number of roots per explant and the highest mean length of roots (Table 3; Figure 1D,E). Root development was also the highest in this treatment, with 93% of microshoots producing roots. IBA at 0.5 and 1.5 μM concentrations, and 1.0 μM NAA, also produced high numbers of roots, but the roots in these treatments were much shorter than in the optimum 1 μM NAA concentration. In general, the number of roots produced in IAA-supplemented media was lower than in IBA- and NAA-supplemented media.

Figure 1. Micropropagation of *Lagerstroemia speciosa* L. (**A**) Multiple shoot induction from nodal segments cultured for 4 weeks using MS medium supplemented with 5.0 µM *m*eta-Topolin (*m*T) (Bar = 0.86 cm); (**B**) mass multiplication of shoots after 8 weeks of being cultured using MS medium supplemented with 5.0 µM *m*T (Bar = 0.89 cm); (**C**) multiplication and elongation of shoots after 12 weeks using the optimal concentration of *m*T (5.0 µM) + NAA (0.5 µM) (Bar = 1.70 cm); (**D**) rooting of in vitro shoots after 4 weeks using MS medium supplemented with 1 µM indole-3-butyric acid (Bar = 0.60 cm); (**E**) rooted microshoot ready for transplantation (Bar 0.71 cm); and (**F**) a 4-week-old acclimatized plant (Bar = 0.71).

Table 3. In vitro rooting response of *Lagerstroemia speciosa* L. microshoots in basal medium supplemented with indole-3-acetic acid (IAA), indole-3-butyric acid (IBA), and α-naphthaleneacetic acid (NAA) after 4 weeks in the culture.

Plant Growth Regulators (μM)			% Shoots with Roots	Number of Roots Per Microshoot	Root Length (cm)
IAA	IBA	NAA			
0.0			00	0.00 ± 0.00^{k}	0.00 ± 0.00^{h}
0.5			85	3.44 ± 0.14^{g}	2.46 ± 0.06^{b}
1.0			85	4.36 ± 0.14^{f}	2.51 ± 0.22^{b}
1.5			70	3.84 ± 0.04^{gh}	2.02 ± 0.12^{d}
2.0			65	3.24 ± 0.06^{h}	1.98 ± 0.12^{cd}
2.5			60	2.60 ± 0.10^{ij}	1.60 ± 0.10^{ef}
	0.5		65	7.36 ± 0.16^{c}	2.20 ± 0.20^{bc}
	1.0		93	10.3 ± 0.12^{a}	3.56 ± 0.05^{a}
	1.5		70	8.20 ± 0.20^{b}	1.76 ± 0.06^{de}
	2.0		60	6.52 ± 0.21^{d}	1.54 ± 0.04^{ef}
	2.5		50	3.20 ± 0.20^{h}	1.36 ± 0.16^{f}
		0.5	70	5.69 ± 0.22^{e}	1.78 ± 0.12^{de}
		1.0	75	7.24 ± 0.26^{c}	2.24 ± 0.07^{bc}
		1.5	65	6.10 ± 0.04^{de}	2.08 ± 0.20^{cd}
		2.0	65	2.72 ± 0.13^{i}	1.54 ± 0.05^{ef}
		2.5	55	2.20 ± 0.00^{j}	0.84 ± 0.04^{g}

Values represent mean ± SE of three repeated experiments, each with 20 replications. Means in a column with different letters (superscript) are significantly different according to Duncan's multiple range test ($p \leq 0.05$). Plant growth regulators: IAA—Indole-3-acetic acid, IBA—Indole-3-butyric acid and NAA—α-naphthaleneacetic acid.

2.2. Acclimation and Assessment of Photosynthetic Pigments, Antioxidative Enzymes and Genetic Fidelity

Overall, 85% plants survived the acclimation process and transfer to garden soil (Figure 1F). The concentration of chlorophyll a and b behaved similarly over the acclimation period; the concentration declined in the first week and quickly recovered by the third and fourth weeks (Figure 2). This difference was statistically significant for both photosynthetic pigments. On the other hand, the photosynthetic rate and carotenoid concentration displayed a more stable increase without any decline after exflasking (Figure 2). The concentrations of all three assayed antioxidant enzymes increased over the period of acclimation, reaching their maximum in the fourth week, when the plants were transferred to the greenhouse (Figure 3). MDA production, an indication of lipid peroxidation, was low in the rooted plantlets at the time of exflasking, but increased almost threefold in the first week of acclimation, after which it decreased gradually until the plants were transferred to the greenhouse (Figure 4).

RAPD and ISSR markers were used for testing the genetic fidelity of micropropagated plants grown in the greenhouse. The mother plant was used as the control. The screening of 18 RAPD primers generated a total of 127 amplified bands with an average of 7.0 bands/primer (Table 4, Figure 5A). Among the 13 ISSR primers screened, 9 primers showed clear, reproducible bands with an average of 7.8 bands/primer (Table 5, Figure 5B). All the bands were monomorphic among the tested plants and the mother plant.

Table 4. Randomly amplified polymorphic DNA primers (RAPD) used to screen ten micropropagated plantlets of *Lagerstroemia speciosa* L.

S. No.	Kit A		
	Primers	Primers Sequences (5′-3′)	No. of Bands
1	OPA 01	GTTTCGCTCG	9
2	OPA 02	TGATCCCTGG	11
3	OPA 03	CATCCCCCTG	8
4	OPA 04	GGACTGGAGT	10

Table 4. *Cont.*

S. No.	Kit A		
	Primers	Primers Sequences (5′-3′)	No. of Bands
5	OPA 05	TGCGCCCTTC	6
6	OPA 06	TGCTCTGCCC	3
7	OPA 07	GGTGACGCAG	10
8	OPA 08	GTCCACACGG	11
9	OPA 09	TGGGGGACTC	9
10	OPA 10	CTGCTGGGAC	9
11	OPA 11	GTAGACCCGT	8
12	OPA 12	CCTTGACGCA	5
13	OPA 13	TTCCCCCGCT	3
14	OPA 14	TCCGCTCTGG	Nil
15	OPA 15	GGAGGGTGTT	3
16	OPA 16	TTTGCCCGGA	3
17	OPA 17	AGGGAACGAG	9
18	OPA 18	CCACAGCAGT	7
19	OPA 19	ACCCCCGAAG	3
20	OPA 20	GGACCCTTAC	Nil

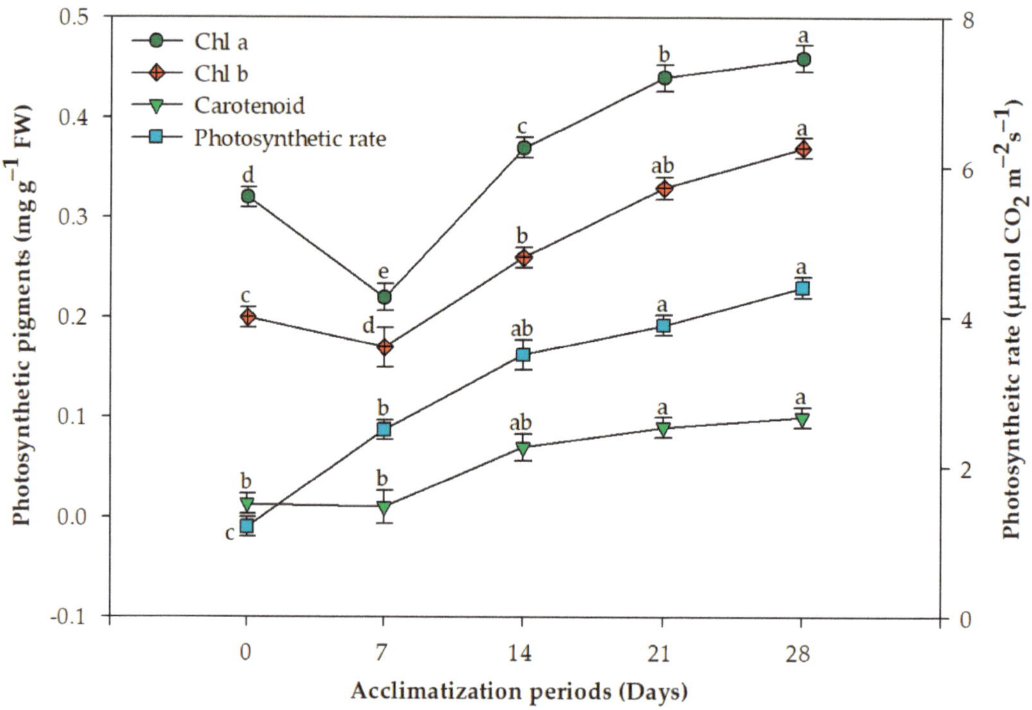

Figure 2. Changes in chlorophyll a, chlorophyll b, carotenoid content, and net photosynthetic rate during the period of acclimation for *Lagerstroemia speciosa* L. plantlets. Line graphs represent mean ± SE of three repeated experiments, each with 20 replications. Means within a category with different letters are significantly different according to Duncan's multiple range test ($p \leq 0.05$).

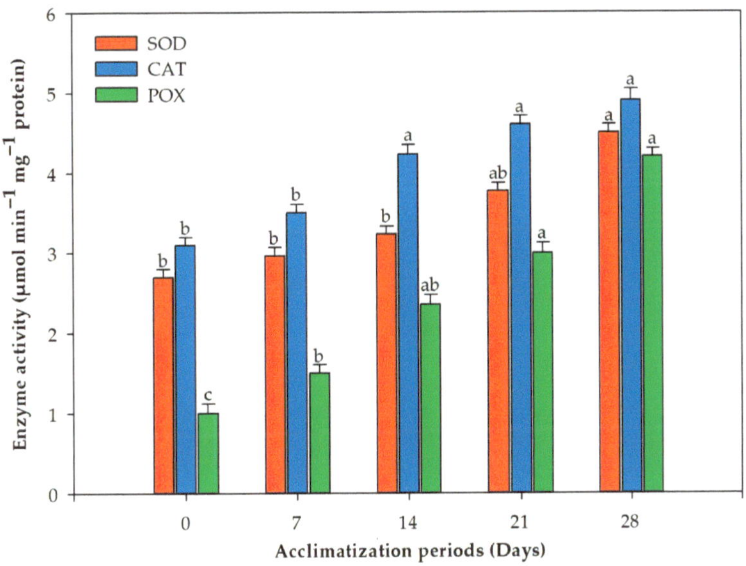

Figure 3. Changes in catalase (CAT), peroxidase (POX), and superoxide dismutase (SOD) activity in leaves of *Lagerstroemia speciosa* L. during the acclimation period. Bars represent mean ± SE of three repeated experiments, each with 20 replications. Means of an enzyme activity with different letters are significantly different according to Duncan's multiple range test ($p \leq 0.05$).

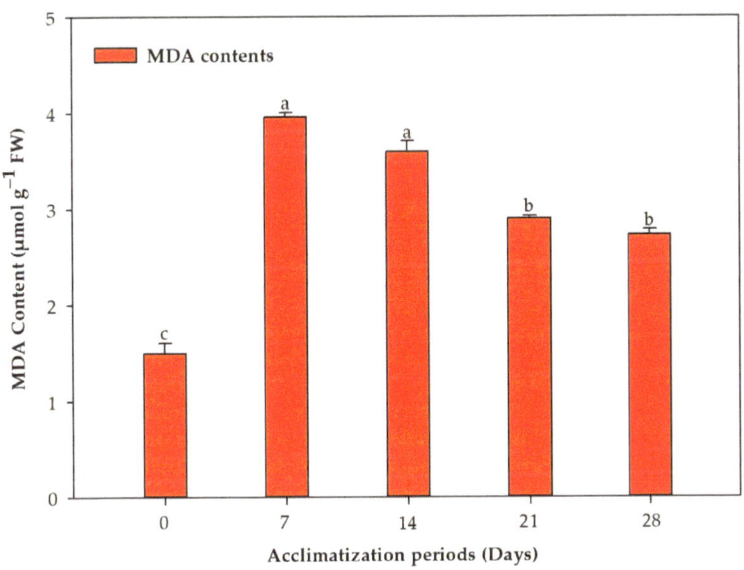

Figure 4. Changes in malondialdehyde (MDA) concentration in *Lagerstroemia speciosa* L. leaves during the acclimation period. Bars represent the mean ± SE of three repeated experiments each with 20 replications. Means in a column with different letters are significantly different according to Duncan's multiple range test ($p \leq 0.05$).

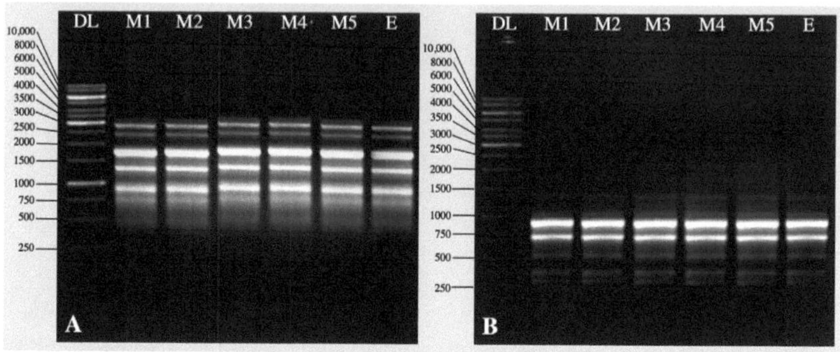

Figure 5. Random amplified polymorphic DNA (RAPD) and inter-simple sequence repeat (ISSR) profiles of *Lagerstroemia speciosa* L. (**A**) Amplified profile from RAPD primer OPA-01; (**B**) amplified profile from ISSR primer UBC-880. DL—DNA ladder; lanes 1–5 randomly selected in vitro plants; lane E—Donor plant.

Table 5. Inter-simple sequence repeats (ISSR) primers used to validate the genetic fidelity of micropropagated plantlets of *Lagerstroemia speciosa* L.

S. No.	Primers	Primers Sequences (5′-3′)	Annealing Temperature (°C)	No. of Bands
1	UBC-801	$(AT)_8T$	35.0	3
2	UBC-811	$(GA)_8C$	49.0	5
3	UBC-825	$(AC)_8T$	45.7	Nil
4	UBC-827	$(AC)_8G$	49.0	Nil
5	UBC-834	$(AG)_8YT$	49.0	9
6	UBC-841	$(GA)_8YC$	49.0	Nil
7	UBC-855	$(AC)_8YT$	49.0	6
8	UBC-866	$(CTC)_6$	55.4	Nil
9	UBC-868	$(GAA)_6$	45.7	2
10	UBC-880	$(GGGGT)_3G$	49.0	12
11	UBC-889	$DBDA(CA)_6C$	45.7	13
12	UBC-891	$HVHT\ (GT)_6G$	45.7	10
13	UBC-900	ACTTCCCCACAGGTTAACAC	58.0	11

3. Discussion

In this study, we report the development of an efficient micropropagation protocol optimized for shoot proliferation from readily available nodal segments of mature trees of *L. speciosa* using *m*T and NAA in a 10:1 proportion, followed by rooting with IBA in MS medium. Previously, when BAP was used at the proliferation stage in *L. speciosa*, rooting was successful when forced ventilation was used in the containers [19]; however, this is costly and impractical for mass-scale operations. In another experiment with this species, the number of shoots per explant obtained with the optimal BAP concentration after 4 weeks was only two [16], compared with the 12 shoots that were obtained using *m*T alone (Table 1) in our research. Furthermore, Lim-Ho and Lee [20] reported that Kn in any concentration, or increasing the BAP concentration beyond 1 mg/L, resulted in callus induction and poor plant regeneration in *L. speciosa*. They managed to increase shoot proliferation using N6-(2-isopentenyl) adenine (2iP) in MS medium; however, CKs with an isoprenoid side chain such as 2iP have the disadvantage of fast oxidation compared with ArCKs [38].

The superiority of *m*T over BAP in micropropagation has been reported recently in other perennial species with medicinal values such as *Oxystelma esculentum* [31], *Pterocarpus marsupium* [29], *Scaevola taccada* [39], and *Bacopa monnieri* [40], as well as in perennial species used in horticulture such as *Daphne mezereum* [37], *Ribes grossularia* [33], *Actinidia chinensis* [35], and *Franklinia alatamaha* [41]. Working on two *Prunus* rootstocks, Gentile, et al. [42] observed that adventitious regeneration from in vitro leaf explants was possible in leaves from shoots growing in media supplemented with *m*T but not BAP; therefore, our research on the micropropagation of *L. speciosa* focused on *m*T. Our first experiment that used *m*T alone in MS media was promising, with a higher number of shoots proliferating than other CKs in *L. speciosa* propagation, based on what has been reported thus far [17,20,21]. When the optimum *m*T concentration of 5 μM was combined with different auxins, we found that the propagation efficiency and quality of shoots could be further improved in combination with 0.5 μM NAA. The synergistic actions of *m*T in combination with auxins in achieving significantly improved shoot multiplication rates have been documented in *Aloe polyphylla* [43], *Uniola paniculata* [44], and *Pterocarpus marsupium* [29].

The problems described in most of the research on micropropagation using BAP, zeatin, and Kn in perennial species, include the low regeneration potential, callus induction at the basal end of the explant, vitrification, and difficulty in rooting micropropagated shoots [36,45,46]. The improvement in shoot proliferation without callus induction when BAP or other CKs are replaced by *m*T has been attributed to its higher activity because of the presence of a hydroxyl group in the N9-position of the purine ring, leading to the formation of the *O*-glucoside conjugate N6(3-O-β-D-glucopyranosyl) benzyladenine-9-glucoside [22,47]. It has also been shown that substitution at the N9 position of CK's purine ring significantly enhances acropetal transport of the CK, impeding the accumulation of CK glucoside forms in roots [45,46]. This allows for the gradual release of the active form of CK, enhancing the distribution of endogenous CKs in different plant tissues, and thus improving rooting and overall micropropagation efficiency [38]. In our research, as expected, *m*T-induced shoots of *L. speciosa* produced roots with all three auxins tested; however, 1 μM IBA proved to be significantly better in terms of the number of roots, root length, and the percentage of shoots (93%) rooted. This was followed by the successful establishment of a high percentage (95%) of rooted plantlets.

During in vitro culture, plantlets are subjected to diverse stresses that amplify the production of reactive oxygen species (ROS); however, unlike under in vivo conditions, in vitro stress has been studied less in terms of cellular mechanisms to manage stress; namely, constitutive and induced production of radical scavengers, free radical and oxidized-protein enzymatic degradation pathways, and DNA repair mechanisms [48,49]. Oxidative stress in the pathway is associated with the reprogramming of explants to produce fully functional plants results in the peroxidation of lipids, leading to membrane damage, protein degradation, enzyme inactivation, and genetic and epigenetic changes [48,50]. Demonstrating that the elevated amounts of free radicals and their reaction products can lead to morphogenetic recalcitrance in plant tissue culture, Benson [51] emphasized the need for combining studies on oxidative stress with plant development. In our research we focused on the plant acclimation stage, one of the critical stages of micropropagation, particularly because the *m*T -induced proliferation rates were high and produced shoots that were healthy in the optimized media. Furthermore, only a few studies on the oxidative stress in plants during the acclimation period have been conducted.

We combined studies on photosynthetic rates and pigment production with antioxidant enzyme activity during the period of acclimation. Our results revealed that there is a significant decrease in chlorophyll pigments during the first week of acclimation, and this negatively correlated with MDA content in the leaves, indicating high lipid peroxidation at this stage. Activity of all three oxidative enzymes, CAT, POX and SOD, had similar trends throughout the acclimation period; their activity increased as the plants acclimatized. Similar trends in carotenoid accumulation were also observed. Tissue-culture-raised plantlets are exposed to low PPFD. During acclimation, and in the greenhouse, plants are subjected to elevated PPFD that results in photo-inhibition [52]. Carotenoids play a pivotal role in defending the photosynthetic apparatus from photo-oxidative injuries [53], and their elevation along with high activity of ROS scavenging enzymes in *L. speciosa* during acclimation was a result of the plants reacting to the stresses of acclimation, resulting in high survival of our regenerated plantlets. Similar to our results, Goncalves, et al. [54] observed that chlorophyll and carotenoid contents in *Plantago algarbiensis* increased significantly during acclimation compared to the in vitro stage; however, these parameters in *P. almogravensis*, the other *Plantago* species they studied, remained high throughout the in vitro, acclimation, and greenhouse stages [54]. This shows a need to study these processes in the latter stages of micropropagation in different species to understand the outcomes. Similarly, Ahmad, et al. [55] found that the chlorophyll and carotenoid contents of *Decalepis salicifolia* decreased in the first week of acclimation, followed by increases in subsequent weeks; antioxidant enzyme activities also behaved similarly to our results with *L. speciosa*.

The retention of genetic consistency in regenerated plants is critical for any micropropagation system, as well as for conservation programs; therefore, it is important to appraise the genetic consistency of tissue culture-raised plants prior to commercialization or germplasm conservation. In the present experiment, both ISSR and RAPD primers showed high monomorphic patterns, and a way of authenticating the genetic integrity in the micropropagation system was developed. Our results indicate that axillary bud multiplication using *m*T and NAA is a safe method for producing true-to-type plants of *L. speciosa*. The variations observed in the 'Williams' banana when topolins or BAP were used, were attributed to excessive subculture cycles [56]. Moreover, the variation was not different between plants regenerated using BAP and *m*T in the 'Williams' banana [57] and sesame [30], whereas the regenerants were stable in *Salvia sclarea* [34] and *Daphne mezereum* [37]. There are a number of studies confirming the genetic fidelity of regenerated plants when the cytokinin for axillary bud proliferation is optimized for micropropagation in woody species with medicinal values, such as *Withania somnifera* [58], *Rauvolfia serpentina* [59], and *Cassia alata* [60]. Our results also show that somaclonal variation can be avoided when the growth regulators are used in optimal concentrations and subculturing is performed at regular intervals.

4. Materials and Methods

4.1. Plant Materials and Surface Sterilization

Juvenile healthy shoots of mature *L. speciosa* plants growing at the Department of Botany of Aligarh Muslim University, Aligarh, Uttar Pradesh, India was collected and washed under running tap water for about 30 min, then treated with the laboratory detergent "Labolene" (Qualigens, Mumbai, India) 5% (*v*/*v*) for 5 min, followed by three to four washes with sterile distilled water. The washed shoots were surface sterilized with $HgCl_2$ 0.1% (*w*/*v*) for 4–5 min followed by repeated washing with autoclaved distilled water under aseptic conditions in a laminar air flow cabinet. Nodal explants (1–1.5 cm) were isolated and cultured on sterile initiation media of different compositions as described below.

4.2. Basal Media and Culture Environment

The basal growth medium (BM) comprised Murashige and Skoog [61] salts (MS), 3% (*w*/*v*) sucrose, and 0.8% (*w*/*v*) agar (Qualigens Fine Chemicals, Mumbai, India). The medium pH was adjusted to 5.8 prior to adding agar, and the media was then autoclaved at 121 °C for 20 min. The cultures were incubated in a growth room at 25 ± 2 °C with 50 µmol m^{-2} s^{-1} photosynthetic photon flux density (PPFD) provided by white 40 W fluorescent lights (Philips India Ltd., Kolkata, India) at culture container level and 16 h photoperiod.

4.3. Shoot Initiation and Proliferation

In the first experiment, nodal explants (axillary buds) were isolated aseptically from surface sterilized shoots and cultured on BM supplemented with different concentrations of *m*T (0.0, 0.1, 0.5, 1.0, 2.5, 5.0, 7.5, 10.0, 12.5, and 15.0 µM) alone. The number of shoots per explant and the length of shoots were recorded on the 4th, 8th, and 12th week. To test if a combination with auxin would further enhance shoot proliferation, the *m*T treatment that gave the best proliferation of shoots (5 μM) was combined with α-naphthaleneacetic acid (NAA), indole-3-butyric acid (IBA), or indole-3-acetic acid (IAA) at five different concentrations (0.1, 0.5, 1.0, 1.5, and 2.0 μM) for the second shoot initiation/proliferation experiment. For this experiment, the shoots were initiated as described above and the number of shoots per explant and the length of shoots were recorded on the 8th week. The shoots from the best treatment in this experiment were used for the rooting experiment described below.

4.4. In Vitro Rooting and Plantlet Acclimation

Microshoots (3–4 cm) were dissected aseptically from 8-week-old proliferating axillary bud explants and transferred to root induction media consisting of BM supplemented with one of three auxins viz., IAA, NAA or IBA at 0.5. 1.0, 1.5, 2.0, 2.5 μM concentrations. Root growth was scored after 4 weeks in the culture as rooting percentage, mean number of roots per responding microshoot and mean root length. Microshoots grown on BM with no auxins served as the control.

For acclimation, shoots that had produced roots after 4 weeks of being in the culture under the same conditions as described for shoot organogenesis were removed from the culture tubes, washed to remove agar under tap water, and immediately transferred to polystyrene cups holding sterile Soilrite™ (Keltech Energies Ltd., Bangalore, India). The cups holding plantlets were covered in transparent polythene bags and fed with half-strength MS liquid medium without sucrose at 3-day intervals. The polythene bags were opened gradually over a period of 2 weeks for hardening of the regenerated plantlets. After a further 2 weeks, the acclimatized plants were shifted to pots with regular garden soil and were maintained inside a net house providing 40% filtered sunlight. The number of plants with roots, number of roots per plant, and root length were documented prior to transfer to garden soil.

4.5. Photosynthetic Pigments and Net Photosynthetic Rate Assessment

The method described by Mackinney [62] was used to estimate the content of chloro phyll a and b, and the method of Maclachlan and Zalik [63] for carotenoids in leaf tissue. Approximately 100 mg tissue from the interveinal region of leaves in plants at 0 (at the time of exflasking), 7, 14, 21, and 28 (completely developed leaves) days during acclimation were ground in 5 mL acetone (80%) using a mortar and pestle, and the suspension was filtered through Whatman No. 1 filter paper. The supernatant was ground again, filtered, and the entire filtrate was poured into a test tube and the volume adjusted to 10 mL with 80% acetone. For the chlorophyll content, the optical densities (OD) were recorded at 645 and 663 nm using a spectrophotometer (UV-Pharma Spec 1700, Shimadzu, Japan), whereas, for carotenoids, the OD was recorded at 480 and 510 nm.

The net photosynthetic rate in micropropagated plantlets was quantified at 0 (at the time of exflasking), 7, 14, 21, and 28 days after exflasking using an infra-red gas analyzer (IRGA, LICOR 6400, Lincoln, NE, USA). The photosynthetic rate was quantified using the source of net CO_2 exchange among the leaves with the environment, by enfolding leaves in the leaf compartment of the gas analyzer and recording the rate at which the CO_2 concentration altered over a 10–20 sec period. The net photosynthetic rate was expressed in μmol CO_2 m^{-2} s^{-1}.

4.6. Extraction and Analysis of Antioxidant Enzymes

For the assessment of leaf antioxidant enzyme concentrations, approximately 0.5 g of newly formed leaf tissue were homogenized in a 2.0 mL extraction buffer containing 1% polyvinylpyrrolidone, 1% Triton x-100, and 0.11 g ethylenediaminetetraacetic acid (EDTA) using a pre-chilled mortar and pestle. Filtration of the homogenate via four layers of cheese cloth was followed by centrifugation at 15,000 g for 20 min. The supernatant was used for the enzyme assays. The extraction process was undertaken in the dark at 4 °C.

4.7. Assessment of Superoxide Dismutase (SOD) Activity

SOD (superoxide:superoxide oxireductase, EC 1.15.1.1) was quantified according to Dhindsa et al. [64] with minor modifications. SOD activity in the supernatant was assessed by its capability to inhibit photochemical reduction of nitroblue tetrazolium (NBT). The reaction mixture consisted of 0.5 M potassium phosphate buffer, 200 mM methionine, 1 M sodium carbonate, 2.5 mM NBT, 3 mM EDTA, 0.1 mL enzyme extract, 60 μM riboflavin, and 1.0 mL doubled distilled water. The assay mixture in uniform transparent tubes was shaken and placed under a 15 W fluorescent lamp (Phillips, Kolkata, India) for 10 min (25–28 °C). The controls consisted of solutions and enzymes placed in the dark (blank A) and the reaction mixture without enzymes was placed in the light (blank B). Absorbance for samples along with blank B was read at 560 nm against the blank A. A 50% reduction in color was considered as one unit of SOD enzyme activity and expressed in mg protein min^{-1}.

4.8. Catalase (CAT) Activity Assay

The method described by Aebi [65] with slight modifications was used for the CAT (H_2O_2:H_2O_2 oxidoreductase: EC 1.11.1.6) activity assay in leaves. The reaction mixture consisted of 0.5 M potassium phosphate buffer, 3 mM EDTA, 0.1 mL enzyme extract, and 3 mM H_2O_2 (pH 7.0), and the reaction was allowed to run for 5 min. CAT action was verified via screening H_2O_2 loss and quantifying the decline in absorbance at 240 nm. Calculation of CAT activity was carried out by using extinction coefficient (€) 0.036 mM^{-1} cm^{-1} and activity was expressed in enzyme units (EU) mg^{-1} protein min^{-1}. One unit of enzyme was defined as the amount of catalase necessary to decompose 1 μmol of H_2O_2 per min at pH 7.0 at 25 °C.

4.9. Assessment of Peroxidase (POX) Activity

POX (EC 1.11.1.7) content was measured according to Bergmeyer, et al. [66] Fresh leaves (0.2 mg) were collected and homogenized with 5 mL cold phosphate buffer (50 mM, pH 7.8) using a mortar and pestle, and the homogenate was centrifuged at 10,000 g at 4 °C for 20 min. The supernatant collected was maintained at 4 °C until use in the assay. The assay mixture consisted of 0.1 M phosphate buffer (pH 7.8), 4 mM pyrogallol, 3 mM H_2O_2, and crude enzyme extract. The assay mixture was transferred to a cuvette and the absorbance was measured at 420 nm using a spectrophotometer (UV-Pharma Spec 1700, Shimadzu, Japan). The enzyme activity was expressed as μmol (H_2O_2 consumed) mg^{-1} (protein) s^{-1}.

4.10. Measurement of Lipid Peroxides

A thiobarbituric acid-reactive-substance assay for measuring malondialdehyde (MDA) was used to assess lipid peroxidation. MDA is a secondary end product of polyunsaturated fatty acid oxidation and is considered to be a useful index of general lipid peroxidation [67]. The method described by Cakmak and Horst [68] was used with modifications. Leaf tissue (0.5 g) was homogenized with 5 mL 0.1% trichloroacetic acid (TCA) and centrifuged at 15,000 g for 5 min. Subsequently, approximately 1 mL supernatant was mixed with 4 mL 0.5% (*w*/*v*) thiobarbituric acid (TBA) in 20% TCA. The mixture was held in boiling water for 30 min, then instantly cooled on ice to end the reaction. The samples were then centrifuged at 12,000 g for 30 min. The absorbance of the supernatant was read at 532 nm using a UV-VIS spectrophotometer (UV-1700 Pharma Spec), and the reading at 600 nm was deducted to compensate for non-specific absorption. The amount of lipid peroxides was measured in terms of MDA level, with an extinction coefficient of 155 mM^{-1} cm^{-1}, and expressed as nmol g^{-1} fresh weight.

4.11. DNA Extraction and Molecular Marker Techniques

Isolation of genomic DNA from leaf tissues of micropropagated plants, as well as donor plants (mother plant), was conducted using a modified cetyltrimethylammonium bromide (CTAB) technique [69]. Quantification and purity of total DNA was assessed by

Nanodrop Spectrophotometer (Implen, Munich, Germany). All the samples were diluted to 25 ng/μL in Milli-Q water and stored at 4 °C until further use. Twenty random amplified polymorphic DNA (RAPD) primers (Operon Kit A Technologies Inc., Ebersberg, Germany) and 13 inter-simple sequence repeat (ISSR) primers (UBC, Vancouver, BC, Canada) were used to examine the clonal fidelity among the regenerated plantlets.

PCR amplification reactions were made in 20 μL volumes containing 1X PCR buffer, 25 mM $MgCl_2$ (1.2 μL), 10 mM dNTPs (0.4 μL), 2 μM primers, 3 Unit Taq polymerase (0.2 μL), and 40 ng Template DNA. The amplification of DNA for both RAPD and ISSR analyses was performed in a Thermal Cycler (Biometra, Gottingen, Germany). The PCR amplification program consisted of 45 cycles counting a 94 °C denaturation step of 5 min, an annealing step for 2 min at 35 °C (for RAPD analysis) or 35–58 °C (for ISSR analysis), and a 72 °C elongation of 1 min. A concluding extension was tracked at 72 °C for 10 min. PCR products were analyzed by electrophoresis in 0.8% (*w*/*v*) agarose gels with 4 μL ethidium bromide in a TAE buffer (pH 8.0), run at 60 V for 2 h, and visualized on a UV transilluminator (Bio Rad, Hercules, CA, USA). To regulate the uniformity of the experiments, DNA extraction, PCR reactions and electrophoresis were repeated three times. Distinct and reproducible bands were scored. Bands with the identical movement were considered to be homologous fragments, regardless of intensity.

4.12. Statistical Analysis

The two experiments involving *m*T were established with 20 replicates per treatment in a randomized complete block design with six explants per replicate and the experiments were repeated three times. The rooting experiment with three auxins was established with 10 replicates per treatment in a randomized complete block design with a minimum of six microshoots per replicate. Morphological changes among the cultures were observed and documented at regular intervals. The SPSS version 16 (SPSS Inc. Chicago, IL, USA) was employed for statistical analysis of the data. Duncan's multiple range test (DMRT) at $p \leq 0.05$ was carried out to examine evidence relating to significant variation among mean values, and the results were expressed as the means ± SE of three experiments.

Research manuscripts reporting large datasets that are deposited in a publicly available database should specify where the data have been deposited and provide the relevant accession numbers. If the accession numbers have not yet been obtained at the time of submission, please state that they will be provided during review. They must be provided prior to publication.

Interventionary studies involving animals or humans, and other studies that require ethical approval, must list the authority that provided approval and the corresponding ethical approval code.

5. Conclusions

The present research has successfully established a simple, resourceful, and robust regeneration methodology for micropropagation in *L. speciosa* that will also be useful for ex situ germplasm conservation. The use of the novel ArCK, *m*T, provides a broad platform for in vitro propagation of true-to-type plant material suitable for commercialization, conservation strategies, and pharmaceutical needs. *L. speciosa* shows immense potential for the production of corosolic acid, a "future anti-diabetic drug", and the plant propagation platform developed in this research could pave the way for commercial production of the plant and its exploitation for bioactives. To the best our knowledge, this *m*T-based in vitro propagation protocol developed in our research is several times more efficient than existing ones.

Author Contributions: Conceptualization, N.A., N.F. and M.F.; methodology, N.A., N.F., M.F., A.A.A. and R.P.; validation, N.A., M.F. and A.A.A.; formal analysis, N.A., N.F., M.F., A.A.A. and R.P.; investigation, N.A., N.F., M.F., A.A.A. and R.P.; resources, N.A. and M.F.; writing—original draft preparation N.A., N.F. and M.F.; writing—review and editing, N.A., M.F., A.A.A. and R.P.;

supervision, M.F. and A.A.A.; project administration, A.A.A.; funding acquisition, M.F. and A.A.A. All authors have read and agreed to the published version of the manuscript.

Funding: This study was supported by the Researchers Supporting Project (RSP-2021/86), King Saud University, Riyadh, Saudi Arabia.

Data Availability Statement: Not applicable.

Acknowledgments: The authors are thankful to the Researchers Supporting Project (RSP-2021/86), King Saud University, Riyadh, Saudi Arabia.

Conflicts of Interest: Authors declare that there are no conflict of interest.

References

1. Ellis, R.H.; Mai-Hong, T.; Hong, T.D.; Tan, T.T.; Xuan-Chuong, N.D.; Hung, L.Q.; Ngoc-Tam, B.; Le-Tam, V.T. Comparative analysis by protocol and key of seed storage behaviour of sixty Vietnamese tree species. *Seed Sci. Technol.* **2007**, *35*, 460–476. [CrossRef]
2. Sannigrahi, A.K. Biodegradation of leaf litter of tree species in presence of cow dung and earthworms. *Indian J. Biotechnol.* **2009**, *8*, 335–338.
3. Dhakane, M.U.; Yadav, S.G. Pride of India (*Lagerstroemia speciosa* (L.) Pers.) forming a silver bullet for 21st century. *Int. J. For. Crop Improv.* **2015**, *6*, 79–82. [CrossRef]
4. Deshpande, A.M.; Shirsat, M.K.; Jeyabalan, G. A review of *Lagerstroemia speciosa*: Pride of India. *Int. J. Contemp. Res. Rev.* **2018**, *9*, 20181–20185. [CrossRef]
5. Hayashi, T.; Maruyama, H.; Kasai, R.; Hattori, K.; Takasuga, S.; Hazeki, O.; Yamasaki, K.; Tanaka, T. Ellagitannins from *Lagerstroemia speciosa* as activators of glucose transport in fat cells. *Planta Med.* **2002**, *68*, 173–175. [CrossRef]
6. Kotnala, M.; Chakraborthy, G.; Mazumder, A. *Lagerstroemia* species: A current review. *Int. J. PharmTech Res.* **2013**, *5*, 906–909.
7. Esmail, A.S.A. Medicinal value of *Lagerstroemia speciosa*: An updated review. *Int. J. Curr. Pharm. Res.* **2019**, *11*, 18–26. [CrossRef]
8. Yamaguchi, Y.; Yamada, K.; Yoshikawa, N.; Nakamura, K.; Haginaka, J.; Kunitomo, M. Corosolic acid prevents oxidative stress, inflammation and hypertension in SHR/NDmcr-cp rats, a model of metabolic syndrome. *Life Sci.* **2006**, *79*, 2474–2479. [CrossRef]
9. Park, S.W.; Kwon, M.J.; Yoo, J.Y.; Choi, H.-J.; Ahn, Y.-J. Antiviral activity and possible mode of action of ellagic acid identified in *Lagerstroemia speciosa* leaves toward human rhinoviruses. *BMC Complement. Altern. Med.* **2014**, *14*, 171. [CrossRef]
10. Laruan, L.M.V.; Balangod, T.; Balangcod, K.; Patacsil, M.; Apastol, O.; Manuel, J.; Cortez, S.; Vallejn, V. Phytochemical and antibacterial study of *Lagerstroemia* (L.) Pers and its ethno-medicinal importance to indigenous communities of Benquet provinence, Philippines. *Indian J. Od Tradit. Knowl.* **2013**, *12*, 379–383.
11. Pareek, A.; Suthar, M.; Rathore, G.S.; Bansal, V.; Kumawat, T. In vitro antioxidant studies of *Lagerstroemia speciosa* leaves. *Pharmacogn. J.* **2010**, *2*, 357–360. [CrossRef]
12. Unno, T.; Sugimoto, A.; Kakuda, T. Xanthine oxidase inhibitors from the leaves of *Lagerstroemia speciosa* (L.) Pers. *J. Ethnopharmacol.* **2004**, *93*, 391–395. [CrossRef] [PubMed]
13. Chen, L.X.; Schumacher, H.R. Gout: An evidence-based review. *J. Clin. Rheumatol.* **2008**, *14*, S55–S62. [CrossRef] [PubMed]
14. Shi, L.; Zhang, W.; Zhou, Y.-Y.; Zhang, Y.-N.; Li, J.-Y.; Hu, L.-H.; Li, J. Corosolic acid stimulates glucose uptake via enhancing insulin receptor phosphorylation. *Eur. J. Pharmacol.* **2008**, *584*, 21–29. [CrossRef]
15. Judy, W.V.; Hari, S.P.; Stogsdill, W.W.; Judy, J.S.; Naguib, Y.M.A.; Passwater, R. Antidiabetic activity of a standardized extract (Glucosol™) from *Lagerstroemia speciosa* leaves in Type II diabetics: A dose-dependence study. *J. Ethnopharmacol.* **2003**, *87*, 115–117. [CrossRef]
16. Vijayan, A.; Padmesh Pillai, P.; Hemanthakumar, A.S.; Krishnan, P.N. Improved in vitro propagation, genetic stability and analysis of corosolic acid synthesis in regenerants of *Lagerstroemia speciosa* (L.) Pers. by HPLC and gene expression profiles. *Plant Cell Tissue Organ Cult.* **2015**, *120*, 1209–1214. [CrossRef]
17. Sivadas, D.; Pandurangan, A.G. Non-viable seed set in *Lagerstroemia speciosa* (Lythraceae) populations in the Western Ghats, India. *Trop. Ecol.* **2022**, *63*, 141–144. [CrossRef]
18. Hadiuzzaman, S. In vitro cloning from seedling explants of *Lagerstroemia speciosa* Pers. and *L. thorellii* Gagnep. *Bangladesh J. Bot.* **1992**, *21*, 59–64.
19. Zobayed, S.M.A. In vitro propagation of *Lagerstroemia* spp. from nodal explants and gaseous composition in the culture headspace. *Environ. Control Biol.* **2000**, *38*, 1–11. [CrossRef]
20. Lim-Ho, C.L.; Lee, S.K. Micropropagation of *Lagerstroemia speciosa* (L.) Pers. (Lythraceae). *Gard. Bull. Singap.* **1985**, *38*, 175–184.
21. Strnad, M.; Hanuš, J.; Vaněk, T.; Kamínek, M.; Ballantine, J.A.; Fussell, B.; Hanke, D.E. Meta-topolin, a highly active aromatic cytokinin from poplar leaves (*Populus* × *canadensis* Moench., cv. Robusta). *Phytochemistry* **1997**, *45*, 213–218. [CrossRef]
22. Werbrouck, S.P.; Strnad, M.; Van Onckelen, H.A.; Debergh, P.C. Meta-topolin, an alternative to benzyladenine in tissue culture? *Physiol. Plant.* **1996**, *98*, 291–297. [CrossRef]
23. Lata, H.; Chandra, S.; Techen, N.; Khan, I.A.; ElSohly, M.A. In vitro mass propagation of *Cannabis sativa* L.: A protocol refinement using novel aromatic cytokinin *meta*-topolin and the assessment of eco-physiological, biochemical and genetic fidelity of micropropagated plants. *J. Appl. Res. Med. Aromat. Plants* **2016**, *3*, 18–26. [CrossRef]

24. Naaz, A.; Hussain, S.A.; Anis, M.; Alatar, A.A. Meta-topolin improved micropropagation in *Syzygium cumini* and acclimatization to *ex vitro* conditions. *Biol. Plant.* **2019**, *63*, 174–182. [CrossRef]
25. Teklehaymanot, T.; Wannakrairoj, S.; Pipattanawong, N. Meta-topolin for pineapple shoot multiplication under three in vitro systems. *Am.-Eurasian J. Agric. Environ. Sci.* **2010**, *7*, 157–162.
26. Westhuizen, A.V.D. Use of Meta-topolin as an alternative cytokinin in the tissue culture of *Eucalyptus* species. *Acta Hortic.* **2014**, *1055*, 25–28. [CrossRef]
27. Gentile, A.; Frattarelli, A.; Nota, P.; Condello, E.; Caboni, E. The aromatic cytokinin meta-topolin promotes in vitro propagation, shoot quality and micrografting in *Corylus colurna* L. *Plant Cell Tissue Organ Cult.* **2017**, *128*, 693–703. [CrossRef]
28. Chauhan, R.D.; Taylor, N.J. Meta-topolin stimulates de novo shoot organogenesis and plant regeneration in cassava. *Plant Cell Tissue Organ Cult.* **2018**, *132*, 219–224. [CrossRef]
29. Ahmad, A.; Anis, M.; Khanam, M.N.; Alatar, A.A. Direct shoot organogenesis from shoot tip explants of a highly medicinal valued tree *Pterocarpus marsupium* Roxb. *Vitr. Cell. Dev. Biol.-Plant* **2020**, *56*, 670–681. [CrossRef]
30. Elayaraja, D.; Subramanyam, K.; Vasudevan, V.; Sathish, S.; Kasthurirengan, S.; Ganapathi, A.; Manickavasagam, M. Meta-Topolin (mT) enhances the in vitro regeneration frequency of *Sesamum indicum* (L.). *Biocatal. Agric. Biotechnol.* **2019**, *21*, 101320. [CrossRef]
31. Jayaprakash, K.; Manokari, M.; Badhepuri, M.K.; Raj, M.C.; Dey, A.; Shekhawat, M.S. Influence of meta-topolin on in vitro propagation and foliar micro-morpho-anatomical developments of *Oxystelma esculentum* (L.f.)Sm. *Plant Cell Tissue Organ Cult.* **2021**, *147*, 325–337. [CrossRef]
32. Manokari, M.; Priyadharshini, S.; Jogam, P.; Dey, A.; Shekhawat, M.S. Meta-topolin and liquid medium mediated enhanced micropropagation via *ex vitro* rooting in *Vanilla planifolia* Jacks. ex Andrews. *Plant Cell Tissue Organ Cult.* **2021**, *146*, 69–82. [CrossRef]
33. Kucharska, D.; Orlikowska, T.; Maciorowski, R.; Kunka, M.; Wojcik, D.; Pluta, S. Application of meta-Topolin for improving micropropagation of gooseberry (*Ribes grossularia*). *Sci. Hortic.* **2020**, *272*, 109529. [CrossRef]
34. Erisen, S.; Kurt-Gur, G.; Servi, H. In vitro propagation of *Salvia sclarea* L. by meta-Topolin, and assessment of genetic stability and secondary metabolite profiling of micropropagated plants. *Ind. Crops Prod.* **2020**, *157*, 112892. [CrossRef]
35. Saeiahagh, H.; Mousavi, M.; Wiedow, C.; Bassett, H.B.; Pathirana, R. Effect of cytokinins and sucrose concentration on the efficiency of micropropagation of 'Zes006' *Actinidia chinensis* var. *chinensis*, a red-fleshed kiwifruit cultivar. *Plant Cell Tissue Organ Cult.* **2019**, *138*, 1–10. [CrossRef]
36. Debenham, M.; Pathirana, R. Establishment and management of an in vitro repository of kiwifruit (*Actinidia* spp.) germplasm. In *Meta-Topolin: A Growth Regulator for Plant Biotechnology and Agriculture*; Ahmad, N., Strnad, M., Eds.; Springer Nature: Cham, Switzerland, 2020; pp. 279–291.
37. Nowakowska, K.; Pacholczak, A. Comparison of the effect of meta-Topolin and benzyladenine during *Daphne mezereum* L. micropropagation. *Agronomy* **2020**, *10*, 1994. [CrossRef]
38. Plíhal, O.; Szüčová, L.; Galuszka, P. N9-substituted aromatic cytokinins with negligible side effects on root development are an emerging tool for in vitro culturing. *Plant Signal. Behav.* **2013**, *8*, e24392. [CrossRef]
39. Shekhawat, M.S.; Priyadharshini, S.; Jogam, P.; Kumar, V.; Manokari, M. Meta-topolin and liquid medium enhanced in vitro regeneration in *Scaevola taccada* (Gaertn.) Roxb. *Vitr. Cell. Dev. Biol.-Plant* **2021**, *57*, 296–306. [CrossRef]
40. Pramanik, B.; Sarkar, S.; Bhattacharyya, S.; Gantait, S. meta-Topolin-induced enhanced biomass production via direct and indirect regeneration, synthetic seed production, and genetic fidelity assessment of *Bacopa monnieri* (L.) Pennell, a memory-booster plant. *Acta Physiol. Plant.* **2021**, *43*, 107. [CrossRef]
41. Gladfelter, H.J.; Johnston, J.; Wilde, H.D.; Merkle, S.A. Adventitious shoot-based propagation of *Franklinia alatamaha* for commercial horticulture and restoration. *Vitr. Cell. Dev. Biol.-Plant* **2020**, *56*, 857–864. [CrossRef]
42. Gentile, A.; Gutierrez, M.J.; Martinez, J.; Frattarelli, A.; Nota, P.; Caboni, E. Effect of *meta*-Topolin on micropropagation and adventitious shoot regeneration in *Prunus* rootstocks. *Plant Cell Tissue Organ Cult.* **2014**, *118*, 373–381. [CrossRef]
43. Bairu, M.W.; Stirk, W.A.; Dolezal, K.; Van Staden, J. Optimizing the micropropagation protocol for the endangered *Aloe polyphylla*: Can meta-topolin and its derivatives serve as replacement for benzyladenine and zeatin? *Plant Cell Tissue Organ Cult.* **2007**, *90*, 15–23. [CrossRef]
44. Valero-Aracama, C.; Kane, M.E.; Wilson, S.B.; Philman, N.L. Substitution of benzyladenine with meta-topolin during shoot multiplication increases acclimatization of difficult- and easy-to-acclimatize sea oats (*Uniola paniculata* L.) genotypes. *Plant Growth Regul.* **2009**, *60*, 43–49. [CrossRef]
45. Podlesakova, K.; Zalabak, D.; Cudejkova, M.; Plihal, O.; Szucova, L.; Dolezal, K.; Spichal, L.; Strnad, M.; Galuszka, P. Novel cytokinin derivatives do not show negative effects on root growth and proliferation in submicromolar range. *PLoS ONE* **2012**, *7*, e39293. [CrossRef]
46. Vylíčilová, H.; Bryksová, M.; Matušková, V.; Doležal, K.; Plíhalová, L.; Strnad, M. Naturally Occurring and Artificial N9-Cytokinin Conjugates: From Synthesis to Biological Activity and Back. *Biomolecules* **2020**, *10*, 832. [CrossRef]
47. Aremu, A.O.; Bairu, M.W.; Dolezal, K.; Finnie, J.F.; Van Staden, J. Topolins: A panacea to plant tissue culture challenges? *Plant Cell Tissue Organ Cult.* **2012**, *108*, 1–16. [CrossRef]
48. Cassells, A.C.; Curry, R.F. Oxidative stress and physiological, epigenetic and genetic variability in plant tissue culture: Implications for micropropagators and genetic engineers. *Plant Cell Tissue Organ Cult.* **2001**, *64*, 145–157. [CrossRef]
49. Şen, A. Oxidative stress studies in plant tissue culture. *Antioxid. Enzym.* **2012**, *3*, 59–88.

50. Bednarek, P.T.; Orłowska, R. Plant tissue culture environment as a switch-key of (epi)genetic changes. *Plant Cell Tissue Organ Cult.* **2020**, *140*, 245–257. [CrossRef]
51. Benson, E.E. Do free radicals have a role in plant tissue culture recalcitrance? *Vitr. Cell. Dev. Biol. Plant* **2000**, *36*, 163–170. [CrossRef]
52. Wang, Y.; Zhang, H.; Hou, P.; Su, X.; Zhao, P.; Zhao, H.; Liu, S. Foliar-applied salicylic acid alleviates heat and high light stress induced photoinhibition in wheat (*Triticum aestivum*) during the grain filling stage by modulating the psbA gene transcription and antioxidant defense. *Plant Growth Regul.* **2014**, *73*, 289–297. [CrossRef]
53. Jeon, M.-W.; Ali, M.B.; Hahn, E.-J.; Paek, K.-Y. Effects of photon flux density on the morphology, photosynthesis and growth of a CAM orchid, *Doritaenopsis* during post-micropropagation acclimatization. *Plant Growth Regul.* **2005**, *45*, 139–147. [CrossRef]
54. Goncalves, S.; Martins, N.; Romano, A. Physiological traits and oxidative stress markers during acclimatization of micropropagated plants from two endangered *Plantago* species: *P. algarbiensis* Samp. and *P. almogravensis* Franco. *Vitr. Cell. Dev. Biol.-Plant* **2017**, *53*, 249–255. [CrossRef]
55. Ahmad, Z.; Shahzad, A.; Sharma, S.; Parveen, S. *Ex vitro* rescue, physiochemical evaluation, secondary metabolite production and assessment of genetic stability using DNA based molecular markers in regenerated plants of *Decalepis salicifolia* (Bedd. ex Hook.f.) Venter. *Plant Cell Tissue Organ Cult.* **2018**, *132*, 497–510. [CrossRef]
56. Aremu, A.O.; Bairu, M.W.; Szucova, L.; Dolezal, K.; Finnie, J.F.; Van Staden, J. Genetic fidelity in tissue-cultured 'Williams' bananas-The effect of high concentration of topolins and benzyladenine. *Sci. Hortic.* **2013**, *161*, 324–327. [CrossRef]
57. Bairu, M.W.; Stirk, W.A.; Doležal, K.; van Staden, J. The role of topolins in micropropagation and somaclonal variation of banana cultivars 'Williams' and 'Grand Naine' (*Musa* spp. AAA). *Plant Cell Tissue Organ Cult.* **2008**, *95*, 373–379. [CrossRef]
58. Fatima, N.; Ahmad, N.; Anis, M.; Ahmad, I. An improved in vitro encapsulation protocol, biochemical analysis and genetic integrity using DNA based molecular markers in regenerated plants of *Withania somnifera* L. *Ind. Crops Prod.* **2013**, *50*, 468–477. [CrossRef]
59. Faisal, M.; Alatar, A.A.; El-Sheikh, M.A.; Abdel-Salam, E.M.; Qahtan, A.A. Thidiazuron induced in vitro morphogenesis for sustainable supply of genetically true quality plantlets of Brahmi. *Ind. Crops Prod.* **2018**, *118*, 173–179. [CrossRef]
60. Ahmed, M.R.; Anis, M.; Alatar, A.A.; Faisal, M. In vitro clonal propagation and evaluation of genetic fidelity using RAPD and ISSR marker in micropropagated plants of *Cassia alata* L.: A potential medicinal plant. *Agrofor. Syst.* **2017**, *91*, 637–647. [CrossRef]
61. Murashige, T.; Skoog, F. A Revised Medium for Rapid Growth and Bio Assays with Tobacco Tissue Cultures. *Physiol. Plant.* **1962**, *15*, 473–497. [CrossRef]
62. Mackinney, G.R. Absorption of light by chlorophyll solutions. *J. Biol. Chem.* **1941**, *140*, 315–322. [CrossRef]
63. Maclachlan, S.; Zalik, S. Plastid structure, chlorophyll concentration, and free amino acid composition of a chlorophyll mutant of barley. *Can. J. Bot.* **1963**, *41*, 1053–1062. [CrossRef]
64. Dhindsa, R.S.; Plumb-Dhindsa, P.; Thorpe, T.A. Leaf senescence: Correlated with increased levels of membrane permeability and lipid peroxidation, and decreased levels of superoxide dismutase and catalase. *J. Exp. Bot.* **1981**, *32*, 93–101. [CrossRef]
65. Aebi, H. Catalase in vitro. *Methods Enzymol.* **1984**, *105*, 121–126. [PubMed]
66. Bergmeyer, H.U.; Gawehn, K.; Grassl, M. *Methods of Enzymatic Analysis*; Bergmeyer, H.U., Ed.; Verlag Chemie: Wienheim, Germany, 1974; Volume 1, pp. 481–482.
67. Hodges, D.M.; DeLong, J.M.; Forney, C.F.; Prange, R.K. Improving the thiobarbituric acid-reactive-substances assay for estimating lipid peroxidation in plant tissues containing anthocyanin and other interfering compounds. *Planta* **1999**, *207*, 604–611. [CrossRef]
68. Cakmak, I.; Horst, W.J. Effect of aluminium on lipid peroxidation, superoxide dismutase, catalase, and peroxidase activities in root tips of soybean (*Glycine max*). *Physiol. Plant.* **1991**, *83*, 463–468. [CrossRef]
69. Doyle, J. DNA Protocols for Plants. In *Molecular Techniques in Taxonomy*; Hewitt, G.M., Ed.; NATO AS1 Series; Springer: Berlin/Heidelberg, Germany, 1991; pp. 283–293.

plants MDPI

Article

Treatment of Chrysanthemum Synthetic Seeds by Air SDBD Plasma

Nikola Škoro [1,*], Suzana Živković [2], Slađana Jevremović [2,*] and Nevena Puač [1]

[1] Institute of Physics—National Institute of Republic of Serbia, University of Belgrade, Pregrevica 118, 11080 Belgrade, Serbia; nevena@ipb.ac.rs
[2] Institute for Biological Research "Siniša Stanković"—National Institute of Republic of Serbia, University of Belgrade, Despot Stefan Boulevard 142, 11000 Belgrade, Serbia; suzy@ibiss.bg.ac.rs
* Correspondence: nskoro@ipb.ac.rs (N.Š.); sladja@ibiss.bg.ac.rs (S.J.)

Abstract: Herein, we present the effect of surface dielectric barrier discharge (SDBD) air cold plasma on regrowth of chrysanthemum synthetic seeds (synseeds) and subsequent plantlet development. The plasma system used in this study operates in air at the frequency of 50 Hz. The detailed electrical characterization of SDBD was shown, as well as air plasma emission spectra obtained by optical emission spectroscopy. The chrysanthemum synseeds (encapsulated shoot tips) were treated in air plasma for different treatment times (0, 5 or 10 min). Plasma treatment significantly improved the regrowth and whole plantlet development of chrysanthemum synseeds under aseptic (in vitro) and non-aseptic (ex vitro) conditions. We evaluated the effect of SDBD plasma on synseed germination of four chrysanthemum cultivars after direct sowing in soil. Germination of synseeds directly sowed in soil was cultivar-dependent and 1.6−3.7 fold higher after plasma treatment in comparison with untreated synseeds. The study showed a highly effective novel strategy for direct conversion of simple monolayer alginate chrysanthemum synseeds into entire plantlets by plasma pre-conversion treatment. This treatment reduced contamination and displayed a considerable ex vitro ability to convert clonally identical chrysanthemum plants.

Keywords: artificial seeds; plasma treatment; chrysanthemum; dielectric barrier discharge; cold plasma

check for updates

Citation: Škoro, N.; Živković, S.; Jevremović, S.; Puač, N. Treatment of Chrysanthemum Synthetic Seeds by Air SDBD Plasma. *Plants* **2022**, *11*, 907. https://doi.org/10.3390/plants11070907

Academic Editors: Anna De Carlo and Waed Tarraf

Received: 7 March 2022
Accepted: 25 March 2022
Published: 29 March 2022

1. Introduction

Synthetic seed technology is one of the most promising tools in plant biotechnology and may represent an innovative method for massive plant production and sustainable agriculture in the future [1]. Synthetic seeds (artificial seeds or synseeds) have been defined as artificially encapsulated somatic embryos or other non-embryogenic vegetative parts of plants, mainly in alginate, that may be used for storage or sowing under in vitro or ex vitro conditions [2]. The term 'synseeds' was described by Murashige in 1977 [3] as 'an encapsulated single somatic embryo', but later, the definition of artificial seeds was extended to any artificially coated micropropagules that have capability to be sown as a seed and converted into a plant [4,5]. There is a growing trend in applications of synseed technology for medium- and long-term storage of plant material under aseptic conditions [6,7] or as an advanced procedure of cryopreservation by encapsulation–dehydration and encapsulation–vitrification method [8,9]. Synseed technology represents an efficient alternative technique for propagation and germplasm conservation of valuable forest, medicinal and vegetable plant species that reproduce mainly vegetatively or have a problem in seed propagation, i.e., plants that produce non-viable seeds or seedless plants [10].

In synseed technology, an alginate capsule has two roles: (i) it acts as physical barrier of shoot tips against mechanical damage, and (ii) it serves as an artificial endosperm, carbon source and reservoir of nutrients for better survival and supply of energy [9]. Alginates are a group of naturally occurring anionic polysaccharides derived from brown algae cell walls (*Macrocystis pyrifera*, *Limnaris hyperborea*, *Ascophyllum nodosum*) and several bacterial strains

(*Azotobacter*, *Pseudomonas*). Sodium alginate is soluble in water, but when the sodium is replaced with calcium, the ionic bond with calcium cross links the polymer chain in alginate, which results in the formation of an insoluble gel. Sodium alginate and calcium salt are reported to be the best combination for encapsulation, representing the most successful and widely accepted approach to synseed production [9]. Alginates can be formed into diverse semisolid or solid structures because of their ability of sol/gel transition and are commonly used as viscosity-increasing agents, thickeners and suspension and emulsion stabilizers in food and the pharmaceutical industry (code E400-E405) [11]. In addition, alginate gels are the basis for a variety of wound dressings that have showed variety of therapeutically effects, such as hemostatic and bacteriostatic properties [12,13]. On the other hand, in plants, sodium alginate is considered a potential elicitor that improves tolerance to plant environmental stresses, such as drought, inhibiting plant infections and reducing the toxic effect of heavy metals [14,15].

Chrysanthemums (*Chrysanthemum morifolium* Ramat. syn. *C. grandiflorum* Kitam) are, besides roses, the most important economically ornamental crop in the world [16]. They originate from east Asia, a center of their biodiversity; however, to date, many horticultural varieties and cultivars of chrysanthemums are produced using different biotechnological tools [17]. The name chrysanthemum means gold flower, but they are also called "autumn roses" because they were, in the past, used as cut flowers during late summer and autumn. Nowadays, there is constant demand on the market for new cultivars that are available during the whole year. Modern biotechnological tools, such as mutation breeding and micropropagation under in vitro conditions, allow for production of hundreds of new chrysanthemum cultivars every year [18]. Chrysanthemum cultivars are commonly propagated vegetatively by cuttings and suckers and stored as field, greenhouse or in vitro collections due to high spontaneous mutation rates and high levels of ploidy and self-incompatibility [19,20]. Micropropagation of chrysanthemum cultivars, as an in vitro way of vegetative multiplication in culture, was reported for the first time more than 50 years ago [21], and numerous reports about plantlet regeneration from various explants of chrysanthemum have been presented [16,22–24].

The application of synseed technology, accompanied by micropropagation, represents a perfect biotechnological approach that could be used for agricultural improvement of year-round plant production of chrysanthemums. There are several advantages of this approach, including large-scale production; easy handling; short and medium storage (4 °C) or low temperature (−196 °C) storage; easy transportation; and the genetically true-to-type nature of the plants produced from synseeds. On the other hand, there are some limitations of wider usage of synseed technology in commercial applications as published to date, such as implementation of labor-intensive procedures, which include double-layer encapsulation or several media changes to derived plantlets with well-developed shoot and roots at the same time. To date, the application of synseed technology of chrysanthemum cultivars has been investigated for in vitro storage and ex vitro planting (summarized in Table 1). In addition, synseed technology in chrysanthemums is widely used as a part of encapsulation–dehydration and encapsulation–vitrification protocols for long-term storage of chrysanthemum cultivars by cryopreservation in liquid nitrogen [25]. Considering the fact that chrysanthemums are susceptible to mutations, meristem explants (i.e., nodal segments or shoot tips) proved to be the best explant choice for the plant propagation of chrysanthemums, with a high degree of clonal fidelity as mother plants, as well as for synseed production [26–28].

Chrysanthemum synseeds are mainly produced under sterile conditions for short- and long-term storage [19], but sowing of chrysanthemum synthetic seeds under non-aseptic conditions has been also reported [29,30]. Chrysanthemum synseeds easily regrow from Na-alginate beads under sterile conditions, whereas for complete germination and whole-plantlet development (shoot and root), it is necessary to add indole-acetic acid to the encapsulation matrix [29] or, as separate step, in the medium for rooting after shoot regrowth [19]. For sowing under unsterile conditions, the results showed that presence of organic compounds in the gelling matrix and commercial substrates caused microbial contamination in all synseeds and complete inhibition of further regrowth of the shoots or whole plantlet development. In general, difficulties of sowing artificial seeds directly in soil or in commercial substrates, such as compost, vermiculite, etc., under non-sterile conditions are considered to be one of the main limitations for the widescale practical application of synseed technology [2,31]. Some progress has been achieved by using chemical mixtures and antibiotics for preservation of synseeds before sowing [32], but more investigations and novel approaches are still needed to improve the capacity of synseed cultivation under non-sterile conditions.

Table 1. Application of synseed technology for storage and propagation of chrysanthemum cultivars.

Cultivar	Plant Material	Beads	Sowing	Germination (%)	Flowering	References
Clone 'PS 27'	Nodal segments (in vitro)	Monolayered (3% Na-alginate + 0.1 mg/L IAA)	Sterile, water–sand	50	no	[29]
		Double-layered (beads: 3% Na alginate, second layer: water)	Non-sterile, water–perlite	45	no	
Lady group	Shoot tips (in vitro)	Monolayered (3% Na-alginate)	Sterile, agar	52	no	[19]
cv. 'Royal Purple'	Shoot tips (ex vitro)	Monolayered (2.5% Na-alginate, sucrose, vitamin free)	Non-sterile, vermiculite	34	no	[30]

Atmospheric pressure plasma (non-thermal, "cold") systems have been extensively used in biomedical applications for almost two decades [33–36]. In parallel, another field of plasma applications has been growing-plasma agriculture [37,38]. One of the first applications of cold plasmas was the treatments of conventional seeds [39,40]. This includes various applications in seed treatment with several purposes [41–43]. Many authors have shown that plasma treatments can increase seed germination and speed up the whole process of plantlet development [37–40]. The rich plasma chemistry (with reactive oxygen and nitrogen species) interacts with the seed coating and triggers various responses, such as increasing water uptake, changes in the surface of seeds' coating and elimination of pathogens on seed surfaces [44–53]. In this sense, cold plasma treatment can have a multiply positive impact on seed germination and subsequent plant development of conventional seeds without the addition of chemicals that can be harmful for the environment.

Nowadays, there is a plethora of plasma sources that operate at atmospheric pressure [54–57]. They differ in electrode design, type of applied voltage, feeding gas, etc. The variety of atmospheric pressure plasma sources enables a large number of possible applications, but at the same time, comparison between the results of treatment is difficult. Therefore, it is of the utmost importance to obtain detailed characteristics of the plasma device that is used in experiments. One of the first steps that is usually performed includes electrical characterization of the discharge, accompanied by optical emission spectroscopy, which can give insight into the plasma-excited species. These diagnostic techniques represent only the starting point for a detailed description of plasma characteristics and include mass spectrometry, fast imaging, laser-induced fluorescence, etc. [56].

In this paper, we present results of air plasma treatment of *Chrysanthemum* synseeds together with detailed plasma source diagnostics. We performed electrical characterization of the optical emission spectra (OES) of an SDBD system that operates in air at atmospheric pressure. Until now, we are not aware of any research data in the literature about the effect of cold atmospheric plasma treatment on the regrowth (germination) of artificial seeds of any plant species. The objective of this study related to synseed treatment was to:

(1) investigate the influence of cold plasma treatment on chrysanthemum synseed regrowth under in vitro conditions; (2) analyze the impact of cold plasma treatment on regrowth and further plantlet development from synseeds sowed directly in soil (ex vitro); and (3) evaluate the effect of plasma treatment on synseed germination of different chrysanthemum cultivars. In this pioneer work, we present a novel approach in synseed biotechnology and plasma agriculture cold plasma treatment of synseeds (encapsulated shoot tips) prior to sowing to prevent contamination and enhance plant growth.

2. Results

2.1. Regrowth of Plasma-Treated Synthetic Seeds under Aseptic Conditions (In Vitro)

We tested the regrowth rate of chrysanthemum synseeds after plasma treatment cultivated under in vitro (Figure 1) and ex vitro conditions (Figures 2–4). The regrowth and shoot development of plasma-treated and untreated chrysanthemum monolayered, simple synseeds were evaluated for two growing substrates without plant growth regulators under aseptic (in vitro) conditions (Figure 1).

Figure 1. Chrysanthemum synseed germination under in vitro conditions. (**a**) Synseeds before plasma treatment; (**b**,**c**) plasma-treated synseeds grown on solid agar medium—leaf emergence after one week of culture (**b**) and shoot development after four weeks of culture (**c**); (**d**,**f**) plasma-treated synseeds grown on vermiculite + liquid medium—plasma-treated synseeds grown one week (**d**) and four weeks (**e**,**f**). Bars a–f, 5 mm.

After plasma treatment, one set of synseeds was grown on the solid agar medium (AM, Figure 1b,c), and the second set was placed on the vermiculite + liquid medium (VLM, Figure 1d–f). Untreated synseeds served as a control group and were grown on the same medium. Encapsulated chrysanthemum shoot tips cultivated on AM in vitro (Figure 1a) easily continued to grow after plasma treatment, and within the first week, leaf emergence was observed (Figure 1b). Full development of microshoots was established after four weeks of culture (Figure 1c). There was no apparent difference in regrowth between the control (untreated) and plasma-treated synseeds. Additionally, no contamination was observed among treated synseeds. All synseeds, including the control as well as plasma-treated synseeds, were fully developed in microshoots. We further evaluated shoot multiplication after plasma treatment, and no difference between plasma-treated and untreated synseeds was observed (Supplementary Data, Figure S1).

When chrysanthemum synseeds were grown on VLM (sterilized vermiculite moisture with liquid plant regulator free medium) under in vitro conditions (Figure 1d–f), a significant increase in regrowth of plasma-treated synseeds was achieved (Table 2).

Table 2. The effect of plasma treatment on germination of chrysanthemum synseeds grown in vitro on vermiculite + liquid medium.

Plasma Treatment (min)	Leaf Emergence (%) *	Shoot Regrowth (%) **
	1 Week	4 Weeks
0	20 ± 10 [a] ***	0 ± 0 [a]
1	47 ± 13 [a,b]	40 ± 13 [b]
5	67 ± 12 [b]	47 ± 13 [b]
10	67 ± 13 [b]	33 ± 12 [b]

* Leaf emergence is evaluated as first sign of shoot appearance out of alginate beads after one week of growth; ** shoot regrowth was evaluated as fully developed shoot out of beads after four weeks of growth. *** values represent mean ± standard error. The data signed with different letter within the same column are significantly different according to Fisher's LSD test.

After one week of culture, only 20% of untreated synseeds broke out of the alginate capsule, and none of them continued their growth. On the other hand, among plasma-treated synseeds, leaf emergence, as a first sign of synseed germination, was two- to three-fold higher in comparison to untreated synseeds after one week of culture. The best results for shoot regrowth, more than three times higher than that of untreated synseeds, was achieved after plasma treatment for 5 min and 10 min before planting in VLM. After four weeks of cultivation, only plasma-treated chrysanthemum synseeds (33–47%) continued their growth and developed microshoots (Figure 1e,f), whereas the untreated (control) synseeds did not survive on VLM. For further experiments on chrysanthemum synseeds, we used 10 min plasma treatment.

2.2. *Germination of Plasma-Treated Synthetic Seeds under Non-Aseptic Conditions (Ex Vitro)*

We examined the germination of plasma-treated and untreated chrysanthemum synseeds, as well as the subsequent growth and development of plantlets after direct sowing in soil substrate (ex vitro conditions) (Table 3, Figure 2). Plasma-treated synseeds showed more vigorous survival, regrowth and germination in comparison to untreated synseeds (Table 3). After one week of ex vitro cultivation, shoots broke the Na-alginate bead, and the first leaf emerged from the synseeds (Figure 2a,b), which represented the main sign of shoot regrowth. The difference in survival between plasma-untreated (first column) and plasma-treated synseeds (second and third column) was evident (Figure 2a,c). Despite the fact that there was no statistically significant difference in shoot regrowth of plasma-treated (66%) and untreated synseeds (60%), after one week of growth under ex vitro conditions, we noticed that the developed leaves formed from plasma-treated synseeds were wider than those derived from untreated synseeds (Figure 2a). After one week of growth, some untreated synseeds were lost due to desiccation and contamination.

After three weeks of growing under ex vitro conditions, we observed two times higher survival and almost doubled percentage of shoot development of plasma-treated chrysanthemum synseeds in comparison to untreated samples. Namely, only 33.3% of untreated synseeds continued growth and developed very well-formed shoots, whereas more than 60% of plasma-treated chrysanthemum synseeds formed well-developed shoots (Figure 2c). Six weeks after sowing, whole chrysanthemum plantlets with well-formed shoots and roots were developed (Table 3). Only 17% of untreated synseeds fully germinated and developed whole plantlets, whereas 2.5 fold higher synseed germination (over 40%) and whole-plantlet development of plasma-treated synseeds was achieved.

Table 3. The effect of plasma treatment on germination of chrysanthemum synseeds grown ex vitro.

Plasma Treatment (min)	Synseed Germination		
	Leaf Emergence * (%)	Shoot Regrowth * (%)	Plantlet * (%)
0	60 ± 7 [a] **	33 ± 7 [a]	17 ± 5 [a]
10	66 ± 5 [a]	60 ± 5 [b]	41 ± 5 [b]

* Leaf emergence is evaluated as first sign of shoot appearance out of alginate beads one week after sowing; shoot regrowth was evaluated as a fully developed shoot out of the bead three weeks after sowing; plantlet development was recorded as a fully developed plant with well-developed shoot and roots. ** Values represent mean ± standard error. The data signed with a different letter within the same column are significantly different according to Fisher's LSD test.

Figure 2. Chrysanthemum synseed germination under ex vitro conditions, cv. Précocita carna. (**a**) Control (first column) and plasma-treated synseeds (second and third column) after one week of growth; (**b**) first sign of germination (leaf emergence) of plasma-treated synseeds; (**c**) control (first column) and plasma-treated synseeds (second and third column) after three weeks of cultivation ex vitro; (**d**) detail of fully germinated plasma-treated chrysanthemum synseed four weeks after sowing.

We compared the total deterioration (%) of the chrysanthemum plasma-treated and untreated synseeds within the first 6 weeks of growth ex vitro (Figure 3). The values reached almost 85% in the case of untreated synseeds, whereas for the plasma-treated seeds, this percentage was around 60%. According to these results, we can conclude that the plasma treatment of synseeds before sowing under ex vitro conditions enables increased survival of the newly developed plants by 25%.

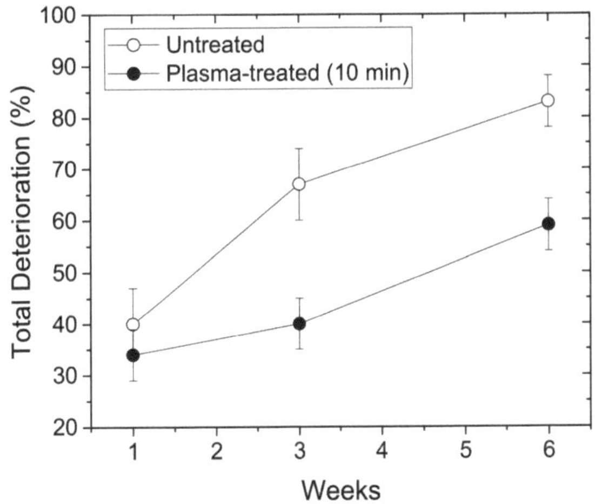

Figure 3. Total deterioration (%) of chrysanthemum synseeds grown ex vitro during 6-week period. Empty symbols: untreated synseeds; full symbols: plasma-treated synseeds.

2.3. Germination of Plasma-Treated Synthetic Seeds of Different Cultivars Ex Vitro

We evaluated the effect of plasma treatment (10 min) on synseed germination of different chrysanthemum cultivars (Table 4, Figure 4). We found that the plasma treatment significantly enhanced the process of synseed germination and conversion to plantlet after direct sowing in soil for all tested cultivars. This effect of plasma was cultivar-dependent (Table 4). Without plasma treatment of chrysanthemum synseeds, frequency of whole plantlet regeneration varied from 6% to 28% depending on the cultivar. On the other hand, plantlet regeneration from plasma-treated synseeds was 22–49%. The highest treatment effects on whole-plantlet development were recorded for cultivars BC and PP (~370% and ~350%, respectively) in comparison to control synseeds, whereas the lowest values were obtained for the PP cultivar (~160%).

Table 4. The effect of plasma treatment on synseed germination and plantlet development of different chrysanthemum cultivars.

Plasma Treatment (min)	Plantlet (%)			
	BC *	Q *	PC *	PP *
0	6 ± 1 [a] **	17 ± 2 [a]	28 ± 1 [a]	14 ± 1 [a]
10	22 ± 6 [b]	40 ± 3 [b]	44 ± 2 [b]	49 ± 3 [b]
Increment (%)	~370	~230	~160	~350

* Chrysanthemum cultivars: Brandsound Liliac (BC), Queens (Q), Précocita Carna (PC) and Précocita Parme (PP); ** values represent mean ± standard error. The data signed with different letters within the same column are significantly different according to Fisher's LSD test.

Chrysanthemum plantlets derived from plasma-treated synseeds continued their growth until full physiological maturity and flowering (Figure 4). During further growth of the plantlets under greenhouse conditions, no morphological differences were observed between the plants developed from untreated (Figure 4a,c) and plasma-treated synseeds (Figure 4b,d). Some of the plants originated from synseeds flowered after six months of ex vitro growth (Figure 4f); during the next flowering season, no morphological or color changes in flowers were observed (Figure 4g–i).

Figure 4. Chrysanthemum plantlets derived from synseeds during growth under greenhouse conditions (ex vitro). (**a–d**) Plants derived from untreated (**a**,**c**) and plasma-treated (**b**,**d**) synseeds of chrysanthemum cv. PC (**a**,**b**) and Q (**c**,**d**) three months after sowing; (**e**,**f**) Plantlets derived from untreated (**e**) and plasma-treated (**f**) synseeds of cv. PC six months after sowing; (**g–i**) untreated (left) and plasma-treated (right) chrysanthemum plants twelve months after sowing, during next flowering season, cv. BL; (**h**,**i**) flowering of chrysanthemum plants derived from untreated (**h**) and plasma-treated synseeds (**i**), cv. BL.

2.4. Characterization of DBD Plasma Source

Detailed plasma diagnostics was carried out prior to the treatments. After assessment of the range of plasma conditions that could be achieved with the DBD source, we selected one condition for the synseed treatment (part 4.3). The SDBD, previously used in treatments of flour [57], was characterized by using commercially available voltage probes. In order to properly assess the discharge current, one needs to subtract the displacement current. The first step was determination of the capacitance of the SDBD when the discharge was not ignited. This capacitance represents the passive capacitance of the plasma system, and it depends mainly on the geometry of the system, so it was determined for several interelectrode distances. The voltage at the powered electrode and current waveforms for one of the cases is represented in Figure 5. The peak-to-peak value of the voltage is 700 V (black line), whereas the peak-to-peak value of the current (red line) is ~12 μA.

The current measured when there is no discharge ignited is only the displacement current. It is represented by:

$$i_{disp}(t) = C_p \frac{dv(t)}{dt} \quad (1)$$

where C_p represents passive capacitance of the system, and v(t) represents instantaneous voltage measured at the powered electrode. As $i_{nodischarge}(t) = i_{disp}(t)$, the C_p value calculated as a parameter from Equation (1) for different electrode distances is given in Table 5.

An example of the voltage and current waveforms with the discharge ignited is shown in Figure 6. The voltage waveform was measured by HV probe at the powered electrode, whereas the current waveform represents the current through discharge. It was obtained by subtraction of the displacement current from the total current measured in the grounded branch of the electrical circuit. Variation of measured voltage and current were below 2% and 4%, respectively. Consequently, the root mean square (RMS) values were calculated with the same uncertainties. We can see that the voltage waveform is a sine function with a dominant first harmonic (50 Hz). The shape of the current waveform reflects a discharge with microfilaments that can be seen through the appearance of current peaks superimposed on the waveform [58,59].

Table 5. Values of stray capacitance in the SDBD system.

Electrode Distance-d [mm]	Stray Capacitance—*Cp* [pF]
2	20
3	35
4	41
5	45

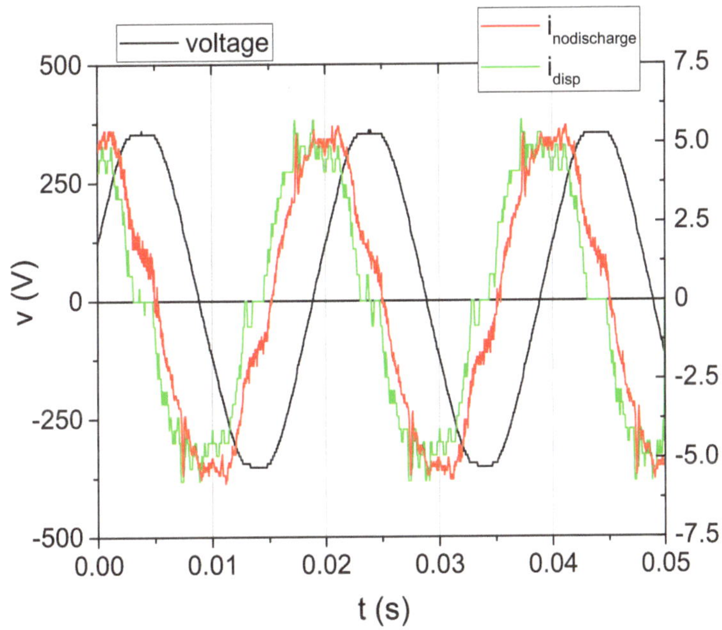

Figure 5. Waveforms of voltage (black line) and current (red line) obtained for d = 5 mm without discharge (input voltage V_{input} = 6 V). Displacement current (green line) is calculated by using voltage signal (Equation (1)).

The dependence of the output RMS voltage measured at the powered electrode on the input voltage is shown in Figure 7, left-hand axis. We can see that the dependence is linear, and it does not change significantly with an increase in the distance between electrodes.

The maximum operating voltage that could be obtained with this power supply system and the plasma source geometries was ~25 $kV_{peak\text{-}to\text{-}peak}$ (~8560 V_{RMS}). The dependence of the measured current on the input voltage is shown in Figure 7 on the right-hand axis. In this case, the dependence was not linear, and it changes with the distance between the powered electrode and the bottom plate, i.e., sample. For all distances presented here, the maximum current was 0.3 m A_{RMS} obtained for d = 2 mm.

The volt–ampere (V-A) characteristics (Figure 8) show that the system is not linear with constant impedance. The complex impedance changes with an increase in the applied voltage, and the non-linearity is the most pronounced for the distance of 2 mm. This can be explained by the number of microfilaments formed and the effective area that they covered. For the interelectrode distances of d = 3, 4 and 5 mm, the V-A dependence is almost the same up to the applied RMS voltage of 6.5 kV. Increasing of the interelectrode distance reduces the number of microfilaments that occur in one half-period, and the ones that are ignited operate totally with low conduction current.

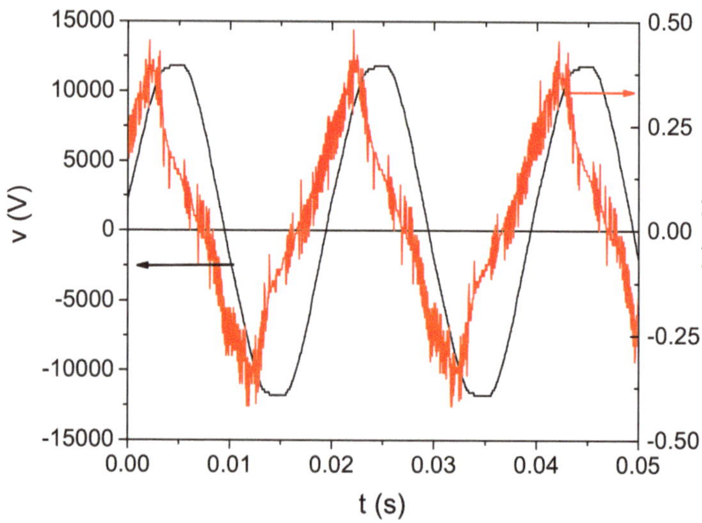

Figure 6. Time dependence of voltage and current signals at V_{input} = 220 V and d = 5 mm for the discharge operating in air.

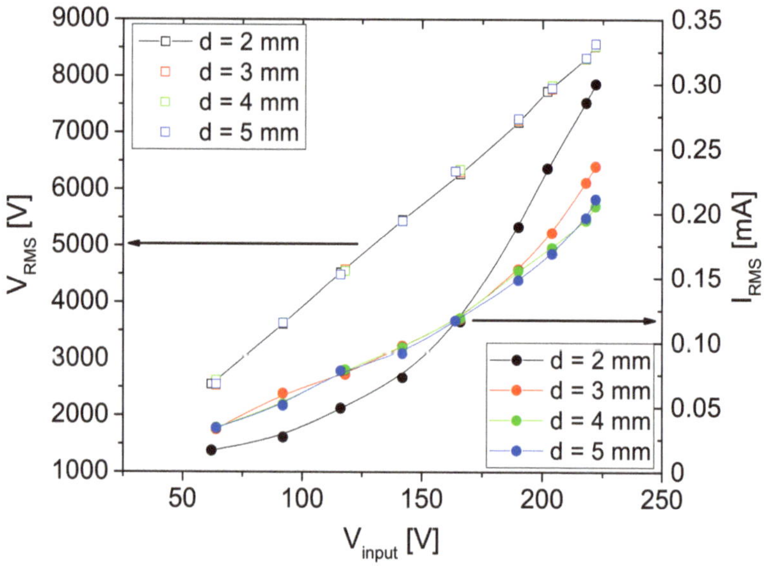

Figure 7. Dependence of the RMS voltage and current values on the input voltage V_{input} for four geometries. The discharge was operating in air.

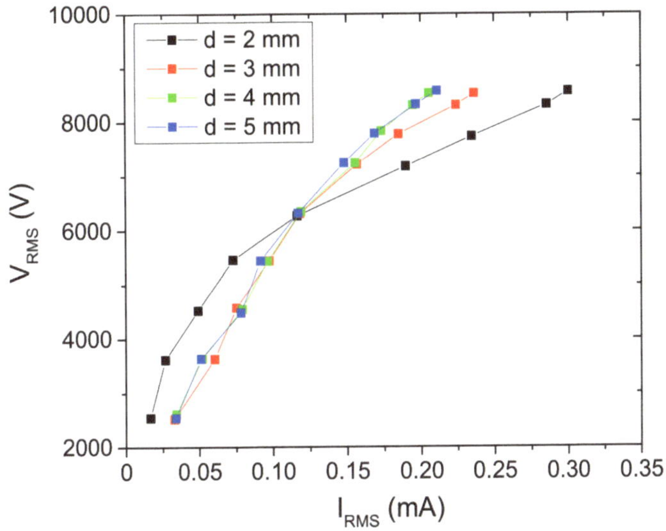

Figure 8. Volt–ampere characteristics of the SDBD obtained for the different distances between the powered electrode and the sample. The discharge was operating in air.

As expected, a similar change can be observed for the power transmitted to the discharge (Figure 9). Power is obtained as a mean value of instantaneous power calculated over several periods. The instantaneous power is obtained by direct multiplication of the waveforms of voltage and discharge current. We can see that the mean power transmitted to the discharge increases with the applied voltage. The maximum power that can be deposited to the discharge is around 1.5 W for the interelectrode distance of 2 mm. For voltages lower than 6.5 kV, the deposited power is similar for distances of d = 3, 4 and 5 mm and lowest for d = 2 mm.

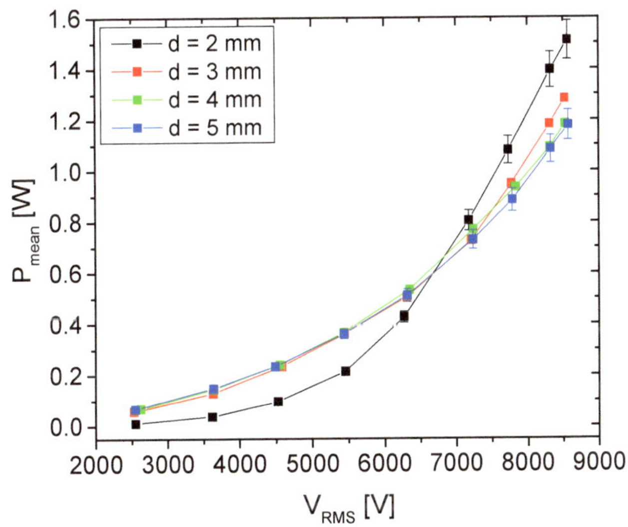

Figure 9. Mean power transmitted to the discharge as a function of the voltage at the powered electrode.

We recorded the emission spectra of the SDBD plasma source operating in air in the wavelength range from 275 nm to 850 nm. In Figure 10, we present the emission spectrum for d = 2 mm up to 500 nm. Almost all lines visible in the spectrum belong to N_2 Second Positive System (SPS): 313.67 nm, 315.93 nm, 337.13 nm (head of the band), 350.05 nm, 353.67 nm, 357.69 nm, 364.17 nm, 375.54 nm, 380.49 nm, 399.84 nm, 405.94 nm, 420.05 nm, 426.97 nm, 441.67 nm, 449.02 nm.

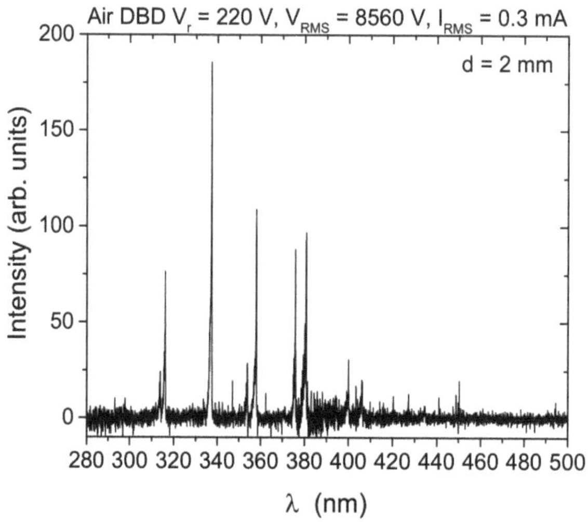

Figure 10. Optical emission spectrum from an air SDBD source obtained from side-on recording of spatially integrated emissions from the whole discharge volume (electrode gap, d = 2 mm). The intensity signal is corrected for spectral efficiency of the optical system.

3. Discussion

First, we will discuss in detail the plasma characteristics used in the study. We used air SDBD because of the plane-parallel geometry, large effective plasma surface and possibility to operate with only air as a feeding gas. When the discharge was ignited, streamers were formed at temporary random points in the form of microfilaments, and they conducted higher discharge current than the rest of the discharge. Because we were not treating plant cells directly but the sodium-alginate-encapsulated plant material, the samples could withstand these local inhomogeneities in the active plasma volume without any damage. Nevertheless, due to the nature of an SDBD source, the total current of filaments was limited, preventing formation of current hot-spots. Another reason for choosing this type of plasma source was its simplicity, both in design and in application. It did not require the addition of feeding gas, and, as an important feature for the future technology, it has a potential for scaling up. Detailed electrical characterization of the SDBD and optical emission spectra was presented where the plasma source, with its plan-parallel geometry, served as a capacitance in the electrical circuit. Regarding this, we used a simple and reliable method to measure the stray capacitance, C_p, for different distances, d, between upper and lower electrode segments. Results showed that with the increase in the distance between the powered electrode and grounded bottom electrode, we had an increase in the system capacitance, which was expected (Table 5). The obtained C_p values allowed for determination of the displacement current of each input voltage. The measured current signals included both the displacement and discharge current. Thus, after subtracting the displacement current, all current waveforms represented only current through the discharge. The capacitance determined for each d enabled calculation of the discharge current for all possible configurations of the plasma source.

With respect to application, one of the most important macro parameters in plasma treatments is power deposited to the discharge. This macro parameter that can be easily monitored; it reflects the electron density and temperature and, to some extent, through these two parameters, plasma chemistry [60]. P_{mean} (Figure 9) depends on the distance between electrodes, as well as formation of microfilaments. Thus, the highest increase and mean power values were achieved for d = 2 mm. For this distance, we chose to treat the synseeds at a power of 1.1 W. This showed to be the optimal value with respect to the effect on the seeds for the three treatment times that we used.

Apart from power, the feeding gas (in this case air) and humidity can also play an important role in plasma chemistry [61]. Although humidity in the treatment environment was not controlled, all experiments were performed in a room with constant humidity. Additionally, measurements of discharge characterization, as well as all treatments, were repeated several times in order to verify the reproducibility of the measurements. All measurements were performed with and without synseed samples. One way to obtain an insight into the chemical reactions occurring in the discharge is optical emission spectroscopy, as it can show the existence of certain excited species. In our experiment, the spectra were recorded for different RMS voltages and interelectrode distances without samples, but all have the same lines belonging only to the N_2 Second Positive System, SPS (Figure 10). Absence of the lines of, e.g., NO, OH and atomic oxygen, from the spectrum of an air DBD has been noted before and is related to dominant excitation and quenching reactions that favor N_2 excitation in filamentary discharges [11,12,61,62]. Moreover, other important reactive species, such as O_3, do not have emissions in the spectral range investigated. Nevertheless, these kinds of plasma sources generate a large amount of ozone and N_2O that is important for treatments of alginate surfaces of synseeds [13,14,63,64].

The most important parameters determining the efficiency of the encapsulation and plant recovery of synseeds are survival, regrowth and capability of initial explants for further plantlet growth to complete plantlets [1]. Chrysanthemum synseeds used in this study (plasma-treated and untreated) easily established regrowth with the first leaf emergence after one week of cultivation. The presence of alginate capsules did not inhibit regrowth of chrysanthemum shoots grown on agar medium. In addition, shoots were developed on plant growth regulator free medium without any callus formation, similar to what was reported for chrysanthemum cultivars by the Lady group [19]. Our results are not surprising because regrowth of encapsulated shoot tips of chrysanthemums is possible on plant growth regulation free medium [19]. In the case of encapsulation of nodal segments, the addition of a small amount of IAA in the encapsulation complex is necessary and obligatory for better regrowth of shoots from synseeds [29]. When synseeds were grown on VLM only, plasma-treated chrysanthemum synseeds continued their regrowth and formed well-developed microshoots. It should be mentioned that microshoots established on VLM were smaller than those developed on standard solid agar medium, which could be related to accessibility of nutrients in this type of medium.

Data about application of synseed technology for short- and long-term storage in vitro or easy transport of valuable genetic resources are available [65,66], whereas data regarding synseed manipulations for ex vitro growth are quite scarce [67,68]. In chrysanthemum, plantlet development from synseeds formed in vitro and sowing ex vitro was successfully achieved from double-layered synseeds [29]. In the current work, we planted untreated and plasma-treated simple, one-layer chrysanthemum synseeds directly in soil substrate. After three weeks of ex vitro cultivation of chrysanthemum synseeds, two-fold higher shoot development was observed in the case of plasma-treated chrysanthemum synseeds in comparison to untreated synseeds. Additionally, after six weeks of growth under ex vitro conditions, plasma-treated chrysanthemum synseeds showed significantly higher plantlet conversion compared to the untreated control. The enhanced survival, regrowth and further complete plantlet formation of plasma-treated chrysanthemum synseeds shown in our work might be explained, besides by antimicrobial effect, by the prolonged effects of chemical changes in the alginate gels after plasma treatment and their antimicrobial properties.

Similar effects were reported for plasma treatment of alginate wound dressings [12]. Plasma treatment of alginate gels inactivated bacterial and fungal infection for a month, which was a long enough period for successful shoot development and complete conversion of synseeds to plantlets. On the other hand, continued growth of untreated chrysanthemum synseeds was significantly reduced due to contamination and lack of adventitious root formation for other plant species [9].

We found that the plasma treatment significantly enhanced synseed germination and complete plantlet development for all investigated chrysanthemum cultivars. The response to plasma treatment of chrysanthemum synseeds was cultivar-dependent. Observed differences could be attributed to the fact that different chrysanthemum cultivars have distinct nutritional requirements, as was reported earlier for other chrysanthemum cultivars [19,30]. According to our results, no morphological disorders were noticed among plantlets derived from untreated and plasma-treated synseeds. The absence of any morphological or flower color alterations in chrysanthemum plants may be explained as a consequence of the regeneration protocol used in this study. First, we used stock shoot cultures derived from one mother plant. In addition, initiation of shoot regeneration was mainly achieved by direct shoot induction on the initial explant, avoiding a callus phase and further shoot multiplication by axillary meristem activation, which minimized possibilities for genetic changes due to somaclonal variations [16,17]. According to available data, this research represents first data about complete chrysanthemum plantlet regeneration from synseeds to flowering plants.

During direct sowing of synseeds, contamination by microorganisms is one of the major hurdles for the commercialization of encapsulation technology for many plant species [9]. Besides that, one of the main limiting factors for plantlet conversion is low-nutrient availability due to inhibition of root growth. Numerous factors are involved in this process, such as poor rooting ability and survival due to the lack of nutrients and oxygen supply. Organic nutrients released by the beads are mainly responsible for severe contamination of synseeds [29,30]. Unfortunately, the depletion of nutritional compounds in beads may cause lower shoot regrowth or complete growth inhibition [28–30]. To date, there are two strategies to overcome this problem in chrysanthemum synseeds. The first strategy is to use double-layered synseeds to restrict contamination, where the second layer is formed by Ca-alginate made with water [29]. A recently reported strategy for both production and sowing of chrysanthemum synseeds in non-aseptic conditions proposes eliminating all carbon sources and organic additives both inside and outside the synseeds [30]. The reported strategy might be promising, but it was applied to one cultivar only, and it is questionable whether it is applicable to other cultivars. Therefore, it is necessary to build up a system that lowers contamination and keeps a nutrient reservoir within the encapsulated plant tissue, which is necessary for successful rooting. Considering the results of the present study, this problem may be successfully solved by plasma treatment of chrysanthemum synseeds before sowing.

There are many research data that demonstrate that cold plasma has a potent general antimicrobial effect through its generation of free radicals, reactive oxygen species (ROS) and reactive nitrogen species, such as hydrogen peroxide, superoxide, singlet oxygen, nitric oxide and ammonia [42]. The generated reactive species or their products are responsible for the antimicrobial effect and, in certain situations, have proven to be non-toxic to eukaryotic cells [69]. The chemical changes produced in the gel are relatively stable, and the anti-microbial properties of such a gel may last for close to one month, as reported after plasma treatment of alginate wound dressings [13]. The treated alginate gels inactivated all of the Gram-negative, Gram-positive and fungal pathogens by generated ROS inside bacterial cells, leading to their rapid death or triggering programmed cell death exhibiting characteristic features of apoptosis [12]. According to our results, we can conclude that treatment with non-thermal plasma generates chemical and physical responses in an alginate gel, producing changes that implicate not only biocidal effects but possibly growth-promoting effects.

4. Materials and Methods

4.1. Atmospheric Pressure Plasma Source

In the experiments, we used a circular surface dielectric barrier discharge (SDBD) source with an outer diameter of 90 mm. Side- and top-view schematics of the source are shown in Figure 11.

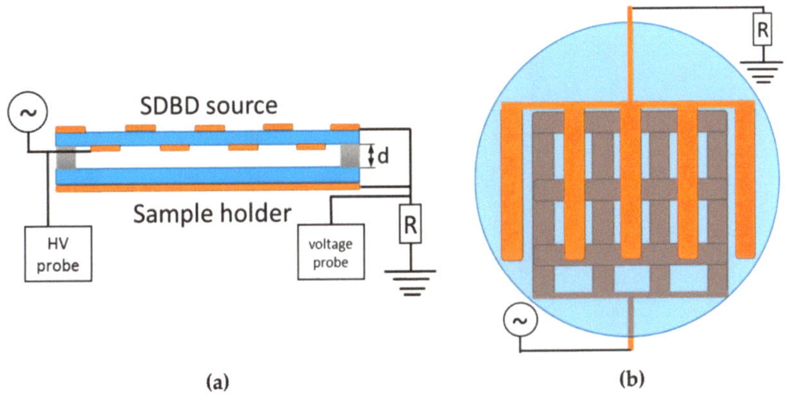

Figure 11. SDBD source schematics: (**a**) side view of the setup and schematics of electrical circuit; (**b**) top view of an upper electrode part. Dimensions of the source are given in the text.

The source consists of the two plane-parallel glass plates (2 mm thickness). The lower glass plate is positioned on a stainless-steel plate and serves as a sample holder. The upper glass plate is covered with conductive strips made of 5 mm wide copper tape. The strips are placed on both sides of the glass dielectric (Figure 11a). On the top side, they form a comb-like structure with gaps between the strips of 12 mm (Figure 11b). On the bottom side, the conductive strips form a grid structure making 12 mm side squares. The strip structures on opposite sides of the glass plate are shifted in such way that longer sides of the conducive strips are not overlapping (Figure 11b). The copper strips fixed to the upper part were grounded, whereas the strips on the bottom part of the top glass plate were powered with 50 Hz high-voltage (HV) sine signal. The outer edges of both powered and grounded electrode structures on the top part form a rectangle. Along the sides of this rectangle, plastic spacers were placed, keeping fixed distance in between the glass plates and bordering the active plasma volume of the SDBD source. Characterization of the device was performed by using spacers with thicknesses (d) of 2, 3, 4 and 5 mm.

The HV signal at the powered electrode was supplied by a homemade HV transformer. The input voltage (V_{input}) to the HV transformer was provided by a variable voltage regulator connected to the electrical power grid. The HV signal was monitored by a high-voltage probe (Tektronix 6015A, North Star High Voltage, Beaverton, OR, USA) connected to the circuit close to the powered electrode. Grounded electrodes on the top and bottom part of the source were connected to the same grounding point. In this grounded line, the probe (Agilent 10076A, Agilent Technologies, Beijing, China) allowed tracing of the voltage drop on the R = 15 kΩ resistor, thus monitoring of the total current in the discharge. Electrical signals were recorded by oscilloscope (Agilent DSO6052A, Agilent Technologies, Beijing, China) and saved on a computer for further processing.

4.2. Synthetic Seed Production

We used four chrysanthemum cultivars as starting plant material for the experiments: Brandsound Liliac (BL), Précocita Carna (PC), Précocita Parme (PP) and Queens (Q). Shoot cultures were established from one branch of the mother plant of each cultivar. Nodal and internodal stem (Figure 12a) and leaf segments (Figure 12b) of the mother plant were, after surface sterilization, grown on Murashige and Skoog mineral solution and vitamins [70] solidified with 7% agar, with the addition of 3% sucrose and 100 mg/L myo-inositol (MS medium) supplemented with α-naphthylacetic acid (0.1 mg/L NAA, Sigma-Aldrich, St. Luis, MO, USA) and 1.0 mg/L 6-benzylpurine (BAP, Sigma-Aldrich, St. Luis, MO, USA) for shoot induction. Initially regenerated shoots were further subcultured on the same medium, and stable stock cultures of chrysanthemum cultivars were established and subcultured in 4-week intervals (Figure 12c).

Figure 12. Chrysanthemum synthetic seed production. (**a**,**b**) Initiation of shoot regeneration in steam segment of cv. PC (**a**) and leaf culture of cv. BL (**b**); (**c**) stock shoot cultures used as primary explants for synthetic seed production; (**d**) sucking of shoot tips with sodium alginate; (**e**) formation of alginate beads in complex solution consisting of 100 mM $CaCl_2 \times 2H_2O$; (**f**) well-formed alginate beads after 20 min in complex solution; (**g**) washed synseeds ready for treatment.

For production of synthetic seeds, we used axillary shoots developed from nodal stem segments (1 cm) grown for 3 weeks on plant regulator free MS medium described above. After 3 weeks of growth, shoot tips (2–3 mm) were excised from same-sized axillary shoots (2 cm) and washed for 20 min in MS liquid medium without $CaCl_2 \times 2H_2O$. Encapsulation was performed according to the following procedure: the explants were plunged into a solution of sodium alginate (2.5%, *w*/*v*, medium viscosity, Carlo Erba, Carnadero, Italy) made with MS liquid medium without $CaCl_2 \times 2H_2O$. Subsequently, droplets of the alginate solution containing one shoot tip were sucked into pipettes with shortened sterile plastic tips (Figure 12d) and dropped into a complex solution (Figure 12e) consisting of 100 mM$CaCl_2 \times 2H_2O$ (Sigma-Aldrich, St. Louis, MO, USA). Formed alginate beads (5–6 mm in diameter) were maintained for 20 min in this solution with continuous slow agitation (Figure 12f). The encapsulated shoot tips (synseeds) were retrieved and rinsed three times in sterile distilled water in order to remove traces of $CaCl_2 \times 2H_2O$ (Figure 12g). Finally, water was decanted, and synseeds were placed on sterile filter paper for a few minutes (Figure 1a).

All nutritional media were adjusted to pH 5.8 before sterilization. All media, sodium alginate and complexing solution were sterilized in an autoclave for 20 min at 114 °C. In vitro cultures were grown in a growth room at 23 ± 2 °C, with a photoperiod of 16 h day/8 h night.

4.3. Plasma Treatment of Synthetic Seeds

Chrysanthemum synseeds (Figure 1a) of the same size (5–6 mm in diameter) were exposed to an air SDBD. Twenty-five synseeds were gently placed onto a sterile glass Petri dish (diameter, 90 mm) and then placed below the powered electrode in unsterile conditions and exposed to the plasma all at the same time. The distance between the powered electrode and the synseed surface was ~2 mm. Due to the fact that seeds are ball-shaped and vary slightly in size, the distance could only be estimated. The treatment times were 1 min, 5 min and 10 min. The applied HV in all treatments was $V_{peak\text{-}to\text{-}peak} = 22$ kV ($V_{RMS} = 7.8$ kV), which corresponded to a mean power of 1.1 W.

4.4. Sowing of Synthetic Seeds

Untreated and plasma-treated synseeds were grown on three different sowing substrates as follows: (i) sterile agar MS medium (AM), grown under in vitro conditions; (ii) sterilized vermiculite + liquid MS medium (VLM), grown under in vitro conditions; (iii) direct sowing into unsterilized soil substrate, grown under ex vitro conditions (greenhouse).

For the in vitro experiment, untreated (control) and cold plasma-treated chrysanthemum synseeds were grown on solid plant regulator free MS medium (30 mL) filled in baby jars. Similarly, untreated and cold plasma-treated synseeds (5 per baby jar) were grown in baby jars (5) filled with 10 g vermiculite and moisture with 10 mL liquid plant regulator free MS medium. Baby jars filled with 10 g of vermiculite were sterilized in an autoclave for 20 min at 114 °C. In vitro cultures with synseeds grown on AM and VLM were grown in a growth room at 23 ± 2 °C, with a photoperiod of 16 h day/8 h night. Germination of the synseeds under in vitro conditions was estimated as a frequencies of shoot regrowth from alginate beads at two steps: leaf emergence after one week and full shoot development after four weeks of cultivation.

For the ex vitro experiment, untreated (control) and cold plasma-treated synseeds were sowed directly in plastic containers with 18 places (3 × 6) filled with commercial substrate (Floradur® Seed, Floradur, Oldenbürg, Germany) during April and May. Two synthetic seeds were placed on the top of the substrate, one in each place inside a plastic container. All containers were covered with transparent foil during the first four weeks of growing. The synseeds were sprayed weekly with unsterile MS mineral solution. Germination of synseeds under ex vitro conditions was estimated as a frequencies of leaf emergence from alginate beads after one week of cultivation, full shoot development was assigned as complete shoot regrowth from initial synseed after three weeks of cultivation and whole-plantlet formation was recorded when germinated synseeds developed roots after 6 weeks of culture. Total deterioration (%) was evaluated as a percentage of seeds that failed to regrow (germinate) and regenerate to plantlets with respect to the total number of planted synseeds. After twelve weeks of growth in containers, each synseed plantlet was potted in individual pots filled with a mix of peat and perlite (3:1) and grown under greenhouse conditions until flowering. Experiments with plasma treatment were repeated two to four times, and 15–25 synthetic synseeds were used per treatment depending on chrysanthemum cultivar. All data were subjected to statistical analysis using STATISTICA and analysis of variance (ANOVA) using least significance (LSD) tests.

5. Conclusions

Synseed technology may be a useful technique for a propagation systems in terms of fast reproduction of seedless plants, preservation of the genetic uniformity of plants, straight delivery to the field and, last but not least, low cost. The difficulties of sowing artificial seeds directly in soil or on commercial substrates under non-sterile conditions are considered to be one of the main limitations for the practical use of this technique. In this paper, we presented, in detail, the results of the electrical characterization of an SDBD that operates in air and its possible application in synthetic seed technology. According to our results, implementation of the SDBD plasma treatment before sowing represents a promising strategy for future investigations and sustainable use of cold plasma in synseed biotechnology. Plasma-treated chrysanthemum synseeds showed a better survival rate and overall plantlet growth under greenhouse conditions in comparison to untreated synseeds.

In conclusion, to the best of our knowledge, this is the first report about the use of SDBD plasma for seed germination of synthetic seeds under aseptic or non-aseptic conditions. This study demonstrated a highly effective strategy for direct conversion of synseeds into entire plantlets by using plasma pre-conversion treatment. This treatment reduced contamination and displayed a considerable ex vitro ability to convert clonally identical chrysanthemum plants.

Supplementary Materials: The following are available online at https://www.mdpi.com/article/10.3390/plants11070907/s1, Figure S1: Shoot multiplication index in a control group (plasma −) and with plasma treated synseeds (plasma +).

Author Contributions: Conceptualization, N.Š., N.P., S.J. and S.Ž.; methodology, N.Š., N.P., S.J. and S.Ž.; software, N.Š., N.P. and S.J.; formal analysis, N.Š., N.P., S.J. and S.Ž.; investigation, N.Š., N.P., S.J. and S.Ž.; resources, N.Š., N.P., S.J. and S.Ž.; writing—original draft preparation, N.Š., S.J. and S.Ž.; writing—review and editing, N.Š. and N.P.; supervision, N.P.; project administration, N.P.; funding acquisition, N.P. All authors have read and agreed to the published version of the manuscript.

Funding: This research was funded by the Ministry of Education, Science, and Technological Development of the Republic of Serbia, grant numbers 451-03-68/2022-14/200007 and 451-03-68/2022-14/200024.

Institutional Review Board Statement: Not applicable.

Informed Consent Statement: Not applicable.

Data Availability Statement: All data are shown in the manuscript, and raw data are available from corresponding authors upon reasonable request.

Acknowledgments: Results represent a part of activity in COST action-Plasma applications for smart and sustainable agriculture (CA19110). The authors would like to thank chrysanthemum breeder Zlatko Nekić, Serbia, for providing plant material for the experiments. N.P. and N.Š. would like to thank Milica Bajić for performing measurements for electrical characterization.

Conflicts of Interest: The authors declare no conflict of interest.

References

1. Rihan, H.Z.; Kareem, F.; El-Mahrouk, E.; Fuller, M.P. Artificial seeds (Principle, aspects and applications). *Agronomy* **2017**, *7*, 71. [CrossRef]
2. Yücesan, B. Synseed: A new trend in seed technology. In *Synthetic Seeds, Germplasm Regeneration, Preservation and Prospects*; Faisal, M., Alatar, A.A., Eds.; Springer Nature: Cham, Switzerland, 2019; pp. 61–75. [CrossRef]
3. Murashige, T. Plant cell and organ culture as horticultural practice. *Acta Hortic.* **1977**, *78*, 17–30. [CrossRef]
4. Standardi, A.; Picconi, E. Recent perspectives on the synthetic seed technology using non–embryogenic in vitro-derived explants. *Int. J. Plant Sci.* **1998**, *159*, 968–978. [CrossRef]
5. Ara, H.; Jaiswal, U.; Jaiswal, V.S. Synthetic seed: Prospects and limitations. *Curr. Sci.* **2000**, *78*, 1438–1444. [CrossRef]
6. Benelli, C.; Micheli, M.; De Carlo, A. An improved encapsulation protocol for regrowth and conservation of four ornamental species. *Acta Soc. Bot. Pol.* **2017**, *86*, 3559. [CrossRef]
7. Phanornchai, S.; Bodhipadma, K.; Noichnda, S.; Leung, D.W.M. Short-term storability of alginate-encapsulated Persian violet microshoots for germplasm exchange. *Plants* **2022**, *11*, 185. [CrossRef]
8. Sakai, A.; Engelmann, F. Vitrification, encapsulation-vitrification and droplet-vitrification: A review. *Cryo Lett.* **2007**, *28*, 151–172.
9. Sharma, S.; Shahza, A.; Teixeira da Silva, J.A. Synseed technology—A complete synthesis. *Biotechnol. Adv.* **2013**, *31*, 186–207. [CrossRef]
10. Saxena, A.; Shukla, M.; Saxena, P. Synthetic seeds: Relevance to endangered germplasm conservation in vitro. In *Synthetics Seeds, Germplasm Regeneration, Preservation and Prospects*; Faisal, M., Alatar, A.A., Eds.; Springer Nature: Cham, Switzerland, 2019; pp. 21–59. [CrossRef]
11. Szekalska, M.; Puciłowska, A.; Szymańska, E.; Ciosek, P.; Winnicka, K. Alginate: Current use and future perspectives in pharmaceutical and biomedical applications. *Int. J. Polym. Sci.* **2016**, *2016*, 7697031. [CrossRef]
12. Wiegand, C.; Heinze, T.; Hipler, U.C. Comparative in vitro study on cytotoxicity, antimicrobial activity, and binding capacity for pathophysiological factors in chronic wounds of alginate and silver-containing alginate. *Wound Repair Regen.* **2009**, *17*, 511–521. [CrossRef]

13. Poor, A.E.; Ercan, U.K.; Yost, A.; Brooks, A.D.; Joshi, S.G. Control of multi-drug-resistant pathogens with non-thermal-plasma-treated alginate wound dressing. *Surg. Infect.* **2014**, *15*, 233–243. [CrossRef] [PubMed]
14. Salachna, P.; Grzeszczuk, M.; Miller, E.; Soból, M. Oligo-alginate with low molecular mass improves growth and physiological activity of Eucomisa utumnalis under salinity stress. *Molecules* **2018**, *23*, 812. [CrossRef] [PubMed]
15. Golkar, P.; Taghizadeh, M.; Noormohammadi, A. Effect of sodium alginate on secondary metabolites and antioxidant activity of safflower genotypes under in vitro salinity stress. *In Vitro Cell Dev. Biol.-Plant* **2019**, *55*, 527–538. [CrossRef]
16. Teixeira da Silva, J.A. Ornamental chrysanthemums: Improvement by biotechnology. *Plant Cell Tissue Organ Cult.* **2004**, *79*, 1–18. [CrossRef]
17. Teixeira da Silva, J.A.; Kulus, D. Chrysanthemum biotechnology: Discoveries from the recent literature. *Folia Hortic.* **2014**, *26*, 67–77. [CrossRef]
18. Zalewska, M.; Lema-Rumińska, J.; Miler, N. In vitro propagation using adventitious buds technique as a source of new variability in chrysanthemum. *Sci. Hortic.* **2007**, *113*, 70–73. [CrossRef]
19. Kulus, D.; Zalewska, M. In vitro plant recovery from alginate-encapsulated chrysanthemum x grandiflorum (Ramat.) Kitam. shoot tips. *Propag. Ornam. Plants* **2014**, *14*, 3–12.
20. Zalewska, M.; Tymoszuk, A.; Miler, N. New Chrysanthemum cultivars as a result of in vitro mutagenesis with the application of different explant types. *Acta Sci. Pol. Hort. Cult.* **2011**, *10*, 109–123.
21. Hill, G.P. Shoot formation in tissue cultures of chrysanthemum *Chrysanthemum morifolium* Ramat. *JPN J. Breed.* **1968**, *42*, 386–389.
22. Jevremović, S.; Radojević, L. In vitro plant regeneration from stem segments of several cultivars of chrysanthemum (*Chrysanthemum morifolium* Ramat.) *Bull. Inst. Jard. Bot. Univ. Belgrade* **1995**, *29*, 107–114.
23. Jevremović, S.; Subotić, A.; Miljković, D.; Trifunović, M.; Petrić, M.; Cingel, A. Clonal fidelity of chrysanthemum cultivars after long term micropropagation by stem culture. *Acta Hortic.* **2012**, *961*, 211–216. [CrossRef]
24. Jevremovic, S.; Subotic, A. Micropropagation of chrysanthemum cultivars in Serbia. In Proceedings of the IX International Scientific Agriculture Symposium, Agrosym 2018, Jahorina, Bosnia and Herzegovina, 4–7 October 2018; Kovacevic, D., Ed.; University of East Sarajevo, Faculty of Agriculture: Sarajevo, Bosnia and Herzegovina, 2018; pp. 408–413.
25. Teixeira da Silva, J.A.; Kim, H.; Engelmann, F. Chrysanthemum low-temperature storage and cryopreservation: A review. *Plant Cell Tissue Organ Cult.* **2015**, *120*, 423–440. [CrossRef]
26. Rout, G.R.; Das, P. Recent trends in the biotechnology of Chrysanthemum: A crucial review. *Sci. Hortic.* **1997**, *69*, 239–257. [CrossRef]
27. Teixeira da Silva, J.A. Chrysanthemum: Advances in tissue culture, cryopreservation, postharvest technology, genetics and transgenic biotechnology. *Biotechnol. Adv.* **2003**, *21*, 715–766. [CrossRef]
28. Kulus, D. Application of synthetic seeds in propagation, storage, and preservation of Asteraceae plant species. In *Synthetic Seeds, Germplasm Regeneration, Preservation and Prospects*; Faisal, M., Alatar, A.A., Eds.; Springer Nature: Cham, Switzerland, 2019; pp. 155–179. [CrossRef]
29. Pinker, I.; Abdel-Rahman, S.S. Artificial seeds for propagation of *Dendrantema* × *grandiflora* (Ramat.). *Propag. Ornam. Plants* **2005**, *5*, 186–191.
30. Hung, C.D.; Dung, C.D. Production of chrysanthemum synthetic seeds under non-aseptic conditions for direct transfer to commercial greenhouses. *Plant Cell Tissue Organ Cult.* **2015**, *122*, 639–648. [CrossRef]
31. Reddy, M.C.; Rama Murthy, K.S.; Pullaiah, T. Synthetic seeds: A review in agriculture and forestry. *Afr. J. Biotechnol.* **2012**, *11*, 14254–14275. [CrossRef]
32. Rihan, H.Z.; Al-Issawi, M.; Al-Swedi, F.; Fuller, M.P. The effect of using PPM (Plant preservative mixture) on the development of cauliflower microshoots and the quality of the artificial seed produced. *Sci. Hortic.* **2012**, *141*, 47–52. [CrossRef]
33. Adamovich, I.; Baalrud, S.D.; Bogaerts, A.; Bruggeman, P.J.; Cappelli, M.; Colombo, V.; Czarnetzki, U.; Ebert, U.; Eden, J.G.; Favia, P.; et al. The Plasma Roadmap: Low temperature plasma science and technology. *J. Phys. D Appl. Phys.* **2017**, *50*, 323001. [CrossRef]
34. Weltmann, K.-D.; von Woedtke, T. Plasma medicine—current state of research and medical application. *Plasma Phys. Control. Fusion* **2017**, *59*, 014031. [CrossRef]
35. Laroussi, M.; Kong, M.G.; Morfill, G.; Stolz, W. *(Eds). Plasma Medicine Applications of Low-Temperature Gas Plasmas in Medicine and Biology*; Cambridge University Press: Cambridge, UK, 2012.
36. Lazović, S.; Puač, N.; Miletić, M.; Pavlica, D.; Jovanović, M.; Bugarski, D.; Mojsilović, S.; Maletić, D.; Malović, G.; Milenković, P.; et al. The effect of a plasma needle on bacteria in planktonic samples and on peripheral blood mesenchymal stemcells. *New J. Phys.* **2010**, *12*, 083037. [CrossRef]
37. Puac, N.; Gherardi, M.; Shiratani, M. Plasma agriculture: A rapidly emerging field. *Plasma Proc. Polym.* **2018**, *15*, 1700174. [CrossRef]
38. Attri, P.; Ishikawa, K.; Okumura, T.; Koga, K.; Shiratani, M. Plasma agriculture from laboratory to farm: A review. *Processes* **2020**, *8*, 1002. [CrossRef]
39. Živković, S.; Puač, N.; Giba, Z.; Grubišić, D.; Petrović, Z.L. The stimulatory effect of non-equilibrium (low temperature) air plasma pretreatment on light-induced germination of *Paulownia tomentosa* seeds. *Seed Sci. Tech.* **2004**, *32*, 693–701. [CrossRef]
40. Sera, B.; Stranák, V.; Serý, M.; Tichý, M.; Spatenka, P. Germination of *Chenopodium album* in response to microwave plasma treatment. *Plasma Sci. Technol.* **2008**, *10*, 506. [CrossRef]
41. Lu, X.; Reuter, S.; Laroussi, M.; Liu, D. *Non Equilibrium Atmospheric Pressure Plasma Jets: Fundamentals, Diagnostics, and Medical Applications*, 1st ed.; CRC Press: Boca Raton, FL, USA, 2019; pp. 1–426.

42. Waskow, A.; Avino, F.; Howling, A.; Furno, I. Entering the plasma agriculture field: An attempt to standardize protocols for plasma treatment of seeds. *Plasma Process Polym.* **2022**, *19*, e2100152. [CrossRef]
43. Motyka-Pomagruk, A.; Dzimitrowicz, A.; Orlowski, J.; Babinska, W.; Terefinko, D.; Rychlowski, M.; Prusinski, M.; Pohl, P.; Lojkowska, E.; Jamroz, P.; et al. Implementation of a non-thermal atmospheric pressure plasma for eradication of plant pathogens from a surface of economically important seeds. *Int. J. Mol. Sci.* **2021**, *22*, 79256. [CrossRef]
44. Lu, X.; Naidis, G.V.; Laroussi, M.; Reuter, S.; Graves, D.B.; Ostrikov, K. Reactive species in non-equilibrium atmospheric-pressure plasmas: Generation, transport, and biological effects. *Phys. Rep.* **2016**, *630*, 1–84. [CrossRef]
45. Bradu, C.; Kutasi, K.; Magureanu, M.; Puač, N.; Živković, S. Reactive nitrogen species in plasma-activated water: Generation, chemistry and application in agriculture. *J. Phys. D Appl. Phys.* **2020**, *53*, 223001. [CrossRef]
46. Matthes, R.; Bender, C.; Schlüter, R.; Koban, I.; Bussiahn, R.; Reuter, S.; Lademann, J.; Weltmann, K.D.; Kramer, A. Antimicrobial efficacy of two surface barrier discharges with air plasma against in vitro biofilms. *PLoS ONE* **2013**, *8*, 70462. [CrossRef]
47. Bormashenko, E.; Grynyov, R.; Bormashenko, Y.; Driori, E. Cold radiofrequency plasma treatment modifies wettability and germination speed of plant seeds. *Sci. Rep.* **2012**, *2*, 741. [CrossRef] [PubMed]
48. Sivachandiran, L.; Khacef, A. Enhanced seed germination and plant growth by atmospheric pressure cold air plasma: Combined effect of seed and water treatment. *RSC Adv.* **2017**, *7*, 1822–1832. [CrossRef]
49. Mildažienė, V.; Aleknavičiūtė, V.; Žūkienė, R.; Giedrė, P.; Naučienė, Z.; Filatova, I.; Lyushkevich, V.; Haimi, P.; Tamošiūnė, I.; Baniulis, D. Treatment of Common Sunflower (*Helianthus annus* L.) Seeds with radio-frequency electromagnetic field and cold plasma induces changes in seed phytohormone balance, seedling development and leaf protein expression. *Sci. Rep.* **2019**, *9*, 6437. [CrossRef] [PubMed]
50. Zahoranová, A.; Henselová, M.; Hudecová, D.; Kaliňáková, B.; Kováčik, D.; Medvecká, V.; Černák, M. Effect of cold atmospheric pressure plasma on the wheat seedlings vigor and on the inactivation of microorganisms on the seeds surface. *Plasma Chem. Plasma Process.* **2016**, *36*, 398–414. [CrossRef]
51. Waskow, A.; Betschart, J.; Butscher, D.; Oberbossel, G.; Klöti, D.; Büttner-Mainik, A.; Adamcik, J.; von Rohr, P.R.; Schuppler, M. Characterization of efficiency and mechanisms of cold atmospheric pressure plasma decontamination of seeds for sprout sroduction. *Front. Microbiol.* **2018**, *9*, 3164. [CrossRef]
52. Liu, B.; Honnorat, B.; Yang, H.; Arancibia, J.; Rajjou, L.; Rousseau, A. Non-thermal DBD plasma array on seed germination of different plant species. *J. Phys. D Appl. Phys.* **2019**, *52*, 025401. [CrossRef]
53. Chen, H.H.; Chang, H.C.; Chen, Y.K.; Hung, C.L.; Lin, S.Y.; Chen, Y.S. An improved process for high nutrition of germinated brown rice production: Low-pressure plasma. *Food Chem.* **2016**, *191*, 120–127. [CrossRef]
54. Brandenburg, R. Dielectric barrier discharges: Progress on plasma. *Plasma Sources Sci. Technol.* **2017**, *26*, 053001. [CrossRef]
55. Winter, J.; Brandenburg, R.; Weltmann, K.D. Atmospheric pressure plasma jets: An overview of devices and new directions. *Plasma Sources Sci. Technol.* **2015**, *24*, 064001.
56. Bruggeman, P.; Brandenburg, R. Atmospheric pressure discharge filaments and microplasmas: Physics, chemistry and diagnostics. *J. Phys. D Appl. Phys.* **2013**, *46*, 464001. [CrossRef]
57. Janić Hajnal, E.; Vukić, M.; Pezo, L.; Orčić, D.; Puač, N.; Škoro, N.; Milidrag, A.; Šoronja Simović, D. Effect of atmospheric cold plasma treatments on reduction of alternaria toxins content in wheat flour. *Toxins* **2019**, *11*, 704. [CrossRef] [PubMed]
58. Brandenburg, R.; Navrátil, Z.; Jansky, J.; Stahel, P.; Trunec, D.; Wagner, H.E. The transition between different modes of barrier discharges at atmospheric pressure. *J. Phys. D Appl. Phys.* **2009**, *42*, 085208. [CrossRef]
59. Wang, C.; Zhang, G.; Wang, X. Comparison of discharge characteristics of a dielectric barrier discharge with different electrode structures. *Vacuum* **2012**, *86*, 960. [CrossRef]
60. Sakiyama, Y.; Graves, D.B.; Chang, H.-W.; Shimizu, T.; Morfill, G.E. Plasma chemistry model of surface microdischarge in humid air and dynamics of reactive neutral species. *J. Phys. D Appl. Phys.* **2012**, *45*, 425201. [CrossRef]
61. Tang, Q.; Jiang, W.; Cheng, Y.; Lin, S.; Lim, T.M.; Xiong, J. Generation of Reactive Species by Gas-Phase Dielectric Barrier Discharges. *Ind. Eng. Chem. Res.* **2011**, *50*, 9839–9846. [CrossRef]
62. Kozlov, K.V.; Brandenburg, R.; Wagner, H. E.; Morozov, A.M.; Michel, P. Investigation of the filamentary and diffuse mode of barrier discharges in N_2/O_2 mixtures at atmospheric pressure by cross-correlation spectroscopy. *J. Phys. D Appl. Phys.* **2005**, *38*, 518–529. [CrossRef]
63. Shang, K.; Wang, M.; Peng, B.; Li, J.; Lu, N.; Jiang, N.; Wu, Y. Characterization of a novel volume-surface DBD reactor: Discharge characteristics, ozone production and benzene degradation. *J. Phys. D Appl. Phys.* **2020**, *53*, 065201. [CrossRef]
64. Jiang, N.; Guo, L.; Qiu, C.; Zhang, Y.; Shang, K.; Lu, N.; Li, J.; Wu, Y. Reactive species distribution characteristics and toluene destruction in the three-electrode DBD reactor energized by different pulsed modes. *Chem. Eng. J.* **2018**, *350*, 12–19. [CrossRef]
65. Liberatore, C.M.; Rodolfi Beghèa, D.; Fabbri, A.; Ganino, T.; Chianconea, B. Adventitious shoot organogenesis and encapsulation technology in hop (*Humulus lupulus* L.). *Sci. Hortic.* **2020**, *270*, 109416. [CrossRef]
66. Benelli, C. Encapsulation of shoot tips and nodal segments for in vitro storage of "kober 5BB" grapevine rootstock. *Horticulturae* **2016**, *10*, 185. [CrossRef]
67. Hung, C.D.; Trueman, S.J. Encapsulation technology for short term preservation and germplasm distribution of the African mahogany *Khaya senegalensis*. *Plant Cell Tissue Organ. Cult.* **2011**, *107*, 397–405. [CrossRef]
68. Hung, C.D.; Trueman, S.J. Alginate encapsulation of shoot tips and nodal segments for short term storage and distribution of the eucalypt *Corymbia torelliana* × *C. citriodora*. *Acta Physiol. Plant.* **2012**, *34*, 117–128. [CrossRef]

69. Dobrynin, D.; Fridman, G.; Friedman, G.; Fridman, A. Physical and biological mechanisms of direct plasma interaction with living tissue. *New J. Phys.* **2009**, *11*, 115020. [CrossRef]
70. Murashige, T.; Skoog, F. A revised medium for rapid growth and bio assays with tobacco tissue cultures. *Physiol. Plant* **1962**, *15*, 473–497. [CrossRef]

MDPI AG
Grosspeteranlage 5
4052 Basel
Switzerland
Tel.: +41 61 683 77 34

Plants Editorial Office
E-mail: plants@mdpi.com
www.mdpi.com/journal/plants

www.ingramcontent.com/pod-product-compliance
Lightning Source LLC
Chambersburg PA
CBHW041357210926
51850CB00013B/3110
* 9 7 8 3 7 2 5 8 2 6 6 8 1 *